Responsible Innovation 1

Jeroen van den Hoven • Neelke Doorn
Tsjalling Swierstra • Bert-Jaap Koops
Henny Romijn
Editors

Responsible Innovation 1

Innovative Solutions for Global Issues

Springer

Editors
Jeroen van den Hoven
Neelke Doorn
Delft University of Technology
Delft, The Netherlands

Bert-Jaap Koops
Tilburg Law School
University of Tilburg
Tilburg, The Netherlands

Tsjalling Swierstra
Faculty of Arts and Social Sciences
University of Maastricht
Maastricht, The Netherlands

Henny Romijn
Eindhoven University of Technology
Eindhoven, The Netherlands

ISBN 978-94-017-8955-4 ISBN 978-94-017-8956-1 (eBook)
DOI 10.1007/978-94-017-8956-1
Springer Dordrecht Heidelberg New York London

Library of Congress Control Number: 2014943765

Printed on acid-free paper

Springer is part of Springer Science+Business Media (www.springer.com)

Acknowledgments

This volume is based upon work that was originally presented at the First International Conference on Responsible Innovation on April 18–19, 2011, hosted by the Netherlands Organisation for Scientific Research (NWO) and co-organized by the Philosophy Section of the Technical University of Delft. The conference brought together the first results of research projects under the NWO Research Program "Responsible Innovation" (Maatschappelijk Verantwoord Innoveren). The theme of this first conference was "Innovative Solutions for Global Issues."

We wish to thank all people who contributed to the completion of this volume. We particularly acknowledge the referees who provided useful feedback on earlier drafts of the chapters and the manuscript as a whole. We gratefully acknowledge the effort that all authors mustered throughout the entire process of planning, composition, and revision of the manuscript. On behalf of NWO, MVI program coordinator Jasper Roodenburg is acknowledged for making possible the conference.

The Editors

Contents

Contributors

Marlous E. Arentshorst Athena Institute for Research on Innovation and Communication in Health and Life Sciences, VU University Amsterdam, Amsterdam, The Netherlands

Bart H.M. Custers eLaw@Leiden, The Centre for Law in the Information Society, Leiden University, Leiden, The Netherlands

WODC – Ministry of Security and Justice, The Netherlands

Claire B. Blaauw Department of Health Law, University Medical Center Groningen, Groningen, The Netherlands

Marianne Boenink Department of Philosophy, University of Twente, Enschede, The Netherlands

Jacqueline E.W. Broerse Athena Institute for Research on Innovation and Communication in Health and Life Sciences, VU University Amsterdam, Amsterdam, The Netherlands

Yvonne M. Cuijpers Innovation Studies, Copernicus Institute of Sustainable Development, Utrecht University, Utrecht, The Netherlands

Tjard de Cock Buning Athena Institute for Research on Innovation and Communication in Health and Life Sciences, VU University Amsterdam, Amsterdam, The Netherlands

Gerard de Groot Development Research Institute (IVO), Tilburg University, Tilburg, The Netherlands

Job de Haan Tilburg School of Economics and Management, Tilburg University, Tilburg, The Netherlands

Jan De Winter Centre for Logic and Philosophy of Science, Department of Philosophy and Moral Sciences, Ghent University, Ghent, Belgium

Virginia Dignum Information and Communication Technology, Delft University of Technology, Delft, The Netherlands

Neelke Doorn Delft University of Technology, Delft, The Netherlands

Julie Egal Institute Strategy for Innovation and Service, ESSEC Business School, Paris, France

Elisabeth Eppinger Faculty of Economics and Social Sciences, Chair of Innovation Management and Entrepreneurship, University of Potsdam, Potsdam, Germany

Armin Grunwald Karlsruhe Institute of Technology, Karlsruhe, Germany

Jeroen Hasselaar Department of Anesthesiology, Pain and Palliative Medicine, Radboud University Nijmegen Medical Centre, Nijmegen, The Netherlands

Joep H. Hubben Department of Health Law, University Medical Center Groningen, Groningen, The Netherlands

Nguyen Thi Huong School of International Education (SIE), Hanoi University of Science and Technology, Hanoi, Vietnam

Frank Willem Jansen Department of Biomechanical Engineering, Delft University of Technology, Delft, The Netherlands

Department of Gynecology, Leiden University Medical Center, Leiden, The Netherlands

Catholijn M. Jonker Interactive Intelligence, Delft University of Technology, Delft, The Netherlands

Prabhu V. Kandachar Faculty of Industrial Design Engineering, Delft University of Technology, Delft, The Netherlands

Jan Otto Kroesen Faculty of Technology, Policy and Management, Department of Philosophy, Delft University of Technology, Delft, The Netherlands

Annemarie Mink Faculty of Industrial Design Engineering, Delft University of Technology, Delft, The Netherlands

Ellen H.M. Moors Innovation Studies, Copernicus Institute of Sustainable Development, Utrecht University, Utrecht, The Netherlands

Vikram Singh Parmar Faculty of Industrial Design Engineering, Delft University of Technology, Delft, The Netherlands

Center for Innovative Business Design, Ahmedabad University, Gujarat, India

Xavier Pavie Institute Strategy for Innovation and Service, ESSEC Business School, Paris, France

Udo Pesch Faculty of Technology, Management and Policy, Department of Values, Technology and Innovation, Delft University of Technology, Delft, The Netherlands

Wim Ravesteijn Faculty of Technology, Policy and Management, Department of Philosophy, Delft University of Technology, Delft, The Netherlands

Anneloes Roelofsen Dutch Cancer Society, Amsterdam, The Netherlands

Nigel Roome Governance & Ethics Management Domain, Vlerick Leuven Gent School of Management, Leuven, Belgium

Lambèr Royakkers School of Innovation Sciences, Eindhoven University of Technology, Eindhoven, The Netherlands

Bart W. Schermer eLaw@Leiden, The Centre for Law in the Information Society, Leiden University, Leiden, The Netherlands

Henk J. Schers Department of Primary and Community Care, Radboud University Nijmegen Medical Centre, Nijmegen, The Netherlands

Tim Schwanen Transport Studies Unit, School of Geography and the Environment, University of Oxford, Oxford, UK

Tsjalling Swierstra Faculty of Arts and Social Sciences, University of Maastricht, Maastricht, The Netherlands

Peter Tinnemann Charité – Universitätsmedizin Berlin, Campus Charité Mitte, Institute for Social Medicine, Epidemiology and Health Economics, Berlin, Germany

Anya Topolski Institute for Philosophy, KU Leuven, Leuven, Belgium

Irina van Aalst Department of Human Geography and Planning, Faculty of Geosciences, Utrecht University, Utrecht, The Netherlands

John J. van den Dobbelsteen Department of Biomechanical Engineering, Delft University of Technology, Delft, The Netherlands

Jeroen van den Hoven Department of Values, Technology and Innovation, Delft University of Technology, Delft, The Netherlands

Simone van der Hof eLaw@Leiden, Universiteit Leiden, Leiden, The Netherlands

Janneke M. van der Zwaan Information and Communication Technology, Delft University of Technology, Delft, The Netherlands

Jelle Van Gurp Department of Anesthesiology, Pain and Palliative Medicine, Radboud University Nijmegen Medical Centre, Nijmegen, The Netherlands

Evert van Leeuwen Department of IQhealthcare; Ethics, Radboud University Nijmegen Medical Centre, Nijmegen, The Netherlands

Harro van Lente Innovation Studies, Copernicus Institute of Sustainable Development, Utrecht University, Utrecht, The Netherlands

Ilse van Liempt Department of Human Geography and Planning, Faculty of Geosciences, Utrecht University, Utrecht, The Netherlands

Martine Van Selm Amsterdam School of Communication Research (ASCOR), University of Amsterdam, Amsterdam, The Netherlands

Kris Vissers Department of Anesthesiology, Pain and Palliative Medicine, Radboud University Nijmegen Medical Centre, Nijmegen, The Netherlands

Jaap Voeten Development Research Institute (IVO), Tilburg University, Tilburg, The Netherlands

René von Schomberg European Commission, Brussels, Belgium

Katinka Waelbers Faculty of Arts and Social Sciences, University of Maastricht, Maastricht, The Netherlands

Hans Christian Wilms Max Planck Research Group "Democratic Legitimacy of Ethical Decisions: Ethics and Law in the Areas of Biotechnology and Biomedicine", Max Planck Institute for Comparative Public Law and International Law, Heidelberg, Germany

Author Biographies

Marlous E. Arentshorst holds a master degree in health and life sciences. In her Ph.D.-research she explores options how a (more) socially responsible research and innovation process of medical neuroimaging technologies can be created in order to establish an appropriate societal embedding of these technologies in the Dutch clinical context. Hereto an interactive, multi-stakeholder approach is applied in which stakeholders from science and society are actively involved.

Bart H.M. Custers is research manager at eLaw, the Institute for Law in the Information Society at Leiden University, the Netherlands. His research is focused on discrimination and privacy issues of new technologies, particularly data mining and profiling, the most important tools to deal with Big Data. In 2013 dr. Custers published the book "Privacy and Discrimination in the Information Society". He has published his work, over 60 publications, in both scientific journals and newspapers.

Claire B. Blaauw obtained her degree in law at the University of Nijmegen. From 2008 to 2012 she worked as a Hospital Lawyer and as a lecturer Health Law at UMCG (Groningen). From 2012 she works as a lawyer at Medirisk with specialization in medical liability.

Marianne Boenink is an assistant professor at the Department of Philosophy, University of Twente. Her research focuses on philosophy and ethics of biomedical technologies. She is particularly interested in the role of science and technology in visions of the future of medicine. She also investigates methodologies for ethical and societal reflection on emerging technologies. She is PI of the project 'Responsible innovation of early diagnostics for Alzheimer's Disease', funded by the 'Responsible Innovation'-program of the Dutch organization for Scientific Research (NWO), 2009–2014.

Jacqueline E.W. Broerse is professor of innovation and communication in the health and life sciences. She holds a master degree in biomedical sciences and obtained her Ph.D. degree on the development of an interactive approach in research agenda setting processes. Her current research is focused on methodology development for facilitating a dialogue between science and society, and management of systemic change processes, to contribute to more equitable and inclusive innovation processes. Many of her research projects are in the health sector.

Yvonne M. Cuijpers has a background in Philosophy of Science Technology and Society at the University of Twente and has worked on research projects about system innovation in animal husbandry and alternatives for animal testing. Currently she is affiliated to the Utrecht University, where she is preparing a doctoral thesis on the subject of responsible early diagnostics for Alzheimer's Disease.

Tjard de Cock Buning is professor of applied ethics in life sciences. He was trained as bio-physicist in the neurosciences and as philosopher of science. His current research focuses on complex multi-stakeholder problems in the life sciences. By engaging patients as well as industry, government and academia in agenda setting and decisions processes he develops and investigates methodologies of various forms of dialogue. The last years he dealt with topics as health system reforms, alcohol & drugs policy, genetic modified food, neuroimaging and predictive medicine.

Gerard de Groot is retired director and senior economist at the Development Research Institute of Tilburg University. Throughout his career, he carried out a series of research projects in the fields of economic development, innovation, poverty alleviation in developing countries. Moreover, he was responsible for in various capacity building projects of African Universities financed by Dutch and international donors. With these programs, he enabled a substantial number of Ph.D. candidate from Africa to graduate at Tilburg University.

Job de Haan is associate professor on international production management at Tilburg University's School of economics and Management. His research interests include lean logistics, international supply chains, corporate social responsibility and innovation. Dr. de Haan was involved in the responsible innovation research in Vietnam and particularly brought in the qualitative research and case studies methodologies of international value chains.

Jan De Winter studied moral sciences at Ghent University. He works as a Ph.D. fellow of the Research Foundation – Flanders (FWO) at the Centre for Logic and Philosophy of Science (Ghent University). His research focuses on issues in social epistemology and research ethics. He also published some papers on explanation.

Virginia Dignum is an associate professor at the Faculty of Technology, Policy and Management, Delft University of Technology. She got her Ph.D. in 2004 at Utrecht University. Her research focuses on agent based models of organizations, and the interaction between people and intelligent systems and teams. In 2006, she was awarded the prestigious Veni grant from NWO (Dutch Organization for Scientific Research). She has organized many international conferences and workshops, and has more than 120 peer-reviewed publications and books.

Neelke Doorn has a background in civil engineering and philosophy. She is currently affiliated to Delft University of Technology and the 3TU. Centre for

Ethics and Technology. For her Ph.D. research, she was involved in an ethical parallel research on Ambient Intelligence Technology. Neelke Doorn has published on topics within the field of applied ethics, notably engineering ethics and medical ethics, responsible innovation, and water governance. Her current research focuses on distributive issues in technology regulation and water governance.

After graduating from the French Business School ESSEC and two years of work experience in a major consulting company, **Julie Egal** has joined the ESSEC ISIS in 2009 as a research associate. She contributed to several research projects, focused on innovation and responsibility in the service industry.

Elisabeth Eppinger is a researcher and lecturer at the University of Potsdam, Germany, Chair for Innovation Management and Entrepreneurship. She holds a degree in Science and Technology Studies from Maastricht University and University of Strasbourg and a degree in engineering. Her research focus is on innovation, intellectual property rights and technology transfer. Currently, she is the project leader of a research project on innovation and business models in the pharmaceutical industry.

Armin Grunwald is physicist by education. After occupations in industry and research institutes he is now full professor of philosophy and ethics of technology at Karlsruhe Institute of Technology (KIT), director of the Institute for Technology Assessment and Systems (ITAS) at KIT and director of the Office of Technology Assessment at the German Parliament at Berlin. His main research areas are theory and methodology of Technology Assessment, theory and methodology of sustainable development, and ethics of technology

Jeroen Hasselaar (Ph.D.) is assistant professor in palliative care at RadboudUMC Nijmegen and project leader for palliative care on behalf of the Dutch federation of UMCs. He is vice-chair of the pain and palliative care center at RadboudUMC Nijmegen. He graduated in health sciences (Rotterdam, cum laude) and Applied ethics (Utrecht) and holds a Ph.D. in Medicine (Nijmegen).

Professor **Joep H. Hubben** is professor of health law at the University Medical Centre and Faculty of Law of the University of Groningen and lawyer at Nysingh Lawyers & Notaries NV. He was previously employed as public health inspector and counselor to the Arnhem Court of Appeal.

Nguyen Thi Huong is researcher and staff member of the School of International Education of Hanoi University of Science and Technology. She graduated from University of Leipzig in entrepreneurial competencies and their impact on the performance of small enterprises. Mrs. Huong acted as researcher and Vietnamese counterpart in the NWO Responsible Innovation research project at Tilburg University. Within Hanoi University of Science and Technology she continues with research in the fields of informal institutions, sustainable business and innovation.

Frank Willem Jansen is a gynaecologist at the Leiden University Medical Center. He is appointed as professor at the Leiden University and the Technical University

Delft for Minimally Invasive Surgery (MIS). His research line is on quality control in MIS and prevention of complications. Herewith guidance is obtained to implement this high technological innovation in a patient safe way. At this moment he is the president of the Dutch Society of MIS.

Catholijn M. Jonker is full professor of the Interactive Intelligence group at the Faculty of Electrical Engineering, Mathematics and Computer Science of the Delft University of Technology. Her recent publications address cognitive processes and concepts such as trust, negotiation, and the dynamics of individual agents and organisations. In Delft, she works with an interdisciplinary team to engineer human experience through multi-modal interaction between natural and artificial actors in a social dynamic context.

Prabhu V. Kandachar is extensively involved in projects involving students and businesses to identify opportunities as well as to design & prototype products and services for the Base-of-the-Pyramid (BoP) and emerging markets. Issues covered include water, healthcare, energy, housing, etc., in countries like India, Indonesia, China, Brazil, Ghana, Tanzania, Honduras, Philippines, Pakistan, Madagascar, etc. He was also directing research work on some healthcare issues of the poor in developing countries. He has several keynote lectures and publications including an edited book on this topic.

Bert-Jaap Koops is Professor of Regulation & Technology at the Tilburg Institute for Law, Technology, and Society (TILT), the Netherlands. From 2005 to 2010, he was a member of *De Jonge Akademie*, a young-researcher branch of the Royal Netherlands Academy of Arts and Sciences. His main research fields are cybercrime, cyber-investigation, forensics, privacy, and data protection. He is also interested in identity, digital constitutional rights, techno-regulation, and regulatory implications of human enhancement, genetics, robotics, and neuroscience.

Jan Otto Kroesen is Assistant Professor at the Technological University Delft. He teaches ethics, language philosophy and intercultural communication. His research is focused on the co-development of society, institutions, values and technology, especially in view of future strategies for developing countries, and the technology transfer and the institutional transition required to reach that objective.

Annemarie Mink is investigating the implications of Amartya Sen's capability approach on product development for rural people in emerging markets. In 2006 she re-designed a silk reeling machine for rural women in eastern India. After graduation she developed this machine further, in collaboration with NGO's. The machine was later patented in her name and named Anna Charkha. Currently 156 machines are operational running on solar power. Besides doing research on responsible design, she teaches master students on research methods and guides graduation students.

Ellen H.M. Moors is Professor of Sustainable Innovation. Her research in the field of innovation studies focuses on the dynamics and governance of emerging technologies in science-based sectors, such as agro-food, life sciences and health

& ageing. Responsible innovation is an important topic in her work. The theoretical focus of her work is on the role of user innovations and user-producer interactions in emerging technological innovation systems and changing institutional governance arrangements in emerging technology fields.

Vikram Singh Parmar has been working in the base of the pyramid context since the past 10 years in the domain such as medical devices for affordable health, agriculture, and education. He is interested in persuasive technology and how it can be used to change social beliefs and attitudes. In VentureStudio, his goal is to offer a platform to entrepreneurs and grassroots innovators to refine their innovation by building prototypes, providing design and engineering support, and determining their viability. The objective of the Studio is to demonstrate a working model for an innovation eco-system which can be replicated in other parts of the country.

Xavier Pavie, Ph.D. is the Director of the Institute for Strategic Innovation & Services (ISIS) of ESSEC Business School. He is also research-associate at IREPH (Institut Recherche Philosophique) – Université Paris Ouest. For many years his publications have emphasized philosophical approaches to innovation management and particularly the notion of 'responsible-innovation' as a source of innovation and performance. He has published several articles and a dozen of books on philosophy and on innovation and has developed courses, conferences and workshops around the topic of responsible innovation.

Udo Pesch is assistant professor at Delft University of Technology. He is affiliated to the department of Values & Technology of the department of Technology, Policy and Management. His disciplinary interests include science and technology studies, environmental politics, sustainable innovation, public administration, and philosophy.

Wim Ravesteijn is a Senior Lecturer & Researcher at TU Delft and affiliated to both Harbin (China) Institute of Technology and Beijing Information Science & Technology University as a Visiting/Contract Professor. He teaches Technology Dynamics, Innovation Management & Impact Assessment from historical, international, socio-cultural and ethical perspectives. His present research is focused on Responsible Port Innovation in China and North-West Europe. He has published, among others, on water resources development & management in Europe, Indonesia and China.

Anneloes Roelofsen has a background in medical biology. She obtained her Ph.D. degree on the development of a constructive technology assessment (CTA) approach that aims to facilitate broad societal reflection on emerging technologies, and to guide research activities into societal desirable directions. In her current position at the Dutch Cancer Society she focuses on patient participation in health research.

Henny Romijn is an Associate Professor of Technology & Development Studies in the School of Innovation Sciences of Eindhoven University of Technology. Her research focuses on sustainable pro-poor innovation in the global South, using evolutionary economics & transition studies as analytical perspectives. Before turning into an academic, she worked for the International Labour Organisation on employment promotion projects in East Africa and South Asia. She is currently coordinating a programme funded by the Dutch Science Organisation's programme on Responsible Innovation which addresses the drivers and consequences of the emergence of contentious biofuel projects in Tanzania and India.

Nigel Roome is professor of governance, corporate responsibility and sustainable development at Vlerick Business School (Leuven University Belgium). His research interests examine the relationship between business strategies, technological and management innovation and systemic changes arising from sustainable development. He also advised various multi-lateral organizations, including the European Commission, and governments in developing innovation and sustainable development policies. Prof. Roome acted as Project leader of the NWO responsible innovation research project in Vietnam of Tilburg University.

Lambèr Royakkers is associate professor in Ethics and Technology at Eindhoven University of Technology, and associate professor in Military Ethics at Netherlands Defense Academy. He is also chairman of the Ethics Advisory Board of the European FP7-project SUBCOP (Suicide Bomber Counteraction and Prevention, 2013–2016). He is co-author of the book *Ethics, Engineering and Technology* (Wiley-Blackwell, 2011, with professor Ibo van de Poel). His current research involves the ethical aspects of social robots, value sensitive design, and the formalization of (collective) responsibility.

Bart W. Schermer Ph.D., LLM is an assistant professor at the Institute for Law in the Information Society (eLaw@Leiden) at Leiden University, the Netherlands. His research is focused on privacy and cybercrime. Apart from his work for the University Bart is partner at Considerati, an IT law and policy consultancy.

Henk J. Schers (MD, Ph.D.), is a general practitioner and a senior researcher in primary health care at RadboudUMC.

Tim Schwanen is Departmental Lecturer in Transport Studies and Human Geography at the University of Oxford. He holds a Ph.D. Degree from Utrecht University and his research can be positioned at the intersection of urban, transport, cultural, political and economic geography. His current research interests revolve around the geographical dimensions of well-being; social inequalities in passenger transport; and transitions towards low-energy and low-carbon societies.

Prof Dr. **Tsjalling Swierstra** is full professor in Philosophy at the University of Maastricht (the Netherlands). The theoretical interest that connects his work in the philosophy of technology is: how can the dynamic interaction between moral and technological development be analyzed, anticipated and evaluated? He has published articles and books on cloning, new reproductive technologies, genomics, technology ethics in general, and on the ethics of New and Emerging Science and Technology (so-called NEST-ethics) in particular.

Dr. **Peter Tinnemann** heads the Charité Universitätsmedizin Berlin global health research at the Institute of Social Medicine, Epidemiology and Health Economics. He is a Medical Doctor with a Master in Public Health from Cambridge University, United Kingdom with substantial work experiences in international humanitarian aid organisations and the United Kingdom and German National Health Services. As a medical doctor he worked in paediatrics, infectious diseases and tropical medicine settings.

Anya Topolski is a FWO (Research Foundation Flanders) postdoctoral fellow at the Centre for Ethics, Social and Political Philosophy of the Higher Institute of Philosophy at the KU Leuven, Belgium. Her current research involves the deconstruction of the discourse of Judeo-Christianity in relation to European identity formation and its symbolic role in propagating Islamophobia. In 2008 defended her Ph.D. thesis entitled: A Political Ethics of Intersubjectivity: Between Hannah Arendt, Emmanuel Levinas and the Judaic (KU Leuven). Her thesis was awarded the 2009 Auschwitz Stichting prize and is being prepared for publication.

Irina van Aalst is Assistant Professor Urban Geography at the Utrecht University, having previously worked as senior researcher at the Institute for Housing and Mobility Studies at Delft University of Technology. Her research covers a wide range of urban topics and can be positioned at the intersection of urban, cultural and economic geography. She has published on urban dynamics and culture, public spaces, creative industries, surveillance and nightlife.

John J. van den Dobbelsteen obtained his degree in psychophysics at the University of Groningen. His Ph.D. degree was obtained in 2003 at the Department of Neuroscience of the Erasmus MC in Rotterdam. From 2005 he works as assistant at the Department of Biomechanical Engineering of the Delft University of Technology. His research projects focus on the study of instrument-tissue interaction in minimally invasive techniques and on the development of workflow monitoring systems for the OR.

Jeroen van den Hoven is professor of Moral Philosophy at Delft University of Technology. Van den Hoven is Vice Dean of the Faculty of Technology, Policy and Management. He is former Scientific Director of the Centre for Ethics and Technology of the Three Technical Universities in The Netherlands (www.ethicsandtechnology.eu) and Editor in Chief of Ethics and Information Technology (Springer). Jeroen van den Hoven is chair of the responsible innovation ("Maatschappelijk Verantwoord Innoveren") programme of the Netherlands Organization of Scientific Research (NWO).

Simone van der Hof is a full professor in law and the information society. Simone's particular academic interest is in the fields of online privacy, data protection and privacy statements, digital identities, digital child rights, (legal, social, technological) regulation of online child safety, and empowerment of individuals through technology.

Janneke M. van der Zwaan is a Ph.D. candidate at the Faculty of Technology, Policy and Management of Delft University of Technology. In her research, Janneke explores how intelligent virtual agents can provide social support to users. She is particularly interested in endowing these agents with the emotional skills required to comfort users. She has developed a virtual buddy prototype system that provides emotional support and practical advice to victims of cyberbullying.

Jelle Van Gurp is a Ph.D. candidate in the department of Anaesthesiology, Pain and Palliative Care, RadboudUMC, Nijmegen, the Netherlands. He is working on the 'Optimization of complex palliative care at home by means of expert teleconsultation'-project. He has a degree in Media Studies and Philosophy, and, by now, an extensive experience with qualitative research with vulnerable people. His main interests are patient-provider communication (through technology), empathy, and the ethics of caring for extremely vulnerable people.

Evert van Leeuwen, Ph.D., is head of the section Ethics, Philosophy and History of Medicine of the Scientific Institute for Quality of Healthcare: IQ healthcare at the RadboudUMC Nijmegen. Evert van Leeuwen studied philosophy and mathematics at the Free University (VU) of Amsterdam. In 1986 he graduated cum laude in philosophy on the Ph.D. thesis "Descartes' Regulae."

Harro van Lente is professor 'philosophy of sustainable development, from a humanistic perspective' at Maastricht University, where he studies the relation between needs and novelty. He is also Associate Professor of 'Emerging Technologies' at the Utrecht University, where his research focuses on the dynamics of expectations in nanotechnology, hydrogen and medical technologies.

Ilse van Liempt gained her Ph.D. (2007) at the Institute for Migration and Ethnic Studies, worked as a Marie Curie Postdoc Fellow (2008–2010) at Sussex University and is currently employed as Assistant Professor at Utrecht University in the Human Geography Department. She has published widely on migration, human rights, surveillance, security, public space and qualitative research methods.

Martine Van Selm is an Associate Professor in the Amsterdam School of Communication Research and director of the College of Communication at the University of Amsterdam. Van Selm combines her gerontological expertise on personal meaning in later life (Ph.D. 1998, Department of Psychogerontology, Radboud University Nijmegen) with researching media content and media use. She has published on the use of new media in health care organizations, portrayals of old age-stereotypes in the media, corporate communication and lifelong employability, and on qualitative research methods.

Kris Vissers MD, Ph.D. is anesthesiologist, professor in Pain and Palliative Medicine and chairman of the Academic Center of Pain and Palliative Medicine of the RadboudUMC in the Netherlands. He is President Elect and member of the Executive Board of the World Institute of Pain, Honorary Secretary of the Benelux

Chapter of the World Institute of Pain. His fields of research are proactive care planning, innovative care solutions and patient follow-up systems in chronic pain patients and patients in a palliative trajectory.

Jaap Voeten is a development economist at the Tilburg School of Economics and Management (TiSEM) of Tilburg University specialised in poor small and household business in developing countries. His past long-term working experience in Vietnam, where he lived and worked closely with small producers in rural areas, provided the basis for his Ph.D. research on 'Responsible Innovation' within NWO's thematic program. He produced a series of articles on innovation in small producers' clusters with regard to sustainable business and poverty alleviation.

Dr. Dr.phil. **René von Schomberg** is an agricultural scientist and philosopher. Author/co-editor of 12 books. He holds Ph.D.'s from: The University of Twente, the Netherlands (Science and Technology Studies) and J.W. Goethe University in Frankfurt am Main, Germany (Philosophy). He has been a European Union Fellow at George Mason University, USA in 2007 and has been with the European Commission since 1998. Before joining the Commission he was teaching at Twente University and Tilburg University in the Netherlands.

Dr. **Katinka Waelbers** studied both Natural Sciences and Philosophy, and worked as a bioethicist at Utrecht University (1998–2005). She went to the University of Twente (2005–2010) to focus on philosophy of technology, and continued this research line at the University of Maastricht (2010–2012), before she decided to focus on her artistic career. She authored dozens of articles, scientific reports and books on bioethics and philosophy of technology, and she is the author of the book "Doing Good with Technologies".

Hans Christian Wilms studied law in Freiburg/Germany and holds a Ph.D. degree in law of the Ruprecht-Karls-University of Heidelberg. He worked as a research fellow in a Max Planck Research Group about "Democratic Legitimacy of Ethical Decisions" at the Max Planck Institute for Comparative Public Law and International Law in Heidelberg. Scientific ethical codes in German, European and International Law are the centerpiece of his research and were general topic of his thesis.

Part I
Methodological and Conceptual Issues

Chapter 1
Responsible Innovation: A New Look at Technology and Ethics

Jeroen van den Hoven

Abstract This is the introductory chapter to the first volume in a series of five conference proceedings on *Responsible Innovation*. The conferences bring together the results of research projects under the Research Program "Responsible Innovation" (*Maatschappelijk Verantwoord Innoveren*) of the Dutch Research Council, while at the same time providing a platform for a broad and rapidly growing community of international researchers – inside and outside academia – interested and involved in research and R&D projects in *Responsible Innovation*. Together, the contributions in this volume show that responsible innovation is a dynamic and promising field of research.

1.1 Introduction

This is the first volume in a series of five conference proceedings on *Responsible Innovation*. The proceedings correspond with a series of five conferences organized by the Dutch Research Council (NWO) in the period 2011–2016 in The Hague. The conferences bring together the results of research projects under the Research Program "Responsible Innovation" (*Maatschappelijk Verantwoord Innoveren*)[1] of the Dutch Research Council. At the same time the conferences provide a platform

This contribution draws upon previously material published in Van den Hoven (2013).

[1] See www.nwo.nl/mvi for a complete description of the program and for descriptions of projects funded under this program.

J. van den Hoven (✉)
Department of Values, Technology and Innovation, Delft University of Technology, Jaffalaan 5, 2600GA Delft, The Netherlands
e-mail: m.j.vandenhoven@tudelft.nl

J. van den Hoven et al. (eds.), *Responsible Innovation 1: Innovative Solutions for Global Issues*, DOI 10.1007/978-94-017-8956-1_1,

for a broad and rapidly growing community of international researchers – inside and outside academia – involved in research and R&D projects in Responsible Innovation.[2]

The idea of a program on Responsible Innovation in The Netherlands emerged out of discussions organized by the Dutch Research Council between 2003 and 2007. The program is the result of a unique collaboration in the applied ethics of technology of the Dutch Research Council with several ministries, private sector partners, university based research groups, representatives of NGO's in The Netherlands. The program published a first round of calls for proposals in 2009 and 33 projects were eventually awarded with a total budget of 12 million Euro.[3] In 2014 a second phase of the program is launched by the Dutch Research Council in close cooperation with public and private parties involved in the implementation of Dutch Innovation Policy. 2014 is also the year of the start of the European Horizon2020 70 billion Euro R&D Program of the EU. Responsible Innovation has been included in this program and much of the research efforts dealing with ethics and ethical legal and social aspects will be funded under the label of Responsible Research and Innovation.[4]

The program took its present shape in early discussions, which started in 2003, about a successor to the Applied Ethics Program – entitled *Ethics and Public Policy* (*Ethiek en Beleid*) – of the Dutch Research Council. This program was quite successful in the late 1990s. One of the main aims in extending the effort on the part of the Research Council was to make applied ethics even more societally relevant in a sequel program. The goal of societal relevance was infused with the firm belief that philosophy, ethics and the humanities more generally, can be highly relevant to policy making and the professions.

In thinking about the outline and design of the envisaged program a number of considerations were articulated by a group of researchers brought together by the Department of the Humanities of the Research Council.[5] First of all the choice of the

---

[2]In 2013 A Taylor and Francis Journal on Responsible Innovation was established, see http://www.tandfonline.com/loi/tjri20. Various other research groups were established: see e.g. http://www.debatinginnovation.org/; http://responsible-innovation.org.uk/frriict/; www.responsibleinnovation.eu.

[3]The scientific quality of proposals was judged on the basis of an extensive international peer review process and overlooked by an international Advisory Board chaired by professor Armin Gruenwald. The societal relevance of proposals was assessed by a separate board that focused on the societal relevance of proposals chaired initially by professor Alexander Rinnooy Kan en later professor Jacqueline Cramer).

[4]Rene van Schomberg has greatly contributed to the development of this line of thinking within the EU. See his contributions in Owen et al. (2013) entitled "A Vision of Responsible Research and Innovation". Van den Hoven chaired an EU expert group that published a report entitled "Options for strengthening responsible research and innovation". http://bookshop.europa.eu/en/options-for-strengthening-responsible-research-and-innovation-pbkKINA25766/.

[5]This group was chaired by professor Jeroen van den Hoven and supported by Marlies van der Meent, at a later stage by Jasper Roodenburg. After a first phase of exploration, a research agenda group met three times in 2007. Representatives of the various sections of the Dutch

domain and scope of a new applied ethics program was recognized to be important to achieve *practical relevance* of moral philosophy. At the beginning of the twenty-first century the obvious observation to make was that technology and engineering would change the lives of people deeply. If ethics could make contributions to the improvement of society and human wellbeing anywhere, then technology, engineering and applied science would be a promising place to start. Through material culture, artefacts and devices, infrastructures, chemicals, pharmaceuticals, new materials, energy systems, food production systems, transport systems and computers, internet, the lives of people change dramatically, sometimes for the better and sometimes for the worse. The daily newspapers contain a panoply of examples of prominent ethical issues on robotics, internet and social networking sites, nano-technology, genetics, chemical safety, cyber security, climate change, nuclear power, renewable energy systems, smart cities, and big data, to mention only a few. So a leading question in thinking about the program was: How could applied ethics research be geared towards technological innovations and applied science and engineering in thinking about practical innovative solutions for important social and global problems so as to make a real difference in public policy and decision making[6]?

Secondly, *participation by stakeholders* was considered another important requirement for an applied ethics of technology research program. Researchers should not lose sight of the real world and disappear in the ivory towers of academia only to find upon completion of their project that the world had changed in the meanwhile. Input from civil society, consumer organisations, NGO's, decision makers and politicians, professionals and market parties, therefore was considered to be important for the envisaged program. Representatives of the real world of innovation and technology were invited to help to identify ethical issues, provide

Research Council were Jeroen van den Hoven (Humanities) Bert Jaap Koops (Social Sciences) Arie Rip (Technology Foundation STW), Guido de Wert (The Netherlands Organisation for health Research and Development), Michiel Korthals (WOTRO Science for Global Development). Also representatives of ministries were part of these discussions (Foreign Affairs, Home Office, Defence, Economic Affairs and Agriculture, Education, and Health).

[6]The academic climate in The Netherlands is conducive to this approach. The Netherlands is one of the most innovative countries in the world and it has an internationally recognized and excellent research tradition in the study of Science, Technology and Society. Internationally prominent research groups in Science and Technology Studies, were led in the recent past by Wiebe Bijker, Arie Rip and Hans Achterhuis. Historians of technology Johan Schot and Harry Lintsen have established well regarded research programs in the history of technology and innovation. Law and technology, especially in the field of ICT have done well as a result of the work of Hans Franken (Leiden) en Bernd Hugenholtz (Amsterdam) en Corien Prins (Tilburg). Also the technical universities at Delft, Eindhoven and Twente have produced large research programs and built up considerable research capabilities in this field. They have joined forces in a collaborative 3TU. Ethics Centre in 2007 initiated by Jeroen van den Hoven, Anthonie Meijers en Philip Brey. The former Dutch Office of Technology Assessment – The Rathenau Institute – initially lead by Jose van Eindhoven and later Jan Staman, is a very active contributor in this field and adds to a strong presence in public debates about technology and society in Netherlands and in Europe. An applied ethics of technology program thus is situated in a stimulating intellectual context in Dutch Academia.

input and even form part of the selection process of the grant applications. So-called "Valorization Panels" would provide real world feed-back during the execution of the research project "to keep research real".

In order to be relevant to thinking about innovation and allow interested and affected parties take part in discussion about technology, "technology" should be construed in a broad sense, in terms of 'systems of socio-technical systems' and it should be acknowledged that the social context of technology, the regulatory frameworks, incentive structures, institutional arrangements and governance are of equal importance to understanding technology as the engineering aspects. Reactors of whatever type could be very innovative, but without laws, safety norms, security policies, governance and inspection regimes, they may not be acceptable. Non-technical requirements and constraints can make them acceptable and can make all the difference.

Furthermore, a lesson learned from previous applied ethics programs, which the Dutch Research Council had initiated, was concerned with the *timing of applied ethics research*. The research sometimes was not only disappointing as far as its societal relevance was concerned, but also because results were sometimes delivered at such a late stage in the development of the issues that it could no longer usefully be employed to make a difference. In the case of technology and engineering the "too late" of ethical guidance is particularly problematic, because innovations in the field of infrastructures and technical systems have their own development trajectories, investment cycles and path dependencies. Once technology has been developed or has been introduced in society it is extremely difficult or prohibitively expensive to modify it. The problems need to be tackled in *anticipation* and up-stream engagement. This challenge is related to a problem that is inherent in studying the social aspects of technology called *Collingridge Dilemma*: at the time when we can still make changes to the technology, one lacks the information about effects which only the introduction and use of the technology in society could provide, but at the moment that the technology has been introduced in society and information about its effects and morally salient characteristics starts to become available, it is often very hard to still make changes. We should aim to have results of ethical discussions available at a moment when it can still be used to inform the design, implementation or utilization decisions.

Another aspect of practical adequacy is that these suggestions need to have a form that makes it easy to utilize ethical and social science research and make it bear upon technical and engineering work. Insights from research on values and design (value sensitive design) suggested that value considerations could be construed as "requirements" among other "functional requirements" in design of new technology and systems (Van den Hoven 2005, 2007; Van den Hoven and Manders Huits 2009; Van den Hoven et al. 2012, 2014). This consideration together with the lessons learned about the way material culture, devices, artefacts, technical systems, infrastructures and computer code may contain moral ideas, values, norms, or ideals that were inscribed into them and as such can be carriers or barriers of ethics, give the program a distinctive 'design character' (see Friedman et al. 2002; Friedman 2004; Cummings 2006; for an early proposal see Whitbeck 1996). *Articulation* of values, ideals, norms and rules in the context of innovation is important for number of reasons: first to evaluate technical innovations and new institutional designs, and secondly to expose their often hidden value assumptions, and finally to construe values as requirements for design. These uses of value considerations were all seen as important features of the program.

The Responsible Innovation Program originated in the division of Humanities of the Dutch Research Council and grew into a multidisciplinary collaborative scheme, but the desideratum of empirically informed research with explicit normative purchase was retained throughout the discussion and shaping of the program. The desired normative and ethical purchase of the research needs to draw upon an analysis of the problems that is empirically informed by a number of other disciplines. Real world problems in their complexity almost always require multidisciplinary approaches and hardly ever have their solution in one particular specialism or discipline. Solutions to the UN Millennium Problems or Grand Challenges are bound to require expertise from the natural sciences and engineering sciences, the social and behavioural sciences and the humanities. The research program therefore made it a necessary condition for receiving funding that humanities, social science (law and sociology, psychology, economics), applied science and engineering and technological perspectives were all well represented in each project.

The Netherlands has learned some interesting lessons about societal and ethical aspects of innovation in the first decade of the twenty-first century. A first instructive case was the attempt to introduce smart electricity meters nation- wide. In order to make the electricity grids more efficient and meet the EU CO_2 reduction targets by 2020, every household in The Netherlands would have to be transformed into an intelligent node in the electricity network. Each household could thus provide detailed information about electricity consumption and help electricity companies to predict peaks and learn how to "shave off" the peaks in consumption patterns. After some years of R&D, a plan to equip every Dutch household with a smart meter was proposed to parliament. In the meantime however, opposition to the proposal by privacy groups had gradually increased over the years (Abdulkarim 2009). The meter was now seen as a 'spying device' and considered a threat to the personal sphere of life and privacy of families, because it could take snapshots of electricity consumption in the household, store data in a database of the electricity companies for data mining and provide detailed information about what was going on inside the homes of Dutch citizens. By the time the proposal was brought to the upper house of the Dutch parliament for approval, public concern about the privacy aspects was very prominent and the upper house rejected the plan on data protection grounds. The European Commission, being devoted to the development of smart electricity grids in its member states, feared that the Dutch reaction to this type of innovation would set an example for other countries and would jeopardize the EU wide adoption of sustainable and energy saving smart grid solutions in an EU market for electricity (Abdulkarim 2009).

Another story – not very different from that of the smart meter – is the introduction of a nation-wide electronic patient record system in The Netherlands. After 10 years of R&D and preparations, lobbying, stakeholder consultation and debates – and last but not least an estimated investment of 300 million Euro – the proposal was rejected by the upper house in parliament on the basis of privacy and security considerations (Van Twist 2012).

Clearly these innovations in the electricity system and health care system could have helped The Netherlands to achieve cost reduction, greater efficiency, sustainability goals, and in the case of the electronic Patient Record System, higher levels of patient safety. In both cases however privacy considerations were not sufficiently incorporated in the plans so as to make them acceptable. If the engineers had taken privacy and security of patient data more seriously right from the start and if they had made greater efforts to incorporate and express the value of privacy into the architecture at all levels of the system, transparently and demonstrably, then these problems would probably not have arisen.

Two European cases can serve as a contrast with the two aforementioned Dutch failures in innovation. They show that early and serious attention to moral considerations in design and R&D may not only have good moral outcomes, but may also lead to good economic outcomes. Consider the case of so-called 'privacy enhancing technologies'. The emphasis on data protection and the protection of the personal sphere of life is reflected in demanding EU data protection laws and regulation. The rest of the world has always considered the preoccupation with privacy as a typically European political issue. As a result of the sustained and systematic attention to data protection and privacy Europe has become an important cradle of new products and services in the field of Privacy by Design or Privacy Enhancing Technologies. Now the Big Data society is on our doorstep and many computer users– also outside Europe are starting to appreciate products and services that can accommodate user preferences and values concerning privacy, security and identity, Europe has a competitive advantage and is turning out to be an important commercial player in this branch of the IT industry.

A second case concerns Germany's success in development of sustainability technology. Germany is one of the leading countries in the world in sustainability technology. During the twentieth century, in the 1960s and 1970s, the world felt sorry for West Germany. Members of the Green Party seemingly frustrated economic growth by means of their disruptive protests. The conflict between economic growth and sustainability was a genuine value conflict that divided the political landscape and led to tensions in society. But in hindsight the conflict between different value orientations seems to have stimulated innovation instead of having stifled it. The conflict and political tension formed the occasion and trigger for Germany to try to have the cake and eat it. The environmental technology that they felt the need to develop in the past has laid the foundation for commercial successes in the future.

The important lesson to learn from both the two Dutch cautionary tales as well as the two positive European cases is that values and moral considerations (i.e. privacy considerations) should have been taken into account as "non-functional requirements" at a very early stage of the development of the system, alongside with the functional requirements, e.g. storage capacity, speed, bandwidth, compliance with technical standards and protocols. A real innovative design for an Electronic

Patient Record System or a truly smart electricity meter, would thus have anticipated or pre-empted moral concerns and accommodated them into its design, reconciling efficiency, privacy, sustainability and safety. Value – focused thinking at the early stages of development at least might have helped engineers to do a better job in this respect. There is a range of fine grained design features that could have been considered and that could have been presented as choices for consumers. A smart meter is not a given, it is to a large extent what we design and make it to be. Respect for privacy can be built in (Garcia and Jacobs 2011; Jawurek et al. 2011). The question of course immediately presents itself as to which values should be used to inform the design of out technologies and innovations and how exactly? This fundamental ethical problem will not disappear. But two conditions brought about by this Responsible Innovation as envisaged in the RI program of the Dutch Research Council, facilitate debates. First of all, they situate them in a rich context and in a form where they become more amenable to study and informed debate and the other condition is that now explicitly an 'empirical cycle' is introduced in the field of ethics and societal debate, where there was none in the first place. Basic value choices are operationalized, specified and functionally decomposed and result in (non-functional) design requirements. Designs are then proposed that claim to implement and satisfy the requirements. We can then not only check and demonstrate that this is actually the case, but moreover we can see whether the implementation of the values we started out with – supported by independent good moral reasons – have the desired consequences. If not, we can revisit our value vantage points and adjust them in light of our experiences with the innovation and new technologies.

Innovation can thus take the shape of (engineering) design solutions to situations of moral overload (Van den Hoven et al. 2012). One is morally overloaded when one is burdened by conflicting obligations or conflicting values, which cannot be realized at the same time. But as we saw above, conflicts of privacy and national security seem amenable to resolution by design and innovation in the form of privacy enhancing technologies. Conflicts between economic growth and sustainability were resolved by sustainability technology. Some think of these solutions as mere "technical fixes" and not as real solutions to moral problems. I do not take a stance on this issue. I just want to point out that in such cases it seems to me that we have an obligation to bring about the required change by design or innovation (Van den Hoven et al. 2012).

It may seem fairly obvious to claim that we have a higher order moral obligation to innovate when it leads to moral progress, but it requires a considerable shift in our thinking about innovation. First of all we need to learn to think about innovation in light of broad sets of values and moral considerations. Furthermore we have to be able to turn moral values into requirements for design and research and development at an early stage. We also need to involve those who will be affected by the innovations and construe innovations as going beyond quarterly revenues, quick wins and for profit motives. Innovation thus becomes a moral category and is as such concerned primarily with the amplification of the set of obligations we can satisfy. Innovation aims at bringing about changes in the world so that we can fulfil more of our obligations regarding the fellow human beings, the environment, life on the planet, and future generations.

1.2 Structure of the Book

This volume contains a selection of the papers presented at the first conference on Responsible Innovation organized by the Dutch Research Council. The overall theme of this first conference was "Innovative Solutions for Global Problems." This theme is reflected in the different chapters in this volume.

The volume is divided into four parts. Part I is dedicated to methodological and conceptual issues. Following this introductory chapter, Part I contains the two keynote lectures by Armin Grunwald and René von Schomberg respectively. Armin Grunwald (Chap. 2) shows how responsible innovation has its roots in TA – with its experiences on assessment procedures, actor involvement, foresighting – but that it shares an evaluative component with engineering ethics, in particular under the framework of responsibility. Based on his work at the European Union, René von Schomberg (Chap. 3) proposes a framework for Responsible Research and Innovation, operationalizing the general consensual normative anchor points derived from the European Treaties. He argues that, in order to drive innovations towards the 'Grand Challenges' of our time, innovation governance should move far beyond the means of solely market-driven innovations.

Part II is dedicated to governance issues and institutional design. Part II starts with a contribution by Xavier Pavie and Julie Egal (Chap. 4), in which the concept of innovation is elaborated and related to the concept of responsibility. Eppinger and Tinnemanns (Chap. 5) discuss equitable licensing and patent pools to improve technology dissemination of publicly funded research results. Hans Christian Wilms (Chap. 6) discusses how the legal validity of ethical codes can be improved. On the basis of an analysis of recurring epistemic, moral, and socio-economic failures in current biomedical research, Jan De Winter (Chap. 7) evaluates some policy proposals for biomedical research. Similar to Eppinger and Tinnemanns, De Winter stresses the importance of making available the outcomes of publicly funded research.

Values are the common denominator in Part III of the book. The chapters in this part discuss the role of values in a globalizing world and this may force us to rethink our notion of innovation. The concept of responsible innovation is developed in a western context. Several of the contributions in Part III take up the challenge to see whether the concept of responsible innovation can also be applied to context of developing countries. Annemarie Mink et al. (Chap. 8) look at responsible product innovation in India. They show how the capability approach, initially developed by economics Nobel laureate Amartya Sen, can support product designers in their efforts to attune their design to the needs of the poor. Jaap Voeten et al. (Chap. 9) conceptualize responsible innovation in craft villages in Vietnam.

They found that, at the village level, it is better to model responsible innovation as a dynamic societal process. The key question is to what extent innovators assume responsibility for the harmful outcomes of innovation and the resolution of the negative ones. Otto Kroesen and Wim Ravesteijn (Chap. 10) look at the relation between culture and values and they argue that value reorientations should be an integral and explicit part of the development agenda if sustainable results are to be attained. Udo Pesch (Chap. 11) relates the notion of innovation to learning and responsibility. His analysis shows that effective responsibility arrangements require the restoration of institutional domains. In the last contribution in Part III, Katinka Waelbers and Tsjalling Swierstra (Chap. 12) show how technologies may lead to moral change. Responsible innovation should therefore not focus one-sidedly on risks, but also on how new technologies may adversely affect and shape our life.

Part IV in the book deals with concrete technological developments. The first half of Part IV is dedicated to case studies and applications in healthcare and the medical sector. These include contributions on Alzheimer's disease (Yvonne Cuijpers et al.), neuroimaging (Marlous E. Arentshorst et al.; Chap. 14), teleconsultation in palliative care (Jeroen Hasselaar et al.; Chap. 15), video recording in the operating room (Claire B. Blaauw et al.; Chap. 16), and Ambient Assisted Living technologies (Neelke Doorn; Chap. 17). On the basis of a study of the scientific and clinical uncertainties in Alzheimer's disease, Yvonne Cuijpers et al. describe responsible innovation in terms of six 'quandaries': problematic, difficult and ambiguous conditions that somehow require fundamental and practical decisions. These six quandaries may help both researchers and policy makers in becoming aware of the available options and in making their choices more explicit. Chapter 14 is dedicated to neuroimaging technologies. These technologies are expected to provide more insight in both the healthy brain and brain disorders, which will accordingly lead to improved prevention, diagnosis and treatment options. Marlous E. Arentshorst et al. analyze what is required to make neuroimaging technologies live up to these promising expectations. Jeroen Hasselaar et al. present the results of a randomized control trial on the effectiveness of tele-consultation in complex palliative homecare. They explain how collaboration between primary care and hospital care at the "digital" work floor may optimize continuity of care and, in the ideal case, even improve patient participation. In their contribution on a video monitoring system in health care, Claire B. Blaauw et al. show how promising technologies – technologies of which the value has been widely acknowledged – may prompt important legal questions, in this case on the privacy of both patients and medical professionals. They emphasize that these legal questions should be solved prior to implementation of the technologies. Neelke Doorn (Chap. 17) shows how technical researchers tend to make a sharp distinction between the technology they develop and its application; the former supposedly being "neutral." Responsible innovation requires that the gap between applications and technologies be bridged, Doorn argues.

The second half of Part IV is dedicated to case studies and applications in ICT and military technology. Irina van Aalst et al. (Chap. 18) discuss the use of a video surveillance system (CCTV) in urban nightlife districts. The authors show that the benefits of CCTV tend to be overestimated and that the people affected hardly experience an enhanced feeling of safety and wellbeing. They argue for more in-depth investigations of the ambiguous relationships between surveillance and policing, and between wellbeing and exclusion in urban nightlife districts. Bart Custers and Bart Schermer (Chap. 19) discuss the topic of data mining and profiling tools. They show that previous attempts to protect privacy and prevent discrimination in data mining, focused on the wrong question. They show that the important question in data mining and profiling tools should be the question how data can and may be used rather than how access to data can be limited. The contribution by Lambèr Royakkers and Anya Topolski (Chap. 20) is on the role of military robots in modern warfare. They argue that the minimal criteria for ethical decision making in military ethics are twofold: non-binary thinking and reflexivity. In order to respond to the moral questions and dilemmas that will be faced by future military operations using robots, these two criteria are the threshold criteria that need to be fulfilled. In the last contribution to this book, Janneke van de Zwaan et al. (Chap. 21) discuss how technologies can be used to regulate anti-social online behavior such as cyber bullying. The authors develop a tentative set of criteria to assess the effectiveness of internet safety technologies.

Together, the contributions in this volume show that responsible innovation is a dynamic field where still a lot of work needs to be done. Many of the issues explored here require further conceptual investigations and new methodologies. With this first book, we hope to have made a valuable contribution to the fastly growing body of literature on responsible innovation.

References

Al Abdulkarim, Layla O., and Z. Lukszo. 2009. Smart metering for the future energy systems in the Netherlands. In *Proceedings of the fourth international conference on critical infrastructures, 2009 (CRIS 2009)*, March 27–April 30 (2009), 1–7. Linköping University, Sweden. [s.l.]: IEEE. ISBN 978-1-4244-4636-0.

Cummings, M.L. 2006. Integrating ethics in design through the value-sensitive design approach. *Science and Engineering Ethics* 12: 701–715.

Friedman, B. 2004. Value sensitive design. In *Berkshire encyclopedia of human-computer interaction*, ed. W.S. Bainbridge. Great Barrington: Berkshire Publishing Group.

Friedman, B., P. Kahn, and A. Borning. (2002). *Value sensitive design: Theory and methods*. Technical Report 02-12-01. University of Washington, Seattle, USA

Garcia, F., and B. Jacobs. 2011. Privacy friendly energy metering via homomorphic encryption. *Lecture Notes in Computer Science* 6710: 226–238.

Jawurek, M., et al. 2011. Plug-in privacy for smart metering billing. *Lecture Notes in Computer Science* 6794: 192–210.

Owen, Richard, et al. (eds.). 2013. *Responsible innovation*. Oxford: Wiley Blackwell.

Van den Hoven, Jeroen. 2005. Design for values and values for design. *Information Age +, Journal of the Australian Computer Society* 7(2): 4–7.

Van den Hoven, Jeroen. 2007. ICT and value sensitive design. In *The information society: Innovation, legitimacy, ethics and democracy*, ed. Philippe Goujon et al., 67–73. Dordrecht: Springer.

Van den Hoven, Jeroen. 2013. Responsible innovation and value sensitive design. In *Responsible Innovation*, ed. Richard Owen, et al., 75–85. Oxford: Wiley Blackwell.
Van den Hoven, Jeroen, and Noemi Manders Huits. 2009. Value sensitive design. In *A companion to the philosophy of technology*. eds. Jan Kyrre Berg Olsen, et al., Chichester: Blackwell.
Van den Hoven, M.J., G.J. Lokhorst, and I.R. Van de Poel. 2012. Engineering and the problem of moral overload. *Science and Engineering Ethics* 18(1): 143–155.
Van den Hoven, Jeroen, et al. (eds.). 2015, Forthcoming. *The design turn in applied ethics*. Cambridge: Cambridge University Press.
Van Twist, M.J.W. 2012. *Het EPD Voorbij*. Den Haag, februari 2012. https://www.eerstekamer.nl/eu/overig/20120207/rapport_het_epd_voorbij_evaluatie
Whitbeck, Caroline. 1996. Ethics as design: Doing justice to moral problems. *The Hastings Center Report* 26(3): 9–16.

Chapter 2
Technology Assessment for Responsible Innovation

Armin Grunwald

Abstract The ideas of 'responsible development' in the scientific-technological advance and of 'responsible innovation' in the field of new products, services and systems have been discussed for some years now with increasing intensity. Some crucial ideas of Technology Assessment (TA) are an essential part of these debates which leads to the thesis is that TA is one of the main roots of Responsible Innovation. This can be seen best in the effort which has recently been spent to early and upstream engagement at the occasion of new and emerging science and technology. However, Responsible innovation adds explicit ethical reflection to TA and merges both into approaches to shaping technology and innovation: Indeed, the field of the ethics of responsibility and its many applications to the scientific and technological advance is the second major root of Responsible Innovation. Responsible Innovation brings together TA with its experiences on assessment procedures, actor involvement, foresighting and evaluation with engineering ethics, in particular under the framework of responsibility. The chapter describes both, TA and engineering ethics, as origins of 'Responsible Innovation'.

2.1 Introduction and Overview

The advance of science and technology has for decades been accompanied by debates in society and science on issues of risks and chances, potentials and side effects, control and responsibility. Approaches such as Technology Assessment (Decker and Ladikas 2004; Grunwald 2009), social shaping of technology (Yoshinaka et al. 2003), science and engineering ethics (Durbin and Lenk 1987) and Value Sensitive Design (van de Poel 2009) have been developed and are practiced to a

A. Grunwald (✉)
Karlsruhe Institute of Technology, Karlstr. 11, 76133 Karlsruhe, Germany
e-mail: armin.grunwald@kit.edu

J. van den Hoven et al. (eds.), *Responsible Innovation 1: Innovative Solutions for Global Issues*, DOI 10.1007/978-94-017-8956-1_2,

certain extent. All of them have a specific focus, particular theoretical foundations, different rationales, and have been conceptualised for meeting differing challenges and context conditions. All of them also show strengths and weaknesses and specific limitations to application. Therefore, the search for new and better concepts is ongoing – and will, probably, never come to an end. The field of interest – scientific and technological advance – continuously creates new developments with new challenges to analysis, assessment and debate leading to the demand for new conceptual and methodological approaches.

The ideas of 'responsible development' in the scientific-technological advance and of 'responsible innovation' in the field of new products, services and systems have been discussed for some years now with increasing intensity. The technology field in which most of this development took place has been nanotechnology. One of the many examples where responsible development and innovation in this field are postulated is:

> Responsible development of nanotechnology can be characterized as the balancing of efforts to maximize the technology's positive contributions and minimize its negative consequences. Thus, responsible development involves an examination both of applications and of potential implications. It implies a commitment to develop and use technology to help meet the most pressing human and societal needs, while making every reasonable effort to anticipate and mitigate adverse implications or unintended consequences. (National Research Council 2006, p. 73)

This request takes up formulations well known from the history of Technology Assessment (TA) (Grunwald 2009). However, there are new accentuations, shifts of emphasis and some new aspects. My thesis is that TA is one of the main roots of Responsible Innovation (Sect. 2.2). Based on earlier experiences with new technologies such as genetic engineering and with corresponding moral and social conflicts, a strong incentive is to 'get things right from the very beginning' (Roco and Bainbridge 2001).

Early engagement has received increasing awareness in TA over the past decade mainly at the occasion of debates on new and emerging science and technology (NEST) such as nanotechnology, nano-biotechnology and synthetic biology. These fields of development show a strong "enabling character" and will probably lead to a manifold of applications in different areas which are extremely difficult to anticipate. This situation makes it necessary – from a TA perspective – to shape TA as an *accompanying process* reflecting on the ethical, social, legal and economic issues at stake. This process should start in early stages of research and development in order to deal constructively with the Control Dilemma (Collingridge 1980). The notion of "real-time TA" partially refers to this challenge (Guston and Sarewitz 2002).

Responsible innovation adds explicit ethical reflection to this "upstream movement" of TA and includes both into approaches to shaping technology and innovation: The field of the ethics of responsibility and the many applications to the scientific and technological advance is the second major root of Responsible Innovation (see Sect. 2.3). Responsible Innovation brings together TA with its experiences on assessment procedures, actor involvement, foresighting and evaluation with engineering ethics, in particular under the framework of responsibility. Ethical reflection and technology assessment, until recently undertaken more at a distance from R&D and innovation, are increasingly taken up as integrative part of R&D programmes (Siune et al. 2009). Science institutions, including research funding agencies, have started taking a pro-active role in promoting integrative research and development. Thus, the governance of science and of R&D processes is changing which opens up new possibilities and opportunities for involving new actors and new types of reflection.

This paper aims at unfolding the theses briefly outlined above. Short introductions into TA (Sect. 2.2) and the notion of responsibility (Sect. 2.3) are required to characterize Responsible Innovation and to identify its innovative aspects (Sect. 2.3).

2.2 Technology Assessment – Roots and Concepts[1]

Technology Assessment (TA) emerged in the 1970s as a science-based and policy-advising activity (Bimber 1996). In its first period technology was regarded to follow its own dynamics (technology determinism) with the consequence that the main task of TA was seen in its early-warning function in order to enable political actors to undertake measure to, for example, compensate of prevent anticipated negative impacts of technology. The dimension of research and development at the lab level was not addressed at all at that time. This changed completely during the 1980s following the social constructivist paradigm leading to the slogan "shaping of technology" (Bijker et al. 1987; Bijker and Law 1994). By following this framework the approach of Constructive Technology Assessment (CTA) was developed (Rip et al. 1995). CTA began to consider activities at the lab level and in innovation processes (Smits and den Hertog 2007). TA for orientating giving shape to new technology and possibly resulting innovations is since then part of the overall TA portfolio reaching from the political, in particular parliamentarian, level far away from the lab up to concrete intervention in engineering, design and development at the level of research programmes and the concrete work at the lab.

2.2.1 The Demand for TA and Its Development Over Time

In the twentieth century, the importance of science and technology in almost all areas of society (touching on economic growth, health, the army, etc.) has grown

---

[1]This Section summarizes the description of TA to be published in the Handbook "Design for Value" (ed. Ibo van de Poel, forthcoming) focusing on its relevance to Responsible Innovation. For a general and more detailed introduction into TA see Grunwald (2009).

dramatically. Concomitant with this increased significance, the consequences of science and technology for society and the environment have become increasingly serious. Technological progress alters social traditions, fixed cultural habits, relations of humans and nature, collective and individual identities and concepts of the self while calling into question traditional moral norms. Decisions concerning the pursual or abandonment of various technological paths, regulations and innovation programs, new development plans, or the phasing-out of lines of technology often have far-reaching consequences for further development. They can influence competition in relation to economies or careers, trigger or change the direction of flows of raw materials and waste, influence power supplies and long-term security, create acceptance problems, fuel technological conflict, challenge value systems and even affect human nature.

Since the 1960s adverse effects of scientific and technical innovations became obvious some of them were of dramatic proportions: accidents in technical facilities (Chernobyl, Bhopal, Fukushima), threats to the natural environment (air and water pollution, ozone holes, climate change), negative health effects as in the asbestos case, social and cultural side effects (e.g., labour market problems caused by productivity gains) and the intentional abuse of technology (e.g. the attacks on the World Trade Centre in 2001). The emergence of such unexpected and serious negative impacts of technology is central to TA's motivation. Indeed, in many cases, it would have been desirable to have been warned about the disasters in advance, either to prevent them, or to be in a position to undertake compensatory measures.

Early warning in this sense is a necessary precondition to make societal and political *precautionary action* possible: how can a society which places its hopes and trust in innovation and progress, and must continue to do so in the future, protect itself from undesirable, possibly disastrous side effects, and how can it preventatively act to cope with possible future adverse effects? Classic problems of this type are, for example, the use and release of new chemicals – the catastrophic history of asbestos use being a good example (Gee and Greenberg 2002) – and dealing with artificial or technically modified organisms (for further examples, cf. Harremoes et al. 2002). In order to be able to cope rationally – whatever this could mean in a concrete context – with these situations of little or no certain knowledge of the effects of the use of technology, prospective analysis and corresponding procedures for societal risk and chance management are required and have been developed such as the Precautionary Principle (von Schomberg 2005).

Parallel to these developments, broad segments of Western society were confronted with predictions of "Limits of Growth" (Club of Rome) in the 1970s which, for the first time, addressed the grave environmental problems perceived as a side effect of technology and economic growth. The optimistic pro-progress assumption that whatever was scientifically and technically new would definitely benefit the individual and society was challenged. As of the 1960s deepened insight into technological ambivalence led to a crisis of orientation in the way society dealt with science and technology. This (persistent!) crisis forms the most essential motivation of the emergence of TA.

New and additional motivations entered the field of TA over the past decades, leading more and more to a shift from the initial emphasis on early warning towards "shaping technology" according to social values:

- *Concerns of an emerging technocracy*: from the 1960s on there have been concerns that the scientific and technological advance could threaten the functioning of democracy because only few experts were capable of really understanding the complex technologies (Habermas 1970). The technocracy hypothesis was born painting a picture of a future society where experts would make the decisions with respect to their own value systems. One of the many origins of TA is to counteract and to enable and empower society to take active roles in democratic deliberation on science and technology (von Schomberg 1999).
- *Experiences of technology conflicts and of legitimacy deficits*: little acceptance of some political decisions on technology (such as on nuclear power in some countries), doubts about their legitimacy and resulting conflicts motivated TA to think about procedures of conflict prevention and resolution, in particular including participatory approaches (Joss and Belucci 2002).
- *Shaping technology according to social values*: In addition to the idea of procedural approaches to legitimisation issues and conflicts (see above) the approach was born to design technology according to social values – if this would succeed, so the hope, problems of rejection or non-acceptance would no longer occur at all, and a "better technology in a better society" (Rip et al. 1995) could be reached. This line of thought seems to be one of the main sources of Responsible Innovation.
- *Innovation issues*: in the past two decades innovation problems of Western societies became obvious. Related with new political efforts and incentives towards innovation TA was faced with new themes, tasks and motivations. TA was increasingly considered part of regional and national innovation systems (Smits and den Hertog 2007). It also has been expected to contribute to Responsible Innovation (Siune et al. 2009).
- *Shift in the societal communication on new and emerging science and technology (NEST)*: techno-visionary sciences such as nanotechnology, converging technologies, enhancement technologies and synthetic biology entered the arena. The widespread use of visions and metaphors marks the expected revolutionary advance of science in general and became an important factor in societal debates (Grunwald 2007; Selin 2007)

Compared to the initial phase of TA a considerable increase of its diversity and complexity can be observed. In modern TA, it is often not only a question of the consequences of individual technologies, products, or plants, but frequently of complex conflict situations between enabling technologies, innovation potentials, fears and concerns, patterns of production and consumption, lifestyle and culture, and political and strategic decisions (Bechmann et al. 2007; Grunwald 2009; von Schomberg 2012).

2.2.2 TA Approaches and Concepts

Technology Assessment (TA) constitutes an interdisciplinary research field aiming at, generally speaking providing knowledge for better-informed and well-reflected decisions concerning new technologies (Grunwald 2009). Its initial and still valid motivation is to provide answers to the emergence of unintended and often undesirable side effects of science and technology (Bechmann et al. 2007). TA shall add reflexivity to technology governance (Aichholzer et al. 2010) by integrating any available knowledge on possible side effects at an early stage in decision-making processes, by supporting the evaluation of technologies and their impact according to societal values and ethical principles, by elaborating strategies to deal with the uncertainties that inevitably arise, and by contributing to constructive solutions of societal conflicts. There are four partially overlapping branches of TA addressing different targets in the overall technology governance: TA as policy advice, TA as medium of participation, TA for shaping technology directly, and TA in innovation processes:

1. TA has initially been conceptualised as *policy advice* (Bimber 1996; Grunwald 2009). The objective is to support policymakers in addressing the above-mentioned challenges by implementing political measures such as adequate regulation (e.g. the Precautionary Principle), sensible research funding and strategies towards sustainable development involving appropriate technologies. In this mode of operation TA does not *directly* address technology development but considers the *boundary conditions* of technology development and use. *Parliamentary* TA is a sub-category of policy-advising TA presupposing that parliaments play a crucial or at least an important and relevant role in technology governance. In an analysis of the roles of parliamentary TA in technology governance based on a theory of institutions, a variety of possible combinations of different institutional configurations occurs (Cruz-Castro and Sanz-Menendez 2004), which is also enriched by the characteristics of the democratic institutions of a nation state and various political traditions (Vig and Paschen 1999).
2. It became clear during the past decades that citizens, consumers and users, actors of civil society, stakeholders, the media and the public are also engaged in technology governance in different roles. Participatory TA developed approaches to involve these groups in different roles at different stages in technology governance (Joss and Belucci 2002). According to normative ideas of deliberative democracy the assessment of technology should be left neither to the scientific experts (expertocracy) nor to the political deciders alone (decisionism) (see Habermas 1970 to this distinction). Participative TA procedures are deemed to improve the practical and political legitimacy of decisions on technology. The participation of citizens and of those affected is believed to improve the knowledge basis as well as the values fundament on which judgements are based and decisions are made. Participation should make it possible for decisions on

technology to be accepted by a larger spectrum of society despite divergent normative convictions. Several approaches and methods have been developed and applied in the recent years, such as consensus conferences, citizens' juries, and focus groups (Joss and Belucci 2002).

3. Building on research on the genesis of technology made in the framework of social constructivism (Bijker et al. 1987) the idea of *shaping technology* due to social expectations and values came up and motivated the development of several approaches such as Constructive TA (CTA) or Social Shaping of Technology (Yoshinaka et al. 2003). They all aim at increasing reflexivity in technology development and engineering by addressing the level of concrete products, systems and services, going for a "better technology in a better society" (Rip et al. 1995). In the engineering sciences, the challenges with which TA is confronted have been discussed as demands on the profession of engineers. Within the various approaches which can be subsumed under the social constructivist paradigm, the impact of those activities is primarily seen in the field of technology itself: ethical reflection aims to contribute to the technology paths, products and systems to be developed (Yoshinaka et al. 2003).
4. Since the 1990s, new challenges have arisen. In many national economies, serious economic problems have cropped up, which have led to mass unemployment and to the accompanying consequences for the social welfare systems. Increased innovativeness is said to play a key role in solving these problems. On the basis of this analysis, new functions have been ascribed to TA within the scope of innovation research (Smits and den Hertog 2007). Its basic premise is to involve TA in the design of innovative products and processes. This is because innovation research has shown that scientific-technical inventions do not automatically lead to societally relevant and economically profitable innovations. The "supply" from science and technology and the societal "demand" do not always correspond. This means that more attention has to be paid to more pronouncedly orienting towards society's needs within the scientific-technical system, the diffusion of innovations and the analysis of opportunities and constraints. There is a shift of emphasis from "shaping technology" to "shaping innovation".

From its very beginning TA has been confronted with expectations to contribute to research, development and innovation by adding reflexivity, by including perspectives different from those of scientists, engineers and managers, by taking into account (even uncertain) knowledge about consequences and impacts of new science and technologies, and by transforming all these elements into advice to policymakers and society. Responsible innovation draws on the body of knowledge and experience provided by TA's history over decades – but also extends the scope of consideration to ethical issues, in particular to issues of responsibility. In this sense, there is a second major origin of Responsible Innovation: the fields of ethics of responsibility which will shortly been described in the following section.

2.3 Engineering Ethics and the Issue of Responsibility[2]

The broader debate on the ethics of technology and in particular on the responsibility of engineers started in the 1960s, around some issues of non-intended side-effects of technology, primarily in the field of environmental problems. However, it had long been a matter of controversy whether science and engineering have any morally relevant content at all. Until into the 1990s, technology was frequently held to be *value neutral*. Numerous case studies have, however, since recognized the normative background of decisions on technology and made it a subject of reflection (van de Poel 2009). The basic assumption in this transition is that technology should not be viewed solely as a sum of abstract objects or processes, but that the fact should be taken seriously that it is embedded in societal processes (Rip et al. 1995). There is no "pure" technology in the sense of a technology completely independent of this societal dimension. Technology is thus inherently morally relevant, particularly concerning its purposes and goals, the measures and instruments used, and the evolving side effects. Therefore, technology is an appropriate subject for reflections on responsibility (Jonas 1979; Durbin and Lenk 1987).

This is also true of science. The value neutrality of science was postulated in the era of positivism. Since then, there have been many developments that lead one to think about the ethical aspects of science and about science as being subject to human responsibility. Science – analogously to technology – is not operating in an abstract space and does not work by contemplating about how nature works; it is rather involved in societal purposes and strategies: it is science *in* society (Siune et al. 2009). Scientific knowledge not only explains nature but also delivers knowledge for action, manipulation, and intervention. In particular, 'explaining nature' often requires certain types of – mostly technical – intervention.

Consequently, the concept of responsibility has been used repeatedly in connection with scientific and technological progress in the past two to three decades (Durbin and Lenk 1987). It associates ethical questions regarding the justifiability of decisions in and on science and technology with the possible actions of concrete persons and groups and with the challenges posed by uncertain knowledge of the consequences. As a consequence, several commitments of engineering associations to social and moral responsibility were made. Codes of conduct are now established in several associations. On example is the system of engineering values identified by VDI (German Engineering Association) (VDI 1991).

In usages of the notion of responsibility a more or less clear meaning of this notion is mostly simply supposed. "Responsibility" seems to be an everyday word not needing an explanation. However, this might be a misleading assumption, at least in the field of science and technology. A more in-depth view at the concept of responsibility is needed (following Grunwald 1999). Responsibility is result of

---

[2]This brief review of the ethics of responsibility and its role for technology follows my paper to be published in Paslack et al. (2011).

an act of attribution, either if actors attribute the quality to themselves or if the attribution of responsibility is made by others. The attribution of responsibility is itself an act that takes place relative to *rules of attribution* (on this also see Jonas 1979, p. 173). The attribution of responsibility as an active process makes clear that assignments and attributions of responsibility take place in concrete social and political spaces involving and affecting concrete actors in concrete constellations.

The notion of responsibility often is characterized by reconstructions making the places in a sentence explicit which must be filled in to cover the intentions valid in a particular responsibility context (Lenk 1992). A four-place reconstruction seems to be suitable for discussing issues of responsibility in scientific and technical progress:

- *someone* (an actor, e.g. a synthetic biologist) assumes responsibility for
- *something* (such as the results of actions or decisions, e.g. for avoiding bio-safety or bio-security problems) relative to a
- *body of rules* (in general the normative framework valid in the respective situation (Grunwald 2012, Ch. 3; e.g. rules given in a Code of Conduct) and relative to the
- *quality of available knowledge* (knowledge about the consequences of the actions: deterministic, probabilistic, possibilistic knowledge or mere speculative concerns and expectations; cp. von Schomberg 2005 in the context of the Precautionary Principle).

While the first two places are, in a sense, trivial in order to make sense of the word "responsible", the third and fourth places open up essential dimensions of responsibility: the normative rules comprise principles, norms and values being decisive for the judgment whether a specific action or decision is regarded responsible or not – this constitutes the *moral dimension* of responsibility. The knowledge available and the quality of the knowledge including all the uncertainties form its *epistemic dimension*. Reminding the initial observation that the attribution of responsibility is a socially and politically relevant act and influences the governance of the respective field, it comes out as a main result that *all* three dimensions must be considered in prospective debates over responsibility in science and technology:

- the *socio-political dimension* of responsibility mirrors the fact that the attribution of responsibility is an act done by specific actors and affecting others. Attributing responsibilities must, on the one hand, take into account the possibilities of actors to influence actions and decisions in the respective field. On the other, attributing responsibilities has an impact on the *governance* of that field. Relevant questions are: How are the capabilities to act and decide distributed in the field considered? Which social groups are affected and could or should help decide about the distribution of responsibility? Do the questions under consideration concern the "polis" or can they be delegated to groups or subsystems? What consequences would a particular distribution of responsibility have for the governance of the respective field?
- the *moral dimension* of responsibility is reached when the question is posed as to the *body of rules according to which* responsibility *should* be assumed. These rules form the normative context for judging acts to be responsible or

not. Insofar as normative uncertainties arise (Grunwald 2012), e.g., because of moral conflicts, ethical reflection on these rules and their justifiability is needed. Relevant questions are: What criteria allow distinguishing between responsible and irresponsible actions and decisions? Is there consensus or controversy on these criteria among the relevant actors? Can the actions and decisions in question be justified with respect to the rules, values and ethical principles?
- the *epistemic* dimension asks for the quality of the knowledge about the subject of responsibility. This is a relevant issue in debates on scientific responsibility because frequently statements about impacts and consequences of science and new technology show a high degree of uncertainty (von Schomberg 2005). The comment that nothing else comes from "mere possibility arguments" (Hansson 2006) is an indication that in debates over responsibility it is essential that the status of the available knowledge about the futures to be accounted for is determined and is critically reflected from epistemological points of view. Relevant questions are: What is really known about prospective subjects of responsibility? What could be known in case of more research, and which uncertainties are pertinent? How can different uncertainties be qualified and compared to each other? And what is at stake if worse comes to worst?

Debates over responsibility in technology and science frequently are restricted to level (b) and treat exclusively the *ethics* of responsibility. My hypothesis is that the familiar allegations of being simply appellative, of epistemological blindness, and of being politically naïve are related to this approach narrowing responsibility to its moral dimension. The brief theoretical analysis above shows, however, that issues of responsibility are inevitably interdisciplinary. The issue is not one of abstract ethical judgments but of responsible research, development and innovation, which entails the observance of concrete contexts and governance factors as well as of the quality of the knowledge available. Responsible Innovation must be aware of this complex semantic nature of responsibility.

2.4 Responsible Innovation

Responsible Innovation is a rather new element of technology governance. Its emergence (Siune et al. 2009) reflects the diagnosis that available approaches to shape science and technology still do not meet all of the far-ranging expectations. The hope behind the Responsible Innovation movement is that new – or further-developed – approaches could add considerably to existing approaches such as TA and engineering ethics. Indeed, compared to earlier approaches such as SST or CTA there are shifts of accentuation and new focuses of emphasis:

- "Shaping innovation" complements or even replaces the slogan "shaping technology" which characterised the approach by social constructivist ideas to technology. This shift reflects the insight that it is not technology as such which influences society and therefore should be shaped according to society's needs, expectation and values, but it is innovation by which technology and society interact.

- There is a closer look on societal contexts of new technology and science. Responsible Innovation can be regarded as a further step towards taking the demand pull perspective and social values in shaping technology and innovation more serious.[3]
- Instead of expecting distant observation following classical paradigms of science there is a clear indication for intervention into the development and innovation process: Responsible Innovation projects shall "make a difference" not only in terms of research but also as interventions into the "real world".[4]
- Following the above-mentioned issues, Responsible Innovation can be regarded as a radicalisation of the well-known post-normal science (Funtowitz and Ravetz 1993) being even closer to social practice, being prepared for intervention and for taking responsibility for this intervention.

However, what "responsible" in a specific context means and what distinguishes "responsible" from "irresponsible" or less responsible innovation is difficult to identify. The distinction will strongly depend on values, rules, customs etc. and vary according to different context conditions. Difficulties similar to those appearing in applications of the Precautionary Principle (von Schomberg 2005) probably will occur. The notion of Responsible Innovation as such does not give orientation how to deal with these challenges and difficulties. In the following I would like to propose a conceptual framework which might help clarifying the crucial questions and finding answers to them. My reflection starts by thinking about the preconditions of inquiries and thoughts about ethics and responsibility.

Most of our decisions take the form of goal–means deliberations at the action level (Habermas 1973) without any particular reflection on their normative background and responsibility issues. The discourse level, at which the normative background of decision-making and issues of responsibility will explicitly be the subject of matter, is the exception. The great majority of technology-relevant decisions can be classified as "business as usual" or "standard situation in moral respect" in the following sense (Grunwald 2000, 2012): the normative aspects of the basis for the decision including assumptions about responsibility are not made the object of special reflection, but accepted *as given* in the respective situation, thereby also accepting the elements of the normative framework this entails. The reason is that actors *can assume*, in making these decisions, a normative framework – the basis on which the decision can be made – to be given, including assumptions about the distribution of responsibility. Parts of this normative framework are (national and

[3]An expression of this shift was the strong role of the Societal Panel in the application phase of the MVI programme 'Responsible Innovation'.

[4]This is reflected by the foreseen role of the Valorisation Panels in projects the MVI programme "Responsible Innovation".

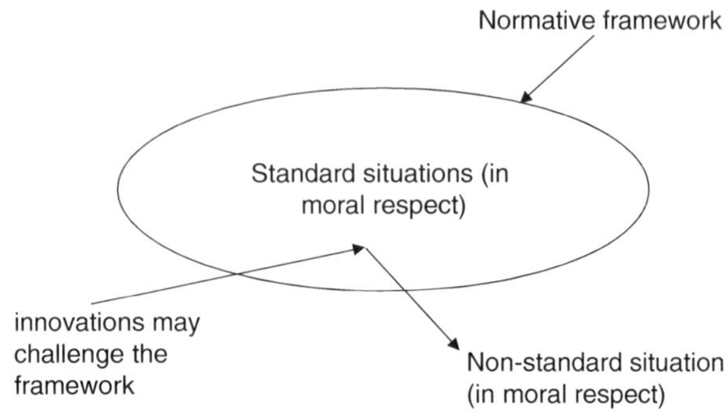

Fig. 2.1 The basic model (Source: Grunwald 2012, Ch. 3)

international) legal regulations, the standard procedures of the relevant institutions (e.g., corporate guidelines), possibly the code of ethical guidelines of the profession concerned, as well as general and un-codified societal customs. The demands on the normative framework which define a business-as-usual situation are formulated more precisely by the following criteria (expanding on Grunwald 2000, 2012):

- *Pragmatic Completeness*
- *Local Consistency*
- *Sufficient Lack of Ambiguity*
- *Acceptance*
- *Compliance*

If these conditions of coherence are satisfied in a specific context, then neither moral conflicts nor ambiguities exist. There is, consequently, no need for explicit ethical reflection and thinking about responsibilities. Participants and others affected by a decision can take information about the normative framework into consideration as axiological information without having to analyze and reflect it. In such "business-as-usual" situations, the *criteria* for making decisions are *a priori* obvious and not questioned (e.g., a cost–benefit analysis in situations in which this is considered an appropriate method according to the accepted normative framework).

However, technical innovations can challenge and possibly "disturb" business as usual situations in moral respect, transform them into non-standard situations and make ethical and responsibility reflection necessary. New scientific knowledge and technological innovation may transform earlier standard situations in a moral respect into *non-standard situations* where one or more of the criteria given above are no longer fulfilled (see Fig. 2.1).

Then, moral ambiguities, conflicts on responsibility and indifferences, as well as new challenges for which moral customs have yet to be established or where there are doubts as to whether established moral traditions apply. In this sense, there is no longer a consensually accepted moral background from which orientation for decision making can be gained. In the following, I will refer to such situations as situations of *normative uncertainty* – then it will be a matter of debate, inquiry or controversy what should be regarded as responsible and what as irresponsible.

Fig. 2.2 Modify innovation (Source: Grunwald 2012, Ch. 3)

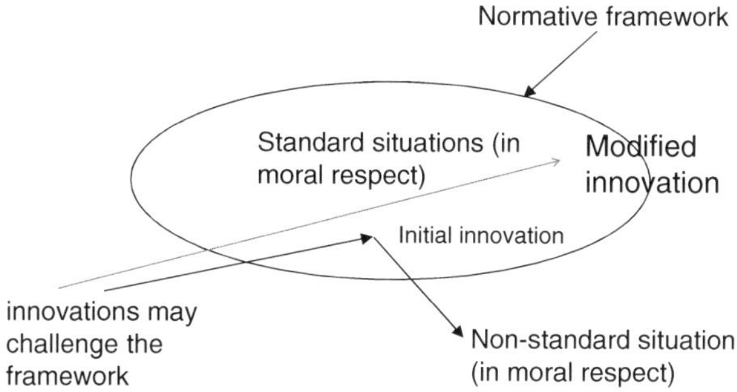

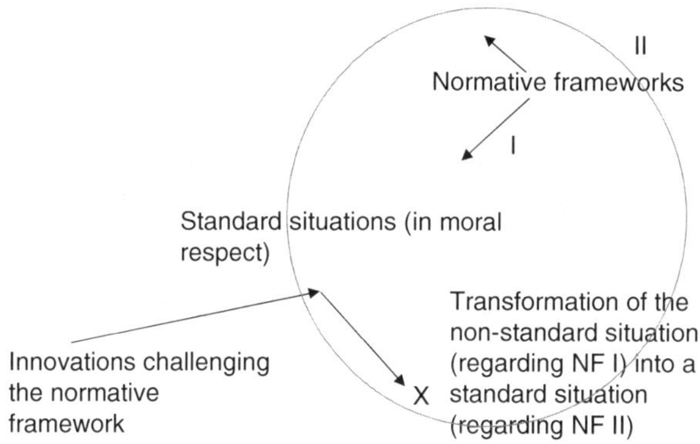

Fig. 2.3 Modification of normative framework (Source: Grunwald 2012, Ch. 3)

In this modified situation, there are simply three options to choose from:

- *The conservative approach*: reject the innovation causing moral trouble – renounce its possible benefits and maintain the initial normative framework.
- *The constructive approach*: Try to modify the properties of the innovation responsible for causing moral trouble (maybe circumstances of its production involving animal experiments or the location of a nuclear waste disposal site in a sacred region of indigenous people) in order to be able to harvest the expected benefits without causing moral trouble (see Fig. 2.2).
- *The techno-optimistic approach*: Modify the normative framework, so that the new technology could be accepted (and the benefits harvested) in a way that would not lead to normative uncertainty and moral conflict (see Fig. 2.3).

Responsibility reflections play a decisive role in determining the criteria of the choice between such alternatives and – in cases 2 and 3 – between different versions and for the concrete consequences. In these cases the reflection is an act of balancing the expected advantages of the innovation or the new technology against the moral or other costs if – as is probably the most common situation – there are no *categorical* ethical arguments for or against. The following can be said about the options:

- *Option 1*: If there would be strong, i.e., categorical, ethical arguments against the new technology then it will probably be rejected. An example is reproductive cloning. Cloning and research on cloning is prohibited in many countries for ethical reasons, and was banned in many Codes of Ethics at the European and international level.
- *Option 2*: The option of shaping technology specifically according to ethical values or principles is behind the approaches of constructive technology assessment (CTA; see Rip et al. 1995), of the social shaping of technology (Yoshinaka et al. 2003), and of value sensitive design (van de Poel 2009, pp. 1001 ff.). The focus is on directing the shaping of technical products or systems along the relevant factors of the normative framework so that the products or systems fit the framework. This would so to speak in itself prevent normative uncertainty from arising.
- *Option 3*: Frequently there are even more complex necessities to balance factors, such as when the (highly promising) use of a new technology or even research on it is not possible except by producing normative uncertainty. Examples are animal experiments undertaken for non-medical purposes (Ferrari et al. 2001) or research in which the moral status of embryos plays a role. The issue is then to examine if and to what extent the affected normative framework can be modified without coming into conflict with the essential ethical principles. Even the handling of technical risks that have to be tolerated in order to utilize an innovation often takes place by means of modifying the normative framework, such as in the implementation of precautionary measures.

Responsibility reflection plays a different role, however, in each of these options. The results of the reflection have to be introduced to the different fields of action (e.g., politics, economics, law). Taking the three dimensions of responsibility mentioned above seriously leads to the conclusion that Responsible Innovation unavoidably requires a more intense inter- and trans-disciplinary cooperation between engineering, social sciences, and applied ethics. The major novelty in this interdisciplinary cooperation might be the integration of ethics (normative reflection on responsibilities) and social sciences such as STS and governance research (empirically dealing with social processes around the attribution of responsibility and their consequences for governance). This integration is at the heart of Responsible Innovation – and a major obstacle might be that applied ethics and social sciences have to deal with deep-ranging controversies and mutual antipathy (Grunwald 1999). It will one of the most exciting challenges in which way these obstacles might be overcome. In the field of technology assessment there are some indications that a constructive cooperation is possible (Grunwald 1999).

The terms of responsible development, responsible research and responsible innovation have been used over the last years to an increasing extent. These terms are highly integrative because they cover issues of engineering ethics, participation, technology assessment, anticipatory governance and science ethics. They include what has been stated in this Chapter about TA: adding reflexivity to technology development and design (see also Voss et al. 2006). In this sense responsible development and innovation might be a new umbrella term (von Schomberg 2012) with new accentuations which may be characterized by:

- involving ethical and social issues more directly in the innovation process by integrative approaches to development and innovation
- bridging the gap between innovation practice, engineering ethics, technology assessment, governance research and social sciences (STS)
- giving new shape to innovation processes and to technology governance according to responsibility reflections in all of its three dimensions mentioned above
- in particular, making the distribution of responsibility among the involved actors as transparent as possible
- supporting "constructive paths" of the co-evolution of technology and the regulative frameworks of society

However, it is important to point out that the model of integrated research including its own ethical and responsibility reflection also harbours problems. The independence of reflection can be threatened especially if the necessary distance to the technical developments and those working on them is lost. Inasmuch as assessment issues becomes part of the development process and would identify itself with the technical success, there might be an accusation that its acceptance was "purchased" or that it was nothing but grease in the process of innovation. Strategies of dealing with such possible developments should be developed and could include means such as careful monitoring activities and a strong role of external review processes. It will be a task for the respective emerging research community around the issue of Responsible Innovation to take care but also the responsible funding agencies should be aware of this challenge.

References

Aichholzer, G., A. Bora, S. Bröchler, M. Decker, and M. Latzer (eds.). 2010. *Technology governance. Der Beitrag der Technikfolgenabschätzung*. Berlin: Edition Sigma.

Bechmann, G., M. Decker, U. Fiedeler, and B.-J. Krings. 2007. Technology assessment in a complex world. *International Journal on Foresight and Innovation Policy* 3: 6–27.

Bijker, W.E., and J. Law (eds.). 1994. *Shaping technology and building society*. Cambridge, MA: MIT Press.

Bijker, W.E., T.P. Hughes, and T.J. Pinch (eds.). 1987. *The social construction of technological systems*. Cambridge, MA: MIT Press.

Bimber, B.A. 1996. *The politics of expertise in congress: The rise and fall of the office of technology assessment*. Albany: State University of New York Press.

Collingridge, D. 1980. *The social control of technology*. New York: St. Martin's Press.

Cruz-Castro, L., and L. Sanz-Menendez. 2004. Politics and institutions: European parliamentary technology assessment. *Technological Forecasting and Social Change* 27: 79–96.

Decker, M., and M. Ladikas (eds.). 2004. *Bridges between science, society and policy. Technology assessment – Methods and impacts*. Berlin: Springer.

Durbin, P., and H. Lenk (eds.). 1987. *Technology and responsibility*. Boston: Reidel Publishing.
Ferrari, A., C. Coenen, A. Grunwald, and A. Sauter. 2001. *Animal Enhancement. Neue technische Möglichkeiten und ethische Fragen*. Bern: Bundesamt für Bauten und Logistik BBL.
Funtowitz, S., and J. Ravetz. 1993. The emergence of post-normal science. In *Science, politics and morality*, ed. R. von Schomberg, 173–188. London.
Gee, D., and M. Greenberg. 2002. Asbestos: From 'magic' to malevolent mineral. In *The precautionary principle in the 20th century. Late lessons from early warnings*, ed. P. Harremoes, D. Gee, M. MacGarvin, A. Stirling, J. Keys, B. Wynne, and S. Guedes Vaz, 49–63. London: Earthscan Publications.
Grunwald, A. 1999. Verantwortungsbegriff und Verantwortungsethik. In *Rationale Technikfolgenbeurteilung*, ed. A. Grunwald, 172–195. Berlin: Springer.
Grunwald, A. 2000. Against over-estimating the role of ethics in technology. *Science and Engineering Ethics* 6: 181–196.
Grunwald, A. 2007. Converging technologies: Visions, increased contingencies of the conditio Humana, and search for orientation. *Futures* 39: 380–392.
Grunwald, A. 2009. Technology assessment: Concepts and methods. In *Philosophy of technology and engineering sciences*, vol. 9, ed. A. Meijers, 1103–1146. Amsterdam: Elsevier.
Grunwald, A. 2012. *Responsible nanobiotechnology. Ethics and philosophy*. Singapore: Pan Stanford Pub.
Guston, D.H., and D. Sarewitz. 2002. Real-time technology assessment. *Technology in Culture* 24: 93–109.
Habermas, J. 1970. *Toward a rational society*. Beacon Press. First publication: Habermas, J. (ed.). 1968. *Technik und Wissenschaft als Ideologie*. Frankfurt.
Habermas, J. 1973. Wahrheitstheorien. In *Wirklichkeit und Reflexion*, ed. H. Fahrenbach, 211–265. Pfullingen: Neske.
Hansson, S.O. 2006. Great uncertainty about small things. In *Nanotechnology challenges – Implications for philosophy, ethics and society*, ed. J. Schummer and D. Baird, 315–325. Singapore: World Scientific.
Harremoes, P., D. Gee, M. MacGarvin, A. Stirling, J. Keys, B. Wynne, and S. Guedes Vaz (eds.). 2002. *The precautionary principle in the 20th century. Late lessons from early warnings*. London: Sage.
Jonas, H. 1979. *Das Prinzip Verantwortung. Versuch einer Ethik für die technologische Zivilisation*. Frankfurt: Suhrkamp.
Joss, S., and S. Belucci (eds.). 2002. *Participatory technology assessment – European perspectives*. London: Westminster University Press.
Lenk, H. 1992. *Zwischen Wissenschaft und Ethik*. Frankfurt: Suhrkamp.
National Research Council. 2006. *A matter of size: Triennial review of the national nanotechnology initiative*. Washington, DC: National Academies Press.
Paslack, R., J.S. Ach, B. Luettenberg, and K. Weltring (eds.). 2011. *Proceed with caution? – Concept and application of the precautionary principle in nanobiotechnology*. Münster: LIT Verlag.
Rip, A., T. Misa, and J. Schot (eds.). 1995. *Managing technology in society*. London: Pinter Publishers.
Roco, M.C., and W.S. Bainbridge (eds.). 2001. *Societal implications of nanoscience and nanotechnology*. Boston: Kluwer.
Selin, C. 2007. Expectations and the emergence of nanotechnology. *Science, Technology and Human Values* 32(2): 196–220.
Siune, K., E. Markus, M. Calloni, U. Felt, A. Gorski, A. Grunwald, A. Rip, V. de Semir, and S. Wyatt. 2009. *Challenging futures of science in society*. Report of the MASIS Expert Group. Brussels: European Commission.
Smits, R., and P. den Hertog. 2007. TA and the management of innovation in economy and society. *International Journal on Foresight and Innovation Policy* 3: 28–52.
van de Poel, I. 2009. Values in engineering design. In *Philosophy of technology and engineering sciences*, vol. 9, ed. A. Meijers, 973–1006. Boston: Elsevier.

VDI – Verein Deutscher Ingenieure 1991. Richtlinie 3780 Technikbewertung, Be-griffe und Grundlagen. Düsseldorf. Available also in English at: www.vdi.de.

Vig, N., and H. Paschen (eds.). 1999. *Parliaments and technology assessment. The development of technology assessment in Europe*. Albany: State University of New York Press.

von Schomberg, R. (ed.). 1999. *Democratizing technology. Theory and practice of a deliberative technology policy*. Hengelo: ICHPA.

von Schomberg, R. 2005. The precautionary principle and its normative challenges. In *The precautionary principle and public policy decision making*, ed. E. Fisher, J. Jones, and R. von Schomberg, 141–165. Cheltenham/Northampton: Edward Elgar.

von Schomberg, R. 2012. Prospects for technology assessment in the 21st century: The quest fort the "right" impacts of science and technology. An outlook towards a framework for responsible research and innovation. In *Technikfolgen abschätzen lehren*, ed. M. Dusseldorp, et al., 43–65. Opladen: Westdeutscher Verlag.

Voss, J.-P., D. Bauknecht, and R. Kemp (eds.). 2006. *Reflexive governance for sustainable development*. Cheltenham: Edward Elgar.

Yoshinaka, Y., C. Clausen, and A. Hansen. 2003. The social shaping of technology: A new space for politics? In *Technikgestaltung: zwischen Wunsch oder Wirklichkeit*, ed. A. Grunwald, 117–131. Berlin: Springer.

Chapter 3
The Quest for the 'Right' Impacts of Science and Technology: A Framework for Responsible Research and Innovation

René von Schomberg

Abstract In this contribution, a framework for 'Responsible Research and Innovation' is proposed. This framework enables to practice Responsible Research and Innovation while addressing both research and innovation processes and research and innovation outcomes and products. The framework operationalizes general consensual normative anchor points derived from the European Treaties in order to drive innovations towards the 'Grand Challenges' of our time for which we share a collective responsibility. This implies an innovation-governance far beyond the means of solely market-driven innovations.

3.1 Introduction

I will outline a framework for Responsible Research and Innovation (RRI). Such a framework builds upon achievements made in the context of 'Science in Society' activities, such as public engagement with science and technology, technology assessment and foresight, and governance and ethics of science and technology. However, RRI reconfigures, redefines and extends these activities with a view on innovation processes and public policy making.

Whereas technology assessments have traditionally addressed the "negative consequences" in terms of risks and adverse effects of technologies, the focus

---

Dr. Dr.phil. René von Schomberg (email: Rene.vonschomberg@ec.europa.eu) is at the European Commission, Directorate General for Research and Innovation. The views expressed here are those of the author and may not in any circumstances be regarded as stating an official position of the European Commission.

R. von Schomberg (✉)
European Commission, Brussels, Belgium
e-mail: Rene.vonschomberg@ec.europa.eu

J. van den Hoven et al. (eds.), *Responsible Innovation 1: Innovative Solutions for Global Issues*, DOI 10.1007/978-94-017-8956-1_3,

of attention within policy is predominantly to demonstrate potentially positive impacts of future outcomes of public policy including research policy. "Negative impacts" are dealt within the context of broader cost-benefit analysis or within specialized fields of policy, such as risk management and risk assessments. The quest for positive or the "right" impacts is a much more overarching feature of public policy. This brings us naturally to the question: what would be the "right" impacts of research and innovation policy? The European Commission has proposed to introduce RRI as a cross-cutting issue under the new Framework Programme for Research and Innovation: Horizon 2020. Horizon 2020 will, among other, address the so called 'Grand Challenges' of our time and RRI could be well linked to driving research and innovation towards particular societal objectives. It is thus important to understand how we can anticipate and assess positive outcomes of science and technology and what type of public policy guidance would be appropriate. In the following, I will answer these questions and how they can be tackled within a new framework for responsible research and innovation.

3.2 The Quest for the Right Impacts and Outcomes of Research

Some philosophers of technology have recently argued that science should move beyond a contractual relationship with society and join in the quest for the common good. In their view, the "good in science, just as in medicine, is integral to and finds its proper place in that overarching common good about which both scientists and citizens deliberate"(Mitcham and Frodeman 2000). This view may sound attractive, but it fails to show how various communities with competing concepts of the "good life", within modern societies, could arrive at a consensus and how this could drive public (research) policy. Moreover, an Aristotelian concept of the good life is difficult to marry with a modern rights' approach, whereby, for instance in the case of the European Union, the European Charter of Fundamental Rights provides a legitimate and actual basis for European Public Policy. Nonetheless, their point of departure remains challenging: "We philosophers believe that publicly funded scientists have a moral and political obligation to consider the broader effects of their research; to paraphrase Socrates, unexamined research is not worth funding" (Frodeman and Holbrook 2007)

European policy however is also increasingly legitimized in terms of public values driving public policies towards positive impacts. The following citations of prominent European policy makers illustrate the case:

- "The defence of human rights and a justice system based on the full respect of human dignity is a key part of our shared European values" Jerzy Buzek, European Parliament President (10 October, 2009)
- "Europe is a community of Values". Van Rompuy, First European Council President, 19 November 2009

- "My political guidelines for the Commission's next mandate stress the idea that Europe's actions must be based on its values". President Barroso, European values in the new global governance, 14 October 2009

Indeed, European public policies are arguably driven towards positive impacts, underlined by common European values. European Environmental policies for example, highlight the European value of maintaining a high level of protection for the environment. Research and Innovation policy seem to have been an exception to the rule and, although we articulate research and innovation policy since recently more and more in terms of public values, research and innovation programme assessments are typically limited to economic terms that "imperfectly take into account these values" (Fisher et al. 2010).

The US National Science Foundation assesses their proposals in terms of "broader impacts" in the framework of considering research proposals worth funding. Under the European Framework Programmes for Research, there is a long tradition of awarding research grants on the basis of anticipated impacts. Indeed, even at the stage of evaluation of research proposals particular impacts are sought. Currently, expected impacts of research topics which are subject to public calls for proposals are listed in the work programmes of the 7th Framework Programme. But are there legitimate, normative assumptions which support these expected impacts that allow an articulation of the 'right impacts' that allow us to steer public research agendas? We can't make an appeal to concepts of the good life, but we can make an appeal to the normative targets which we can find in the Treaty on the European Union. These normative targets have been democratically agreed and provide the legitimate basis for having a public framework programme for research at the European Level. From article 3 of the Treaty on the European Union. European Union (2010) we can derive the following:

- "The Union shall (. . . .) work for the sustainable development of Europe based on balanced economic growth and price stability, a highly competitive social market economy, aiming at full employment and social progress, and a high level of protection and improvement of the quality of the environment. It shall promote scientific and technological advance".
- "It shall combat social exclusion and discrimination, and shall promote social justice and protection, equality between women and men, solidarity between generations and protection of the rights of the child".
- "To promote (. . .) harmonious, balanced and sustainable development of economic activities, a high level of employment and of social protection, equality between men and women, sustainable and non-inflationary growth, a high degree of competitiveness and convergence of economic performance, a high level of protection and improvement of the quality of the environment, the raising of the standard of living and quality of life, and economic and social cohesion and solidarity among Member States".

Rather than pre-empting views and concepts of the "good life", the European Treaty on the European Union provides us then with normative anchor points. These

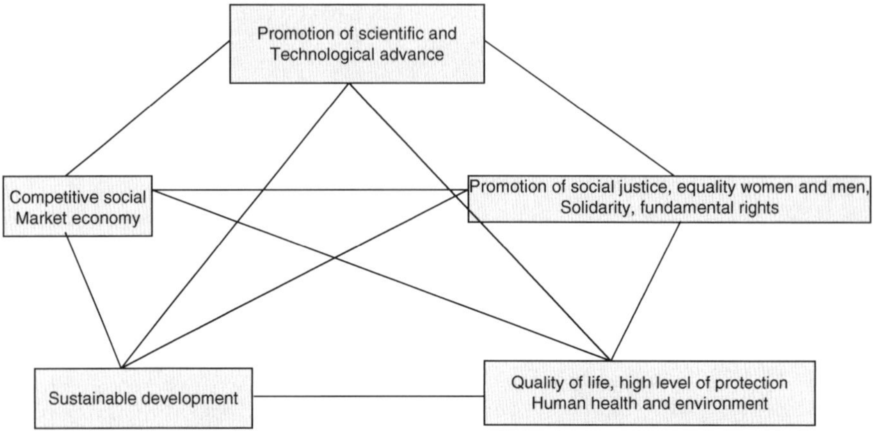

Fig. 3.1 Normative anchor points derived from the Treaty on the European Union

normative anchor points and their mutual relationship thus provide a legitimate basis for defining the type of impacts, or the "right" impacts of research and innovation should pursue. (See Fig. 3.1. above). These are of course normative anchor points which have impacts beyond the EU. The EU's commitment to promote Human Rights and demonstrate solidarity with the poorest on earth is reflected in its international policies. If applied to international Research and Innovation policies, this could invite us to address issue such as "technology divides", ethics free zones and broad benefit sharing from scientific and technological advance (see Ozolina et al. 2012). Research and Innovation policy can also be a form of development policy.

3.3 The Responsible Development of Technologies: A Historical Perspective

The formation of public opinion on new technologies is not a historically or geographically isolated process; rather, it is inevitably linked to prior national and (international) debate on similar topics. Ideally, such debates should enable a learning process – one that allows for the fact that public opinion forms within particular cultures and political systems. It is therefore not surprising that, in the case of nanotechnologies, the nature of public debate and its role in the policy making process is articulated against a background of previous discussion of the introduction of new technologies such as biotechnology, or that specific national experiences with those technologies become important. In particular, the introduction of genetically modified organisms (GMOs) into the environment is a frequent reference point within Europe (whereas more frequently absent in such debates in the USA).

This historical development of policy frameworks can be followed through the ways in which terms are used and defined: initially, definitions are often determined by the use of analogies which, in the initial stages of the policy process, serve to 'normalise' new phenomena. In a number of countries, for instance, GMOs were initially regulated through laws which deal with toxic substances. Subsequently such analogies tend to lose their force as scientific insights on the technology grows and distinct regulatory responses can be made. GMOs, for example, eventually became internationally defined as 'potentially hazardous', and, in the European Union, a case by case approach was adopted under new forms of precautionary regulation. This framework was developed over a period of decades, and thereby took into account the ever-widening realm in which GMOs could have effects: developing from an exclusive focus on direct effects to eventually include indirect and long-term effects. It is not, however, solely the scientific validity of analogies which determines definitions and policy: public interest also plays an important role. Carbon dioxide, for instance, has changed from being viewed as a gas essential to life on earth to being a 'pollutant'. The latest iteration of this evolution came just prior to the Copenhagen summit on climate change in December 2009, when the American Environmental Protection Agency defined greenhouse gases as a "threat to public health" – a definition which has important implications for future policy measures.

In the case of relatively new or emerging technologies, such as nanotechnology policy, then, it seems likely that we are still in the initial phases of development. The process of agreeing on any internationally agreed definitions relating to the technology goes very slow despite repeated announcements of their imminence, and nanoparticles continue to be defined as "chemical substances" under the European regulatory framework REACH. (Analogies are also made with asbestos, as a way to grasp hold of possible environmental and human health effects, but these are contested). There is no certainty that they will become the definitive way to frame risk assessments. To cite one topical example, nanotechnology in food will not start its public and policy life with a historically blank canvas but will be defined as a 'novel food' under a proposal for renewing the Novel Foods regulation. The Novel Foods regulation came into existence in the 1990s with foods containing or consisting of GMO's in mind. Recent proposals for renewing regulation on food additives have made this the first piece of regulation to include explicit reference to nanotechnology.

Public debate that articulates particular interests and scientific debate on the validity of analogical approaches to nanotechnologies will inevitably continue to shape the ways in which nanotechnologies are addressed in regulation and policy. But the governance of the technology, as well as debate around it, has to be seen within its historical context. How did stakeholders behave in previous cases, and what can we learn from these cases with regard to nanotechnology? One answer to this question might point to a learning process around the governance of new technologies, and the development of a consensus that early involvement of both stakeholders and the broader public is of the utmost importance. The European Commission has responded to this with its adoption of a European strategy and action plan on nanotechnologies, which addresses topics from research needs to regulatory responses and ethical issues to the need for international dialogue. This strategy above all emphasizes the "safe, integrated and responsible" development of nanosciences and nanotechnologies – something which several European Research projects has drawn upon in articulating how 'responsible development' might take its course within deliberative fora.[1]

We can conclude that the "safe, integrated and responsible" development gives us *a new anchor point* for making for instance, nanotechnology policy. Obviously, this has to be built on the basic anchor points in the treaty, concerning "a high level of protection of the environment and human health", applying precaution etc.

These normative anchor points, in their mutual interdependency, should guide the impact assessments of technologies, and also the notion of desirable expected impacts of research. This brings us to how we can identify these "right" impacts of research and technologies.

3.4 From Normative Anchor Points Towards Defining 'Grand Challenges' and the Direction of Innovation

Under the prospective framework programme Horizon 2020, a number of 'Grand Challenges' have been defined, which follow the call in the Lund Declaration for a Europe that "must focus on the grand challenges of our time" (Lund Declaration, July 2009). Sustainable solutions are sought in areas such as "global warming, tightening supplies of energy, water and food, ageing societies, public health, pandemics and security" (Lund Declaration, p. 1–2009).

Arguably, the "grand challenges" of our time reflect a number of normative anchor points of the Treaty and thus can be seen as legitimate. The Lund declaration states that in order to be responsive the European Research Area must develop processes for the identification of Grand Challenges, which gain political support and gradually move away from the current thematic approaches, towards a structure where research priorities are based on these 'grand challenges'. It hopes to give direction to research and innovation in the form of "broad areas of issue-oriented research in relevant fields. It calls for amongst other things, broad stakeholder involvement and the establishment of public-private partnerships.

The macro-economic justification of investment in research and innovation emphasizes that innovation is the "only answer" to tackle societal challenges: "returning to growth and higher levels of employment, combating climate change and moving towards a low carbon society" (European Commission 2011, p. 3). This approach implicitly assumes that access to and availability of finance for research and innovation will *automatically* lead to the creations of jobs and economic growth,

---

[1]See the projects contribution in: Rene von Schomberg and Sarah Davies [eds.], Understanding public debate on nanotechnologies. Options for Framing Public Policy. Luxembourg: Publication office of the European Union (2010).

thereby taking on the societal challenges along the way. The more innovation, the better. The faster it becomes available, the better. In this macro – economic model, innovation is assumed to be *steerless but inherently good* as it produces prosperity and jobs and meets societal challenges, addressed through market-demand.

The Lund declaration gives however an alternative justification for investing in research and innovation, primarily framing this in terms of responding to societal Grand Challenges and further stating that "meeting the grand challenges will be a prerequisite for continued economic growth and for improved changes to tackle key issues". Here the assumption is that sustainable economic growth is only possible when particular societal objectives are met, in the form of responding to Grand Challenges. Innovation is neither seen as steerless nor as inherently good. Economic prosperity and the anticipation that h innovation yields positive anticipated impacts (such as the creation of jobs and growth) become *dependent upon the social context*. The Lund Declaration points out those measures are "needed to maximize the economic and *societal* impact of knowledge" (italics by the author). The idea is clear; to steer the innovation process towards societal beneficial objectives. Additional measures that go beyond removing barriers for research and innovation, availability of and access to finance of research and innovation become then necessary. The Lund declaration defines a type of justification of investment in research and innovation towards *particular* positive outcomes. The Lund declaration underlines a justification of research and innovation beyond economic terms and with a view on particular outcomes. Recently, European Commissioner for Research, Innovation and Science, Geoghegan-Quinn stated at a conference on 'Science in Dialogue' that 'research and innovation must responsible to the needs and ambitions of society, reflect its values, and be responsible'.[2]

3.5 A Framework for Responsible Research and Innovation

The following definition for Responsible Research and Innovation is proposed:

Definition: *Responsible Research and Innovation is a transparent, interactive process by which societal actors and innovators become mutually responsive to each other with a view to the (ethical) acceptability, sustainability and societal desirability of the innovation process and its marketable products(in order to allow a proper embedding of scientific and technological advances in our society).*

There is a significant time lag: this can be several decades between the occurrence of technical inventions or planned promising research and the eventual marketing of products resulting from RTD and innovation processes. The societal impacts of scientific and technological advances are difficult to predict. Even major tech-

[2]Conference "Science in Dialogue". Towards a European Model for Responsible Research and Innovation Odense, Denmark 23–25 April 2012.

nological advances such as the use of the internet and the partial failure of the introduction of GMOs in Europe have not been anticipated by governing bodies. Early societal intervention in the Research and Innovation process can help avoid technologies failing to embed in society and/or help that their positive and negative impacts are better governed and exploited at a much earlier stage. Two interrelated dimensions can be identified: the *product* dimension, capturing products in terms of overarching and specific normative anchor points (see discussion above) and a *process* dimension reflecting a deliberative democracy.

The normative anchor points should be reflected in the product dimension. They should be:

(Ethically) acceptable: in an EU context this refers to a mandatory compliance with the fundamental values of the EU charter on fundamental rights [right for privacy etc.] and the safety protection level set by the EU. This may sound obvious, but the practice of implementing ICT technologies has already demonstrated in various cases that the fundamental right for privacy and data protection can and has been neglected. It also refers to the "safety" of products in terms of *acceptable* risks. It goes without saying that ongoing risk assessments are part of the procedure towards acceptable products when safety issues are concerned. However, the issue of safety should be taken in a broader perspective. The United Kingdom's largest public funder of basic innovation research, the Engineering and Physical Science and Research Council asked applicants to report the wider implications and potential risk (environmental, health, societal and ethical) associated with their proposed research in the area of nanosciences (Owen and Goldberg 2010). This highlighted the fact that, often, the risks related to new technologies, can neither be quantified nor a normative baseline of acceptability assumed by scientists (acknowledging that any, particular baseline cannot be assumed to represent *the* baseline of societal acceptance).

Sustainable: contributing to the EU's objective of sustainable development. The EU follows the 1997 UN "definition" of sustainable development, consisting of economic, social and environmental dimensions in mutual dependency. This overarching anchor point can become further materialized under the following one:

Socially desirable: "socially desirable" captures the relevant, and more specific normative anchor points of the Treaty on the European Union, such as "Quality of life", "Equality among men and women" etc.(see above). It has to be noted that a systematic inclusion of these anchor points in product development and evaluation would clearly go beyond simple market profitability, although the latter could be a precondition for the products' viability in market competitive economies. However, it would be consistent with the EU treaty to promote such product development through the financing of research and development actions. In other words, at this point, Responsible Research and Innovation would not need any *new* policy guidelines, but simply would require a consistent application of the EU's fundamentals to the research and innovation process reflected in the

Treaty on the European Union. Perhaps it has been wrongly assumed that these values could not be considered in the context of research and innovation. Since the Lund Declaration, a process to take into account societal objectives in the form of addressing Grand Challenges has been set in motion.

Responsible Research and Innovation features both a product and process dimension:

Product dimension:

Products be evaluated and designed with a view to their normative anchor points: high level of protection to the environment and human health, sustainability, and societal desirability.

Process dimension:

The challenge here is to arrive at a more responsive, adaptive and integrated management of the innovation process. A multidisciplinary approach with the involvement of stakeholders and other interested parties should lead to an inclusive innovation process whereby technical innovators become responsive to societal needs and societal actors become co-responsible for the innovation process by a constructive input in terms of defining societal desirable products. The product and process dimension are naturally interrelated. Implementation is enabled by five mechanisms: technology assessment and foresight, application of the precautionary principle, normative/ethical principles to design technology, innovation governance and stakeholder involvement and public engagement.

Table 3.1 provides a matrix which describes examples of lead questions to be answered by the stakeholder either from a product or process perspective in order to fully implement an RRI scheme (the lead questions with the same shade of grey, represent the alternative emphasis on either the product or process dimension).

3.5.1 Use of Technology Assessment and Technology Foresight

This is done in order to anticipate positive and negative impacts or, whenever possible, define desirable impacts of research and innovation both in terms of impact on consumers and communities. Setting of research priorities and their anticipated impacts needs to be subject to a societal review. This implies broadening the review of research proposals beyond scientific excellence and including societal impacts.[3] Specific Technology Assessment methods also help to identify societal desirable products by addressing the normative anchor points throughout their development.

[3]The Netherlands Organisation for Scientific Research (NWO) has developed a research funding programme on Responsible Innovation under which research proposals are subject to a review in terms of societal relevance. See: http://www.nwo.nl/nwohome.nsf/pages/NWOA_7E2EZG_Eng.

Table 3.1 Responsible research and innovation matrix

Product-dimension / **Process-dimension**	**1. Technology Assessment and Foresight**	**2. Application of the Precautionary Principle**	**3. Normative/ethical principles to design technology**	**4. Innovation governance and stakeholder involvement**	**5. Public engagement**
Technology Assessment and Foresight	x	Development of Procedures to cope with risks	Which design objectives to choose?	Stakeholder involvement in Foresight and TA	How to engage the public?
Application of the Precautionary Principle	Identification of nature of risks	x	Choice and development of standards	Defining proportionality: how much precaution?	How safe is safe enough?
Normative/ethical principles to design technology	"privacy" and "safety" by design	Setting of risk/uncertainty thresholds	x	Which principles to choose?	Which technologies for which social desirable goals?
Innovation governance models and stakeholder involvement	Defining scope and methodology for TA/Foresight by stakeholders	Defining the precautionary approaches by stakeholders	Translating normative principles in technological design	x	How can innovation be geared towards social desirable objective
Public Engagement and Public Debate	Defining/choice of methodology for public engagement	Setting of acceptable standards	Setting of social desirability of RRI outcome	Stakeholders roles in achieving social desirable outcomes	x

The matrix is composed of 10 'twin' issues, representing emphasis on either the process-dimension or the product-dimension. For example, the twin issues 'the identification of nature of risks' and 'the development of procedures to cope with risks' are at the cross-roads of applying the precautionary principle and technology assessment and foresight.

Methodologies to further "script" the future expected impacts of research should be developed (Den Boer et al. 2009). A good example exists in the field of synthetic biology by Marc Bedau et al. (2009). They have identified six key checkpoints in protocell development (e.g. cells produced from non-living components by means of synthetic biology) in which particular attention should be given to specific ethical, social and regulatory issues, and made ten recommendations for responsible protocell science that are tied to the achievement of these checkpoints. Technology Assessment and Technology Foresight can reduce the human cost of trial and error and take advantage of a societal learning process of stakeholders and technical innovators. It creates a possibility for anticipatory governance. This should ultimately lead to products which are (more) societal robust.

3.5.2 *Application of Precautionary Principle*

The precautionary principle is embedded in EU law and applies especially within EU product authorization procedures, e.g. REACH, GMO directives etc. The precautionary principle works as an incentive to make safe and sustainable products and allows governmental bodies to intervene with risk management decisions such as temporary licensing, case by case decision making, whenever necessary, in order to avoid negative impacts. The responsible development of new technologies must be viewed in its historical context. Some governance principles have been inherited from previous cases: this is particularly notable for the application of the precautionary principle to new fields such as that of nanosciences and nanotechnologies.

The precautionary principle is firmly embedded in European policy, and is enshrined in the 1992 Maastricht Treaty as one of the three principles upon which all environmental policy is based. It has been progressively applied to other fields of policy, including food safety, trade and research. The principle runs through legislation for example in the 'No data, no market' principle of the REACH directive for chemical substances, or the pre-market reviews required by the Novel Foods regulation as well as the directive on the deliberate release of GMOs into the environment. More generally, within the context of the general principles and requirements of European food law it is acknowledged that "scientific risk assessment alone cannot provide the full basis for risk management decisions". European Commission (2002) – leaving open the possibility of risk management decision making partly based on ethical principles or particular consumer interests.

In the European Commission's Recommendation on a Code of Conduct Commission of the European Communities (2008) for Nanosciences and Nanotechnologies Research(the principle appears in the call for risk assessment before any public funding of research a strategy currently applied in the 7th Framework Programme for research). Rather than stifling research and innovation, the precautionary principle acts within the Code of Conduct as a focus for action, in that it calls for funding for the development of risk methodologies, the execution of risk research, and the active identification of knowledge gaps.

3.5.3 *Innovation Governance*

3.5.3.1 Multistakeholder Involvement

Multistakeholder involvement in RRI- projects should bring together actors from industry, civil society and research to jointly define an implementation plan for the responsible development of a particular product to be developed within a specific research/innovation field, such as information and communication technology or nanotechnology. Responsible innovation should be materialised in terms of the research and innovation process as well as in terms of (product) outcomes. The advantage is that actors cannot exclusively focus on particular aspects (for instance, civil society organizations addressing only the risk aspects) but have to take a position on all aspects of innovation process as such. Thus allowing a process to go beyond risk governance and move to innovation governance. The company BASF, for example, has established a dialogue forum with civil society organizations and also developed a code of conduct for the development of new products.[4]

3.5.3.2 Use of Codes of Conduct

Codes of Conduct, in contrast to regulatory interventions, allow a constructive steering of the innovation process. They enable the establishment of a proactive scientific community which identifies and reports to public authorities on risks and benefits at an early stage. Codes of Conduct are particular useful when risks are uncertain and when there is uncertain ground for legislative action nanotechnology for example. Codes of Conduct also help to identify knowledge gaps and direct research funds towards societal objectives.

Policy development treads a fine line: governments should not make the mistake of responding too early to a technology, and failing to adequately address its nature, or of acting too late, and thereby missing the opportunity to intervene. A good governance approach, then, might be one which allows flexibility in responding to new developments. After a regulatory review in 2008, the European Commission came to the conclusion that there is no immediate need for new legislation on nanotechnology, and that adequate responses can be developed – especially with regard to risk assessment – by adapting existing legislation.

In the absence of a clear consensus on definitions, the preparation of new nano-specific measures will be difficult and although there continues to be significant scientific uncertainty on the nature of the risks involved, good governance will have to go beyond policy making that focuses only on legislative action. The power of governments is arguably limited by their dependence on the insights and cooperation of societal actors when it comes to the governance of new technologies: the development of a code of conduct, then, is one of their few options for intervening in a timely and responsible manner. The European Commission states in the second implementation report on the action plan for Nanotechnologies that "its effective implementation requires an efficient structure and coordination, and

---

[4]In the BASF Dialogueforum Nano representatives of environmental and consumer organisations, trade unions, scientific institutes and churches. Civil Society Organisations/Non Governmental Organisations) work together with employees of the chemical company BASF SE on various issues related to the subject of nanotechnologies. See for a recent report: http://www.risiko-dialog.ch/component/content/article/507-basf-dialogueforum-nano-final-report-2009-2010.

regular consultation with the Member States and all stakeholders" Commission of the European Communities (2009). Similarly, legislators are dependent on scientists' proactive involvement in communicating possible risks of nanomaterials, and must steer clear of any legislative actions which might restrict scientific communication and reporting on risk. The ideal is a situation in which all the actors involved communicate and collaborate. The philosophy behind the European Commission's code of conduct, then, is precisely to support and promote active and inclusive governance and communication. It assigns responsibilities to actors beyond governments, and promotes these actors' active involvement against the backdrop of a set of basic and widely shared principles of governance and ethics. Through codes of conduct, governments can allocate tasks and roles to all actors involved in technological development, thereby organising collective responsibility for the field (Von Schomberg 2007). Similarly, Mantovani and Porcari (2010) propose a governance plan which both makes use of existing governance structures and suggests new ones, as well as proposing how they should relate to each other.

The European Commission recommendation on a Code of Conduct views Member States of the European Union as responsible actors, and invites them to use the Code as an instrument to encourage dialogue amongst "policy makers, researchers, industry, ethics committees, civil society organisations and society at large"(recommendation number 8 to Member States, cited on page 6 of the Commission's recommendation), as well as to share experiences and to review the Code at European level on a biannual basis. It should be considered that such Codes of Conduct would in the future extend their scope beyond research and also address the innovation process.[5]

3.5.3.3 Adoption of Standards, Certification and Self-Regulation

The adoption of standards and even "definitions" are fundamental requirements to allow for responsible development. The outstanding adoption of a definition for nanoparticles, for example makes legislation and adequate labelling practices difficult, if not impossible (Bush 2010) notes that the use of standards, certifications and accreditations constitute a new form of governance which progressively has replaced and transmuted positive law, as a product of the state, with its market equivalent. Although this form of governance is in need of improvement, we unavoidably have to make productive use of it, as the flood of products and processes coming on to the market will not be manageable through governmental bodies and agencies alone. Yet, the perception and working practice of these standards is significant. In 2005, it was claimed that the EU had forced local authorities to remove see-saws from children's playgrounds. No such EU measures were taken. Some standards were set by the European Committee for Standardisation (CEN), a

[5] The European Project NANOCODE makes this point concerning nanosciences and nanotechnologies, see: http://www.nanocode.eu/.

voluntary organisation made of national standards bodies. CEN sought to limit the height from which children could fall, by specifying the maximum height for seats and stands, and by setting standards for hand supports and footrests. Manufacturers could choose to follow these standards, which carried the advantage of being able to export across Europe, instead of having to apply for certification in each country (European Communities 2006).

The area of data- and privacy protection in the context of the use of ICT and security technologies should also be impacted by forms of self-regulation and standard setting. Data controllers based at operators need to provide accountability, which can be termed as a form of verifiable responsibility (Guagnin et al. 2011). The involvement of third parties which can implement, minimally, a transparent verification practice will be crucial. In other fields, the whole certification can be carried out by a third party. For example, in 1996, the World Wildlife Fund (WWF) and Unilever joined forces and collectively constructed a long-term programme for sustainable fisheries. They founded an *independent* non-profit organisation to foster worldwide fisheries. They also apply "standards of Sustainable Fishing", which is also monitored by independent certifying agencies to control those standards.

Standards will also need to reflect particular ethical considerations and go well beyond mere technical safety issues. Currently, the development of new ISO standards for Nanofood might involve the inclusion of ethical standards. Forsberg (2010).

3.5.4 Ethics as a "Design" Factor of Technology and Increasing Social-Ethical Reflexivity in Research Practices

Ethics should not be seen as being only a constraint of technological advances. Incorporating ethical principles in the design process of technology can lead to well accepted technological advances. As discussed above, in Europe, the employment of Body Imaging Technology at Airports has for example raised constitutional concerns in Germany. It has been questioned whether the introduction is proportional to the objectives being pursued. The introduction of a "smart meter" at the homes of people in the Netherlands to allow for detection of and optimisation of energy use, was rejected on privacy grounds, as it might have allowed third parties to monitor whether people are actually in their homes. These concerns could have been avoided if societal actors had been involved in the design of technology early on. "Privacy by design" has become a good counter example in the field of ICT, by which technology is designed with a view to taking privacy into account as a design principle of the technology itself. Yet, practicing it is still rare. The European project ETICA[6] has

[6]See: http://www.etica-project.eu/.

recommended the introduction of specific governance structures for emerging (ICT) technologies in this regard.

Recently "Midstream Modulation" (Fisher et al. 2006; Fisher 2007) has emerged as a promising approach to increase social-ethical reflexivity within research practices. In the form of laboratory engagement practices, social scientists and ethicists are embedded in research teams of natural scientists. The embedded social scientist engages natural scientists in the wider impact of their work, while doing research in the laboratories. Reports from these practices could feed into schemes on responsible research and innovation.

3.5.5 Deliberative Mechanisms for Allowing Feedback with Policymakers: Devising Models for Responsible Governance and Public Engagement/Public Debate

Continuous feedback from information generated in Technology Assessment, Technology Foresight and demonstration projects to policy makers could allow for a productive innovation cycle. Knowledge assessment procedures should be developed in order to allow assessment of the quality of information within the policy process, especially in areas in which scientific assessments contradict each other or in the case of serious knowledge gaps. (The EC practises this partly with its impact assessments for legislative actions). Knowledge assessment could integrate distinct approaches of cost-benefit analysis and environmental and sustainability impact assessments. In short: models of responsible governance should be devised which allocate roles of responsibility to all actors involved in the innovation process. Ideally, this should lead to a situation in which actors can resolve conflicts and go beyond their traditional roles: companies addressing the benefits and Non-Governmental Organisations the risks. Co-responsibility implies here that actors have to become mutually responsive, thus companies adopting a perspective going beyond immediate market competiveness and NGOs reflecting on the constructive role of new technologies for sustainable product development. In this context, Technology Assessment, as practised, for example, by the Dutch Rathenau Institute, can take up the function of "seducing actors to get involved and act" (Van Est 2010).

On-going public debate and monitoring of public opinion is needed for the legitimacy of research funding and particular scientific and technological advances. Continuous public platforms should replace one-off public engagement activities with a particular technology and, ideally, a link with the policy process should be established. The function of public debate in viable democracies includes enabling policy makers to exercise agenda and priority setting. Public debate, ideally, should have a moderating impact on "Technology Push" and "Policy Pull" of new technologies which sometime unavoidably may occur.

3.6 Outlook for Implementing Responsible Research and Innovation

The most crucial advancement of RRI will be dependent on the willingness of stakeholders to work together toward social desirable products. Up till now, the examples of industry-NGO cooperation has been primarily limited to addressing the risks, e.g. the negative aspects of products. Under the European 7th Framework Programme for Research and Innovation, the 2013 Science in Society Work programme provides an opportunity for a "demonstration project" incentivizing actors from industry, civil society and research institutions to "jointly define an implementation plan for the responsible development of a particular product to be developed within a specific research and innovation field". Responsible Research and Innovation should be shown in terms of the product development process (such as stakeholder involvement, etc.) and the quality of the final product (complying with, among other standards, those relating to sustainability and ethics).

Furthermore, further institutionalizations of technology foresight and technology assessments are necessary within the legislative process. At the European level, now impact assessments have been made mandatory, an opportunity arises to make better and systematic use of assessments. I have argued that we have to go beyond assessing research and innovation beyond their economic impacts. Bozeman and Sarewitz (2011) have proposed a framework for a new approach to assessing the capacity of research programs to achieve social goals. The further development of such frameworks are badly needed as the promises of scientist to address social objectives (regularly leading to a "hype" and corresponding increased levels of research funding) while developing their research is often sharply contrasted with the actual outcomes.

Internationally, a global perspective needs to be developed. Diverging ethical standards at the international level and "ethics-free" zones pose challenges to the introduction of RRI at the global level. Ozolina et al. (2012) have recently addressed the challenges RRI faces at the global level and advocate to advance an international framework for RRI by means of multilateral dialogue.

RRI should become a research and innovation 'design' strategy which drives innovation and gives some "steer" towards achieving societal desirable goals. We can start with this strategy at the level of research funding by public authorities.

References

Bedau, Mark, Emily C. Parke, Uwe Tangen, and Brigitte Hantsche-Tangen. 2009. Social and ethical checkpoints for bottom-up synthetic biology, or protocells. *Systems and Synthetic Biology* 3: 65–75.

Bozeman, Barry, and Daniel Sarewitz. 2011. Public value mapping and science policy evaluation. *Minerva* 49: 1.

Bush, Lawrence. 2010. Standards, law and governance. *Journal of Rural Social Sciences* 25(3): 56–78.

Commission of the European Communities. 2002. *Regulation (EC) no 178/2002 of the European Parliament and of the Council of 28 January2002 laying down the general principles and requirements of food law, establishing the European Food Safety Authority and laying down procedures in matters of food safety.*

Commission of the European Communities. 2008. *Commission recommendation of 7 February 2008, on a code of conduct for responsible nanosciences and nanotechnologies research*, 7 Feb 2008.

Commission of the European Communities. 2009. *Communication from the commission to the council, the European Parliament and the European Economic and Social Committee. Nanosciences and nanotechnologies: An action plan for Europe 2005–2009. Second Implementation Report 2007–2009,* Brussels, 29 Oct 2009, COM (2009) 607 final. Citation on page 10.

Den Boer, Duncan, Arie Rip, and Sylvia Speller. 2009. Scripting possible futures of nanotechnologies: A methodology that enhances reflexivity. *Technology in Society* 31(2009): 295–304.

European Commission. 2011. From Challenges to Opportunities. *Towards a Common Strategic Framework for Research and Innovation Funding*. Green Paper of the European Commission, p. 3.

European Communities. 2006. *Better regulation. Simply explained.* Luxembourg: Office for Official Publications of the European Communities.

European Union. 2010. Consolidated version of the Treaty on European Union. *Official Journal of the European Union* 53: 13, C83 of 30 March 2010, article 3.

Fisher, Erik. 2007. Ethnographic invention: Probing the capacity of laboratory decisions. *NanoEthics* 1(2): 155–165.

Fisher, Erik, Roop L. Mahajan, and Carl Mitcham. 2006. Midstream modulation of technology: Governance from within. *Bulletin of Science, Technology & Society* 26(6): 485–496.

Fisher, Erik, C.P. Slade, D. Anderson, and B. Bozeman. 2010. The public value of nanotechnology? *Scientometrics* 85(1): 29–39.

Forsberg, Ellen Marie. 2010. Safe and socially robust development of nanofood through ISO standards? In *Global food security: Ethical and legal challenges*, ed. C.M. Romeo Casabona, L. Escajedo San Epifanio, and A. Emaldi Cirión. Wageningen: Academic Publishers.

Frodeman, Robert, and J. Britt Holbrook. 2007. Science's social effects. *Issues in Science and Technology*, Spring 2007, pp. 28–30.

Guagnin, Daniel, Leon Hempel, and Carla Ilten. 2011. Privacy practices and the claim for accountability. In *Towards responsible research and innovation in the information and communication technologies and security technologies fields*, ed. Rene von Schomberg. Luxembourg: Publication Office of the European Union.

Lund Declaration. 2009. Conference: new worlds – new solutions. Research and innovation as a basis for developing Europe in a global context, Lund, Sweden, 7–8 July 2009. Online available at: http://www.vr.se/download/18.7dac901212646d84fd38000336/. Accessed 20 May 2014.

Mantovani, Elvio, and Andrea Porcari. 2010. A governance platform to secure the responsible development of nanotechnologies: The Framing Nano Project. In *Understanding public debate on nanotechnologies. Options for framing public policy*, ed. Rene Von Schomberg and Davies Sarah. Luxembourg: Publication office of the European Union.

Mitcham, Carl, and Robert Frodeman. 2000. Beyond the social contract myth: Science should move beyond a contractual relationship with society and join in the quest for the common good. *Issue in Science and Technology Online,* Summer 2000, pp. 15–18.

Owen, Richard, and Nicola Goldberg. 2010. Responsible innovation. A pilot study with the UK Engineering and Physical Science and Research Council. *Risk Analysis* 30(11): 1699.

Ozolina, Zaneta, Carl Mitcham, Doris Schroeder, Emilio Mordini, Paul McCarthy, and John Crowley. 2012. *Ethical and regulatory challenges to science and research policy at the global level.* Expert Group report, Directorate-General for Research and Innovation of the European Commission. Luxembourg: Publication office of the European Union.

Van Est, R. 2010. From techno-talk to social reflection and action. Lessons from public participation in converging Technologies. International workshop "Deliberating converging technologies", IÖW, Berlin, 25–26 Nov 2010.

Von Schomberg, Rene. 2007. *From the ethics of technology towards and ethics of knowledge policy*. Working document of the Service of the European Commission. Obtained through the internet http://ec.europa.eu/research/science-society/pdf/ethicsofknowledgepolicy_en.pdf. Accessed 18 Nov 2012.

Part II
Governance and Institutional Design

Chapter 4
Innovation and Responsibility: A Managerial Approach to the Integration of Responsibility in a Disruptive Innovation Model

Xavier Pavie and Julie Egal

Abstract Progress of modern science and technology provides managers with a very large range of innovation opportunities, which do not necessarily benefit customers and society in the long term, and because they are often primarily concerned with economic value and short-term development, do not take into account the impact and potential threat on society. Because responsibility should not be limited to the scope of social business and micro-projects, we must consider responsibility as a major determinant to innovation, and from a managerial point of view, integrate forecasting and anticipation in the decision-making process. Differing from the traditional approach to responsible innovation often only addressed through an expert perspective, and based on an original survey conducted in companies, this paper aims at providing an insight into managerial decision-making processes regarding the launch of innovation on the market.

4.1 Background: The Challenge of Integrating Responsibility in Traditional Innovation Models

Innovation comes from the Latin *innovationem*, noun of action from *innovare*, *in – novare*: "in" for inside, "novare" for change. Innovation was originally seen as the process that renews something that exists and not, as is commonly assumed, the introduction of something new. Newness often implies uncertainty, regarding consequences and impacts. The consequences of innovation, by nature, simply cannot be predicted despite the many surveys and market studies undertaken by companies prior to launching a new product or service onto the market. In his description of innovation, Schumpeter particularly underlined that innovation only

X. Pavie, Ph.D. (✉) • J. Egal
Institute Strategy for Innovation and Service, ESSEC Business School, Paris, France
e-mail: pavie@essec.edu

J. van den Hoven et al. (eds.), *Responsible Innovation 1: Innovative Solutions for Global Issues*, DOI 10.1007/978-94-017-8956-1_4,

occurs once the product or service has been launched and attracts enough customers to become significantly profitable (Schumpeter 1939).

Anticipating the future consequences of innovation is a major challenge because technological progress and modern science have added complexity to people's lives, and give individual real power over the environment and society, which can eventually threaten the integrity of ecosystems upon which human society depends. Man has the responsibility to protect himself and his sustainability.

"Act only in accordance with that maxim whereby you can at the same time will that it should become a universal law." (Kant 1785) Referring to this version of Kant's categorical imperative, Jonas gives his imperative as follows: "Act so that the effects of your action are compatible with the permanence of genuine human life" or so that they are "not destructive of the future possibility of such life." (Jonas 1979). For Jonas is no logical contradiction in favouring the well-being of the present generation to that of future generations, or in allowing the extinction of the human species by destroying our planet. The imperative of responsibility differs from the ethics of Kant because it relies on the principle that we owe something to the future generations, even if we will never be directly in relationship with them.

Jonas argues that humanity is in a new ethical movement, that the recent scientific, technological and economic developments have raised new challenges for society: Jonas explains that humans now suffer from an ethical gap, created because of the chasm that exists between technological performance and the capacity of individuals for exercising moral responsibility, and that traditional ethics do not provide a clear guidance to the understanding of these issues anymore (Jonas 1979). Since nature now constitutes an important focus of human responsibility, and since many actions undertaken by individuals can have an irreversible effect on nature, this notion of responsibility spreads beyond human relations, and should thus be incorporated in any long-term effects of forecast.

A common acceptance of the term "responsible" within an organizational context is difficult to find. According to François Eswald, "what makes us responsible is the fact that we make decisions when we are responsible for others. This dimension cannot be seized by law because law thinks responsibility in terms of norms and of breaking of those norms. Yet we are not completely feeling responsible when we are submitted to norms. The experiment of responsibility begins with making a decision in which norms had no part" (Eswald 1996). This dimension was the one adopted by Petersen when he underlined the space we implement in responsibility between the 'do no harm' and the 'do good' (Pedersen 2010). The question of submission to norms thus differs from doing good; the latter is defined as going positively beyond norms.

In the recent past, this notion of responsibility has evolved. From consumer credit to the last cellulars, everything has suddenly acquired 'responsible' coating. Following the 'green washing' trend, it appears the next one will be the 'responsibility washing' trend.

There is clearly a need for responsible innovation, but the term is no longer keeping pace with its meaning, too unclear and trivialized. As well as having a passive and defensive coloration, it does not allow to point out the particularities of its object precisely enough; it thus remaining of little use. In this paper we will try to understand what responsibility means as far as innovation is at stake.

First we will try to figure out where responsibility is at stake, by examining Clayton Christensen's innovation model, which stands for a very useful analysis tool regarding differences between disruptive innovation versus incremental innovation (Christensen 2003). Then we will look at catalytic innovation, an attempt by Christensen to provide a societal benefit through disruptive innovation (Christensen 2006). Finally we will see how the results of an original survey targeted at managers show that they choose to innovate whether the future consequences of the launch of an innovation are foreseeable or not. In conclusion, we will try to understand what hinders responsible innovation in companies.

4.2 Introduction to Christensen's Model

Christensen's model distinguishes between two types of innovations: incremental innovation and disruptive innovation (Christensen 2003).

Incremental innovation has a minor impact on the market and does not change conditions of use radically. It usually builds upon existing knowledge and resources within companies: it is competence-/performance- enhancing. This type of innovation is usually pulled by the customers.

In contrast, disruptive innovation consists of designing for a different set of consumers. It has by nature an impact that the market does not expect. It usually modifies conditions of use for customers and usually implies a radical technical or technological change. The personal computer (PC) is an example of disruptive innovation: before PCs, computing was done through expensive mainframe centers, and was therefore not accessible to the mainstream market.

Since a company is able to innovate faster than what customers can "digest", low-end disruption occurs when the rate at which products improve exceeds the rate at which customers can adopt the new performance. At some point, a disruptive technology may enter the market and provide a product which has lower performance than the incumbent but which exceeds the requirements of certain segments. When technology outperforms consumers' expectations, only a niche of "premium" consumers will want to buy the product/services at a high price in a very competitive environment. Other consumers may favour disruptive innovation.

"New market disruption" occurs when a product fits a new or emerging market segment that is not being served by existing industry.

Some disruptive innovations can be hybrid: both low-end and new market (Knopper 2009). For instance, Amazon.com is a low-end disruptive innovation as, since the 1990s, when the music industry phased out the single, many consumers couldn't afford buying music. Amazon put an end to this by enabling "poor" consumers to buy a single song for a cheap amount (0, 99 cents). On the other hand, it eventually became a new market disruption by undermining the sales of physical CD's: total industry sales were about $10 billion last year, down from $14 billion in

2000, according to the Recording Industry Association of America, mainly because of digital music such as music available on amazon.com.

Christensen's disruption model therefore provides a comprehensive and useful insight to understanding innovation (Christensen 2003). We will try to find out where the major risks regarding responsibility stand in this model, starting with disruptive innovation, which seems to be providing more uncertainty than incremental innovation.

4.3 Disruption vs. Responsibility: An Antinomy?

4.3.1 Diffusion of Incremental Innovation

Incremental innovation usually follows a traditional adoption pattern. The traditional adoption curve, as described by Everett Rogers, is a S-shaped curve, showing the rate of adoption of an innovation by four different types of consumers: innovators (2.5 %), early adopters (13.5 %), early majority (34 %), late majority (34 %) and laggards (16 %) (Rogers 2003). The way to develop a market is to follow the curve from left to right, using each captured group as a reference for the next group to adopt the innovation. The early majority naturally follows early adopters, because of learning and adaptability to technological progress. Thus, with incremental innovation, evolution of adoption is rather predictable, and uncertainty is therefore reduced.

Specific challenges arise when disruptive innovation is at stake.

4.3.2 Diffusion of Disruptive Innovation

Being a disruptive innovator sometimes implies "crossing the chasm" of the product/service adoption curve, which is different from following the traditional adoption pattern. According to Moore, for disruptive innovations, adoption does not come in a predictable way: it makes the transition between visionaries (early adopters) and pragmatists (early majority) a difficult and unpredictable step to follow (Moore 1999). Indeed, it is very difficult to convince pragmatists with a totally new product or service. References are very important to them, and they do not necessarily trust early adopters. Pragmatists won't buy until the company and its new offer are established, but in order to establish a company, pragmatists have to be involved … And so if trust is acquired and early majority starts buying, the development can be exponential. But the innovation might as well be rejected by pragmatics and make it have no impact. It is therefore a real challenge for companies to foresee the development of a disruptive innovation in terms of market size (Fig. 4.1).

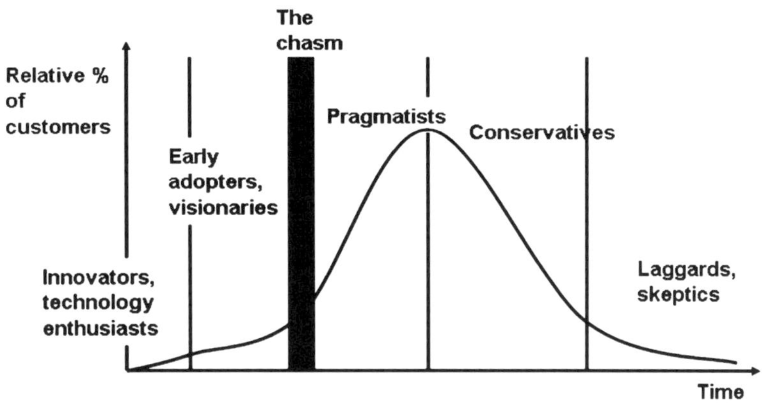

Fig. 4.1 Crossing the Chasm (Moore 1999)

4.3.3 Managing the Life Cycle of Disruptive Innovations

Because of the complexity of anticipating the success and the penetration of disruptive innovation in the market, it is particularly difficult to assess consequences and impacts that arise across the entire life cycle of an innovation, from its production to its withdrawal. Indeed, when the penetration of an innovation surpasses expectations, new challenges can emerge, such as the long-term availability of the resources needed to produce the innovation, the impact of a massive production and use in terms of energy consumption and rejects, and recycling opportunities.

The case study of the CRT versus LCD television market is a good example of what happens when a new technology disrupts a market. Liquid Crystal Displays (LCD) screens are key components in flat panel televisions, but also laptop computers, flat panel monitors, cell phones, PDAs, digital cameras, clocks, watches, GPS receivers, answering machines and other electronic devices. Many of these are in very high demand and LCDs are being incorporated into an increasing number of devices. As far as it is known, Indium Tin Oxide remains the best material for LCD and other flat panel displays applications. It offers the best performances in terms of optical transparency, electrical resistivity, uniformity of transparency and resistivity, chemical and mechanical stability, resistance to corrosion... New material sets could be developed as replacements for ITO, however, this not a likely scenario in the near future. Any change would require significant research and development, life testing, process changes and equipment changes. Indium is expected to disappear in the next decades (forecasts differ slightly from one source to another, but most of them claim that Indium should be extinct by 2025). What is more, at the end of the life cycle, recycling opportunities remain almost inexistent. The use of nitrogen trifluoride (NF3) during the production of LCD screens is another important issue. NF3 is a greenhouse gas, and an important contributor to global warming. As NF3 was not in widespread use by the time Kyoto Protocols were implemented, there is no incentive for companies to reduce their production. The huge success of LCD screens has therefore strong consequences in terms of sustainability of this offer, but also strong environmental impacts.

Another example is the iPod. When more than 20 different versions of iPods are launched by Apple within 8 years, between 2002 and 2012, when more than 275,000,000 iPods were sold by August 2010 making it the first numeric personal stereo far ahead of its competitors, questions arise. Can iPod be easily recycled? How sustainable (regarding natural resources) can such a mass production be?

Even low-end disruptive innovation, which usually relies on lower cost versions of existing solutions, faces similar challenges, since it allows a large number of people to access a product/service they could not previously afford. This generalization of access can become a threat and have a strong impact on society or environment. In India for instance, in addition to increasing the general chaos of the streets, the rise in car ownership, with the development of the Tata Nano and other low-cost cars, worsens air quality and lead to more global warming pollution. Rajendra Pachauri chairman of the Intergovernmental Panel on Climate Change, who got Nobel Prize in 2007, said he had "nightmares" about the impact the Tata car would make to environment. As a consequence of its low cost (120,000 rupees, around 2,500$), the number of Indian families that can afford a car could almost double. Even if the car consumes only at an average of 20 km per gallon – lower than the European average – the impact on environment could be high if about 250,000–300,000 cars were produced per year.

Not knowing the size of potential adoption market makes it difficult for companies to manage the sustainability of the product/service over its entire life cycle. But even if forecasting the adoption was possible, some uncertainty would still remain concerning the consequences of disruptive innovation, following the idea introduced by Hans Jonas that the field of human action is now greater that the field of human knowledge (Jonas 1979), and that models based on risk analysis can fail, because of this knowledge gap.

4.3.4 The Knowledge Gap

Disruptive innovations often rely on new techniques or technologies, for which scientific knowledge is still limited, and for which all consequences cannot always be foreseen.

For instance, the impact of nanotechnologies, which are now used in many consumption products, is still uncertain, and the consequences on health and environment are not precisely known. Nanotechnology is science and engineering at the scale of atoms and molecules. Materials of this size display unusual physical and chemical properties. On the one hand, there are about a 1,000 products with nanotechnologies available on the French market, and in the short term, the greatest advances through nanotechnology should be related to new medical devices and processes, new catalysts for industry and smaller components for computers. The global revenue resulting from nanotechnologies, which was about 40 billion Euros per year in 2001, was estimated at around 700 billion Euros in 2008, and should reach a 1,000 billion Euros in 2015: this would represent the employment of two billion people worldwide. One the other hand, only about 3 % of research publications about nanotechnologies take into consideration the risks on environment and health, despite the fact that it has been proved that nanomaterials can get into the lungs or skin epidermis easier than any other material (INRS 2009). The issue of responsibility in the generalization of nanotechnologies should therefore be discussed.

Nanotechnologies are not the only example of such a dilemma between the economic potential of some scientific developments and the limits of knowledge concerning the consequences of their use. The inability to anticipate the consequences of disruptive innovation, and therefore its consequences on society, ecosystems and the environment, requires the implementation of responsibility as a key element of the model. Understanding that societal stakes are high, because of potential threats created by innovation, is therefore a key challenge for managers, who can no longer only rely on risk analysis models to guide their decisions, and who must also be aware that in case of the emergence of an unanticipated risk, they must preserve society by stepping back and caring for all stakeholders.

The consequences of disruptive innovation are thus led by two major factors of uncertainty: the complexity to anticipate adoption levels and therefore manage the resulting mass effect on the entire life cycle, and the gap created by our limited knowledge and the existence of unpredictable risks. But one cannot describe the importance of responsibility in disruptive innovation without mentioning the Christensen's work on catalytic innovation as a first step towards this issue.

4.4 Catalytic Innovation: A First Step Towards Responsibility?

To a certain extent, Christensen introduced a notion of responsibility, but with a restricted scope.

Admittedly, following Christensen's article on "Disruptive Innovation for Social Change", the disruptive innovation model provides opportunities to create social businesses through catalytic innovation (Christensen et al. 2006). Indeed, disruptive innovations don't meet existing customers' needs for existing products or services. Certain "high tech" features of the established goods, which only appeal to high-end consumers, are not included in disruptive offers, which rely on more basic features and capabilities. Being simpler, these offers are often more convenient, and less expensive, so they appeal to the low-end of the market, who can afford to buy them.

They can provide access to new products and services for people at the Bottom of the Pyramid and therefore contribute to the development of groups of people who are marginalized in society.

"Catalytic innovations", which are a subset of disruptive innovations focusing on social development, can be found in sectors such as education, healthcare, banking. Catalytic innovations are characterized by four important functions, according to Christensen: they create systemic social change, they meet a need that remained unaddressed or over served (when they offer a too high level of performance compared with individual needs) by existing companies, they offer good products that are cheaper and simpler, and they generate donations ("micro-businesses", social funds, volunteer workforce . . .). They are often considered unattractive by competitors but have a dominant position on their market (Christensen et al. 2006).

The example of Eko Bank in India shows how using a very simple interface on mobile phones provides access to basic banking services to a large part of the Indian society. The service is available to the customers on all mobile phones including the most basic models. It provides access to a simple mini savings account. Its functionality range from peer-to-peer money transfers, cash deposit/withdrawal, wage and salary disbursements, to micro-insurance, micro-credit and payments. Mobile technologies provide various opportunities for catalytic innovators, especially in fields like health, or education. These opportunities, when launched successfully on the market, are said to be disruptive innovation, because they provide a simple service to low-end consumers who could not afford it before, based on relatively basic technologies, accessible to the "Bottom of the Pyramid" (Prahalad 2004).

But what is called "Bottom of the Pyramid innovation" or "social entrepreneurship" only accounts for a very small part of responsible innovation. Indeed, it focuses on the present more than the future. What we refer to as "responsible innovation" covers a much larger scope, which is indeed linked not only to the development of society, but which places the individual value at the center of any product or service development. "Social" is not a synonym for "responsible", which embraces a more holistic approach, focused not only on the present but also on the future, not only on society but also on environment and economic sustainability, not only on pragmatic solutions and action but also ethical debates and thinking.

In order to better understand the perception of responsibility by decision-makers in companies, an original online survey was conducted by ESSEC ISIS, in France; among people making decisions regarding innovation in their company. Based on the observation that responsibility had become a major stake for society, and that managers in companies did not seem to make decisions based on this responsibility pre-requisite, this survey aimed at understanding decision-making processes and qualifying the responsibility-sensitiveness of managers. It therefore provides an insight into managerial behaviors towards responsible innovation inside the company, differing from traditional approaches to responsible-innovation which often only address experts' points of view.

4.5 How Decision-Makers Understand and Implement Responsibility: Results of a Survey

4.5.1 Scope of the Survey

The survey was initially sent to a large range of managers, mostly from French companies, and with an interest in innovation. Over a period of 4 months, between January and April 2011, 62 people out of 78 respondents (out of 280 questionnaires sent) completed the entire survey. Among those 78 respondents, 80.8 % came from a service company, 65.4 % were men, and 65.4 % were in a company with more than 2,000 employees. 92.3 % were involved in decisions concerning innovation. Answers from the six people who didn't participate to decision-making processes were not taken into account in the following results.

There were five "profile" questions in the survey: Gender, Company size, Function, Sector of the company, and Innovation decision-making involvement (Yes, always/Yes, sometimes/No). Unfortunately no clear trend related to profile characteristics emerged from the results.

Then, for the decision-makers, six other closed questions were asked (Table 4.1).

4.5.2 About Anticipation and Forecast – Questions 2, 3 & 4

Can you forecast the social/environmental consequences of innovation launched by your company? (Table 4.2)

The results of the survey show that decision-makers have a clearer vision on the potential impacts of innovation launched by their company in the short-term than in the long-term:

- When, in the short-term (3 years to come) 29 % of decision-makers declare they can anticipate precisely the impacts on society, and 23 % the impact on the environment, in the medium term (3–10 years), only 16 % have a precise idea about the social impact and 13 % about the environmental one.
- In the long-term (more than 10 years), only 9 % of decision-makers deem they are able to anticipate precisely the social impacts of innovation, and 8 % the environmental one.
- In the short-term, social impacts are easier to forecast than environmental ones, but in the long-term the results are more balanced.

The survey therefore shows that decision-makers in company are aware of their incapacity to forecast the consequences on innovation on which they decide.

Table 4.1 Survey questions and answers

Question	Possible answers
1. Why do you innovate?	Competition/Compliance to the law/ Customer need/Technological opportunity/Other
2a. Can you forecast the short-term (less than 3 years) consequences and impacts of innovation launched by your company, from a social point of view?	Yes, precisely/Not really, there are still uncertainties/Not at all
2b. Can you forecast the short-term (less than 3 years) consequences and impacts of innovation launched by your company, from an environmental point of view?	Yes, precisely/Not really, there are still uncertainties/Not at all
3a. Can you forecast the medium-term (3–10 years) consequences and impacts of innovation launched by your company, from a social point of view?	Yes, precisely/Not really, there are still uncertainties/Not at all
3b. Can you forecast the medium-term (3–10 years) consequences and impacts of innovation launched by your company, from an environmental point of view?	Yes, precisely/Not really, there are still uncertainties/Not at all
4a. Can you forecast the long-term (more than 10 years) consequences and impacts of innovation launched by your company, from a social point of view?	Yes, precisely/Not really, there are still uncertainties/Not at all
4b. Can you forecast the long-term (more than 10 years) consequences and impacts of innovation launched by your company, from an environmental point of view?	Yes, precisely/Not really, there are still uncertainties/Not at all
5a. If you cannot anticipate the social consequences of innovation, do you choose to innovate anyway?	Yes, for sure/Yes, maybe/No I don't think so/Not at all
5b. If you cannot anticipate the environmental consequences of innovation, do you choose to innovate anyway?	Yes, for sure/Yes, maybe/No I don't think so/Not at all

Table 4.2 Survey results to questions 2, 3 & 4 (61 responses)

	Short term impact		Medium term impact		Long term impact	
	Environmental (%)	Social (%)	Environmental (%)	Social (%)	Environmental (%)	Social (%)
Yes, precisely	23	29	13	16	8	9
Not really, there are still uncertainties	53	65	52	52	34	31
Not at all	24	6	35	32	58	60

4.5.3 About Decision-Making and Responsibility

If you cannot anticipate the social/environmental consequences of innovation, do you choose to innovate anyway? (Table 4.3)

Table 4.3 Survey results to question 5 (61 responses)

	Environmental (%)	Social (%)
Yes, for sure	16	11
Yes, maybe	35	35
No, I don't think so	47	48
Not at all	2	5

Despite the decision-makers' inability to anticipate precisely the social impact, even in the short-term, of innovation, almost 47 % of them choose to innovate anyway. It is even more accurate when the environmental impact is concerned. Indeed, more than 51 % of decision-makers innovate, even if they do not have a clear forecast of the impact of their choice.

Innovation for decision-makers in company remains a necessity, mostly because of market demand (for 82 % of them), and of technological opportunities (60 %). They feel compelled to innovate, or at least they do not feel concerned by a responsibility towards the people and the planet.

Many reasons for this irresponsible behavior can be found in the comments from respondents, from an economic responsibility for the survival of the company ("If I do not innovate first I will lose my competitive advantage"), to the pressure coming from the shareholders. What is more, innovative firms tend to be more decentralized than others, with group projects involving different categories of employees and a flat hierarchy, resulting in a dilution of responsibility in a collective unconsciousness.

It can be noticed that for 52 % of decision-makers, innovation is resulting from a will to comply with the law or anticipate its evolution. Therefore, law can be driving responsibility, even if no ethical conviction lies behind the action.

4.6 Fulfilling Responsible Innovation and the Race for Competitiveness: A Dilemma?

Today, shrinking product life cycle and the race for competitiveness through innovation, because of market pressure, give little time for companies launching new products or services. This "time-based competition", as introduced by Stalk (1988), considers time a resource, an input in the innovation process: since time consumption acts as an opportunity cost, time-based strategy creates competitive advantage for the company. A product 50 % over budget but introduced on time generates higher profit levels than a product brought to market 6 months late (within budget) (Inman 1992). If launched 6 months late, a product with a 5 year life cycle can lose up to 33 % of its total lifetime net profit (INSEAD 2006). The speed of the innovation process therefore often poses a threat to responsibility, since it reduces the time dedicated to research and to the analysis of direct and indirect consequences of new products or services.

As CEOs and managers are rewarded for making quick decisions in complex situations, as they are selected for their ability to "act despite uncertainty", they no longer afford much time for in-depth study and review before making choices, and they tend to rely on quick decisions, which can threaten responsibility in the medium or long run. Many CEOs acknowledge that they "feel overwhelmed by data while still being short on insight" (IBM 2010). But at the same time, they can't wait to act, even in uncertain situations, because if they do not, competitors might consider that taking calculated risks can pay off. The ambiguity is in the notion of "calculated risks". Among the top leadership qualities, creativity is ranked at the top position, followed by integrity and global thinking, but focus on sustainability, humility and fairness stand at the bottom of the list (IBM 2010).

In this context, a framework for responsible innovation should be defined around three axes (Bensaude-Vincent 2009), which usually represent major obstacles to responsibility (Pavie 2012), and which are:

- The unique prism of answering consumer needs.
  Questioning the reasons for developing a particular innovation is of fundamental importance for the firm wishing to integrate responsibility in its strategy. The rise of a consumer need does not mean that it must be automatically met by a new product or service. Nowadays, the market is saturated with products to suit every single consumer's want. If we consider the fact that in less than 10 years, the market has seen a succession of more than 20 generations of iPods, there is, admittedly, a need for the dematerialization of music; but does that necessarily mean that consumers actually need so many different versions of what really is the same product concept and within such a short space of time?
- The innovator's incapability to calculate and predict the consequences of their product/service launch.
  This incapability is enhanced by the endless race to the market for each innovation, thereby generating quick and hasty decision-making processes. Once again, we are faced with the crucial dimension of time. This dimension even concerns an innovation like Facebook. Has the organization attempted to predict the many and risky consequences of a database which may well soon reach its billionth connected member?
- The introduction of new risks with societal and daily consequences on individual lifestyles.
  The consequences of an innovation launched in a particular sector can have knock-on effects in other sectors. This factor is rarely taken into account within innovation projects. The low-energy light bulbs, widely acclaimed for being ecologically-friendly are produced using rare-earths from China. However, it is acknowledged that the extraction of these rare minerals represents an ecological cost so large, that it would be preferable to keep using incandescent light bulbs.

These three axes are essential in the understanding and the possible integration of responsible innovation. They can generate awareness and provide guidance for decision-makers and innovators; they can even stimulate humility, cautiousness, longer-term thinking, and the ability to step back whenever unexpected consequences arise.

These axes must however remain linked to the ultimate objective of a firm to create value, which is a major incentive to innovation but also a condition for economic sustainability and therefore existence. Catalytic innovations, for instance, are successful in combining social benefits and economic sustainability: they illustrate how initiatives with a non-profit drive can create economic value as well. Thus, in order to favour decisions supporting responsible innovation within this framework, further research needs to be done about the relationship between responsibility and economic performance, and the type of organization optimizing the integration of responsibility in business models.

Finally, this framework is one of the many ways which can lead to responsible innovations. Political and legal action, societal debates, and education all need to be combined to progress towards responsible innovation. Company decision-makers are therefore not the only stakeholders involved in this process. Companies' ecosystems, including shareholders, suppliers, and consumers, as well as governments, NGOs, researchers, and other societal and economic influencers need to be aligned to contribute to change the current paradigm, with a drive to achieve responsible innovation.

References

Bensaude-Vincent, B. 2009. *Colloque Innovation responsable du 29 avril 2009*. Collège de France. http://www.universud-paris.fr/sites/default/files/attachments/Innovation_responsable_-_29_avril_-_invitation_et_programme_.pdf

Christensen, C. 2003. *The innovator's solution*. Boston: Harvard Business School Press.

Christensen, C., B. Heiner, R. Ruggles, and T. Sadtler. 2006. Disruptive innovation for social change. *Harvard Business Review* 84(12): 94–101.

Christensen, C.M. 2006. The ongoing process of building a theory of disruption. *The Journal of Product Innovation Management* 23: 39–55.

Ewald, F. 1996. *Histoire de l'Etat-Providence*, 86. Paris: Grasset. Folio.

IBM. 2010. [Online]. Capitalizing on complexity. *IBM's global CEO study 2010*. Available at: http://public.dhe.ibm.com/common/ssi/ecm/en/gbe03297usen/GBE03297USEN.PDF

Inman, R. 1992. Time-based competition: Challenges for industrial purchasing. *Industrial Management* 34(March/April): 31–32.

INRS. 2009. [Online]. (Homepage). *Nanomatériaux*. Available at: http://www.inrs.fr/accueil/produits/mediatheque/doc/publications.html?refINRS=ED%206050

INSEAD. 2006. Understanding and responding to societal demands on corporate social responsibility. RESPONSE Report 2006. http://www.insead.edu/v1/ibis/response_project/documents/Response_FinalReport.pdf. Accessed 15 June 2014.

Jonas, H. 1979. *The imperative of responsibility: In search of ethics for the technological age*. Chicago: University of Chicago Press.

Kant, E. 1785. *Grounding for the Metaphysics of Morals*, 3rd ed. Trans. James W. Ellington Hackett, 1993.

Knopper, S. 2009. *Appetite for self-destruction: The spectacular crash of the record industry in the digital age*. New York: Free Press.

Moore, G. 1999. *Crossing the Chasm*. Rev. ed. New York: HarperCollins Publishers.

Pavie, X. 2012. *L'innovation responsible, levier stratégique pour les organisations*. Eyrolles.

Pedersen, E.R. 2010. Modelling CSR: How managers understand the responsibilities of business towards society. *Journal of Business Ethics* 91(2): 155–166. doi:10.1007/s10551-009-0078-0.

Prahalad, C. 2004. *The fortune at the bottom of the pyramid*. Upper Saddle River: Wharton School Publishing.

Rogers, E. 2003. *Diffusion of innovations*. 5th ed. New York: The Free Press.

Schumpeter, J.A. 1939. *Business cycles. A theoretical, historical and statistical analysis of the capitalist process*. New York: McGraw-Hill Book Company.

Stalk, G. 1988. Time – The next source of competitive advantage. *Harvard Business Review* 41–55.

Chapter 5
Technology Transfer of Publicly Funded Research Results from Academia to Industry: Societal Responsibilities?

Elisabeth Eppinger and Peter Tinnemann

Abstract Publicly funded research aims to serve the public good; hence monopolies introduced by patents are highly debatable in their characteristic to foster innovation and economic growth. However, even with patent protection the research results could be transferred responsibly. This article explores option to enhance technology dissemination of publicly funded research results which are patent protected, using alternative licensing strategies such as *equitable licensing* and *patent pools* instead of exclusive licensing. We found that German research institutes lack incentives to license patents under these schemes and suggest that social responsibilities could be protected by implementing legal frameworks, and through policies of research organizations and research funding organizations. With our analysis we aim to contribute to the responsibility debate of technology transfer from publicly funded research to private industry.

5.1 Introduction

Every year, a substantial amount of tax funded research projects are conducted at universities and public research institutes. Large funding schemes focus on key technologies considered highly relevant for economic development, such as clean technologies, or areas where traditional market incentives fail, such as rare or

E. Eppinger (✉)
Faculty of Economics and Social Sciences, Chair of Innovation Management and Entrepreneurship, University of Potsdam, August-Bebel-Str. 89, 4482 Potsdam, Germany
e-mail: elisabeth.eppinger@ime.uni-potsdam.de

P. Tinnemann
Charité – Universitätsmedizin Berlin, Campus Charité Mitte, Institute for Social Medicine, Epidemiology and Health Economics, Luisenstr. 57, 10098 Berlin, Germany
e-mail: peter.tinnemann@charite.de

J. van den Hoven et al. (eds.), *Responsible Innovation 1: Innovative Solutions for Global Issues*, DOI 10.1007/978-94-017-8956-1_5,

so-called neglected diseases. In order to bring publicly funded research results to the market, they are made available to private industry to allow commercialization of innovations based on these research outcomes.

The traditional concept of publicly funded research is that researchers answer questions and provide solutions to issues which are relevant to society, in response their work is funded by tax payers' money. Publicly funded researchers making their research results accessible to society, in exchange receive reputation and fame. However, this system is changing drastically. Instead of the objective to disseminate the results as wide as possible, researchers at universities and other publicly funded research institutions are increasingly expected to apply free market rules to utilization of their work, to protect and to commercialize their research results.

Since recently, increased intellectual property protection and exclusive licensing of research results is discussed as hindering market competition and for its negative impacts on costs of new technologies, most importantly those of high societal importance (e.g. Anderson 2007; Murray and Stern 2006; Sampat 2009; Straus 2008). Moreover, in today's increasingly commercialized research environment effective transfer of research results from universities and publicly funded research institutes to the highest benefit for society is a matter of ongoing debate. In particular in areas where traditional market incentives fail to foster innovation, such as neglected diseases, the role of publicly funded research and new paths to develop research results towards commercialisation needs to be explored (Tinnemann et al. 2010). Further on, existing technology transfer concepts to address the research and access gap do not yet include social responsibility of academia (Wagner-Ahlfs 2009). While we experience a rise of patenting of publicly funded research, the underlying question remains whether patent protection in general is a suitable tool for technology transfer from academia to private industry. But even with patent protection it is possible to transfer technology responsibly. A few new initiatives started, aiming to achieve a broader dissemination of academia research results which are protected by patents. Amongst them the concept of equitable licensing and the concept of patent pools appear promising.

This paper aims to contribute to the debate on technology transfer and its responsibility in contributing to solve societal problems. By identifying reasons why researchers and technology transfer offices are hesitant to employ new concepts to increase dissemination of patented research outcomes, and by suggesting possible ways to overcome potential obstacles, we want to enhance current technology transfer practices. Furthermore, from a purely scientific perspective, we contribute to current research on the impact of patent rights and innovation, and methods to increase technology dissemination.

The remainder of the paper explores the suitability of two most promising concepts – equitable licensing and patent pools – and reasons why they are not broadly employed by academia yet. Section 5.2 provides a brief introduction on current theory on technology transfer, as well as new technology transfer concepts equitable licensing as developed by Godt (Godt and Marschall 2010), and patent pools. Section 5.3 explains the research question, research design, and the analytical approach – actor-centred institutionalism. Section 5.4 provides an overview of the key findings on incentives to apply exclusive licensing, equitable licensing and patent pools. Moreover, the necessity to include social impact within the effectiveness measurements of technology transfer is discussed. Section 5.5 concludes with a reflection of the results and some thoughts on further research.

5.2 Technology Transfer: Current Rational and Effectiveness Concepts and Alternatives to Exclusive Licensing

It is widely accepted that with our current economic system private industry is not able cover all relevant issues that societies are facing. For example, one major societal challenge is access to medicine in low income countries, and development of new innovative medicines for neglected diseases. While the issue of the existing research and access gaps for medicines for neglected diseases and diseases of poverty is well documented, possible solutions to address those are under debate (Sampat 2009). One possible solution could be publicly funded academic research, whose results are transferred to industry, including to those in low income countries, which provide medicines or patients in these countries. In order to achieve this, patents need to be licensed responsibly. This section explains the background of the current predominant technology transfer system – *exclusive licensing* – and introduces the two alternative concepts which could achieve responsible technology transfer – *equitable licensing* and *patent pools*.

5.2.1 The Rational of Patenting in Academia and Exclusive Licensing for Technology Transfer

Exclusive use of intellectual property rights (IPR) protected research results is intended to foster strategic utilization of market exclusivity for financial gains, hence economic growth. To enhance these strategies for further commercialization of research results, new policies and laws like the Bayh-Dole Act were introduced to allow universities and other publicly funded research institutes to protect the intellectual property (IP) of their research results through patents and other IPR (Mowery et al. 2001). The traditional view that research results should be freely available to everyone interested was succeeded by the rational that private companies are only interested in utilization of publicly funded research results if they will be able to use them exclusively. This policy change was initiated when investigators in the U.S.A. tried to understand why firms do not fully use research results. They concluded that industry would take greater interest if they could obtain, at least some, market exclusivity. Consequently, policy makers advised legislators to allow public research institutes and universities to patent their research results, even when the research was publicly funded.

Since then it is clearly observable that the international technology transfer practices have shifted with the introduction of the Bayh-Dole Act from freely accessible research results towards IPR protected exclusive access for commercial remuneration (Baldini 2009). Policies similar to the US Bayh-Dole Act were recently introduced in European countries such as Germany (Tinnemann et al. 2010) with some regional the differences. However, there are major difference between the legislation in the US and Germany. Whereas the US research institutes remain the owner of the patents, Germany allows the sale of patents owned by public research institutions. The exclusive transfer to firms aims to provide industry with incentives to further develop the research results. But whether the free-market driven concept behind technology transfer of research results is more efficient than the old concept of free access is still yet a matter of debate. To eventually allow answering these questions, suitable instruments and key measures are lacking to obtain data and compare effects.

Two developments of current dominating global economic policies manifest the importance of technology transfer by means of patents: the overall strengthening of the IPR systems through the TRIPS-Agreement (WIPO 1996), and increasing competition for public funding amongst research institutes. The TRIPS-Agreement introduced the concept of IPR also in developing countries and provided greater harmonization of IPR across all WTO members. Accordingly, international technology transfer by means of patents became easier and more interesting for multinational corporations. The increasing competition for public funding entailed the introduction of research results and technology transfer measurements for better comparison, amongst them the number of patent approval and licensing agreements. To ensure that shortcomings of the current system are addressed, it becomes even more important to find alternative ways to overcome disadvantages inherent to exclusive licensing and complete transfer of patent ownership. That does not necessarily mean that the knowledge needs to be brought into the public domain without any protection. Several IPR arrangements exist to shift IP from private property to shared property or common goods. Open source software development is probably the best known concept.

Research analysing technology transfer effectiveness has shown, that simply counting licensing contracts or revenues does not adequately represent the impact of transferred technology. It is much more difficult to define and to measure, because the impacts are numerous, interrelated, and almost impossible to be separated from other aspects that determine the success and failure of a new technology (Astor et al. 2010; Teece 2008). Bozeman (2000) proposed a contingent effectiveness model of technology transfer that aims to capture various effectiveness criteria necessary to assess the transfer outcome (see Fig. 5.1), based on an extensive literature review on empirical and conceptual studies of technology transfer from universities and government laboratories to industry.

His model covers a comprehensive number of relevant aspects influencing the technology transfer process and its effectiveness. According to the model the most important dimensions that impact on effectiveness are the transfer agent, the transfer medium, the transfer object, the transfer recipient, and the demand environment.

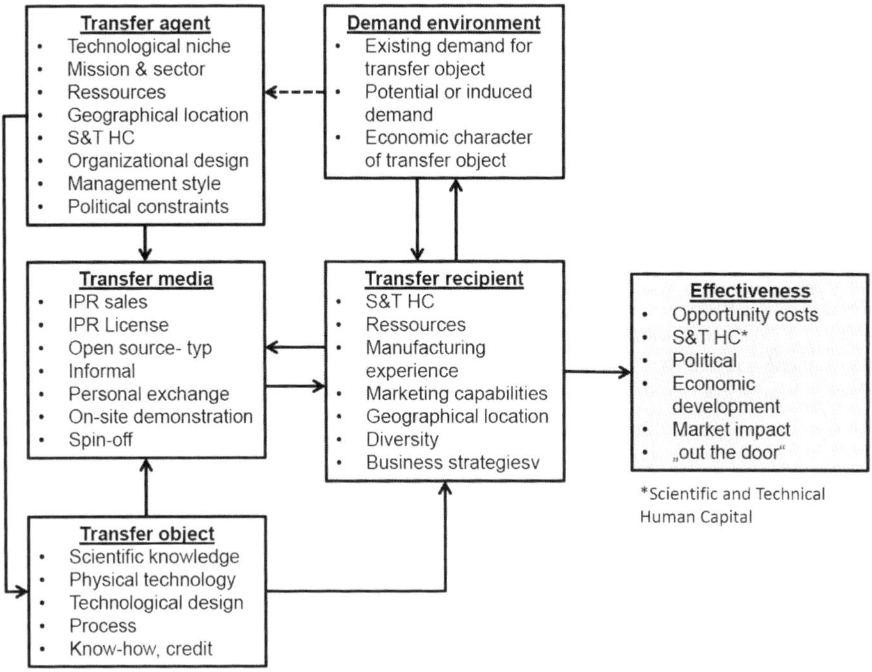

Fig. 5.1 Technology transfer effectiveness model (Source: Based on Bozeman 2000, 369)

Whereas all participating stakeholders and objects have different interests and shape possible transfer modes, and have different criteria to assess the effectiveness. The main effectiveness criteria that Bozeman (2000) defined are:

1. "out-the-door": when the successful transfer process is measured, regardless of further impact,
2. market impact: the commercial impact for the firm, margin, profit, obtained market share and size,
3. economic development: impact on regional or national markets, effects on markets in specific territories,
4. political reward: political reward due to the transfer, e.g. increased funding,
5. opportunity costs: alternative use of resources and impact on other strategic areas,
6. scientific and technical human capital: increase of skills, networks, and infrastructure.

The model represents the combination of different views on technology transfer effectiveness from technology transfer offices (TTOs), policy makers and innovation studies scholars. It subsumes the rational, that the success of technology transfer is dependent on commercial success of firms and economic growth. Patents are designed as an incentive to invest in technological innovations. The predominant economic theory regards patents as goods which are best used, when a single patent owner has full control over the right. This view entails exclusive licensing or sale of patents as most efficient transfer mode. It is challenged by the notion of competition, that competition is important for the public good to decrease prices and incentivize private companies to use their resources including patents most efficiently. Accordingly, non-exclusive licensing to multiple firms could increase the competition amongst them.

More recently developed technology transfer models are equitable licensing and patent pooling, which aim to increase technology dissemination, despite existing patent protection. The concept of equitable licensing, based on humanitarian licensing initiatives of Northern-American universities, is explored and further developed by a German research consortium of the University of Oldenburg, Charité Universitätsmedizin Berlin and BUKO Pharma (Godt 2011; Godt and Marschall 2010). Equitable licensing is of interest to our study, because it is under discussion at German universities, public research institutes and politicians and holds the potential to be implemented in the near future. The concept of a patent pools, broadly employed by various companies in the electronic and IT industries (Grassier and Capria 2003; Verbeure et al. 2006), is adapted by UNITAID ('t Hoen 2011) for the development for neglected tropical diseases and HIV/AIDS medicines. The following two subsections introduce these concepts in more detail.

5.2.2 Equitable Licensing – A Differentiated Transfer Approach

Equitable licensing developed from the initially introduced humanitarian licensing, aims to combine IPR and societal responsible technology transfer by increasing access to research results to achieve higher dissemination and counteract monopolies with high prices (Godt 2011). The discussion about the necessity for new technology transfer concepts started in 2001 when Yale University, holding relevant patents for the HIV-medicine Stavudine (Zerit®), initially provided a single pharmaceutical company with an exclusive license and subsequently renegotiated the licensing contract to allow also other manufacturers bringing the product onto different markets at lower prices (Mimura 2010; Wagner-Ahlfs 2010). However, the concept is not limited to medicines, but it is also enticing for other technologies with high societal impact such as environmental technologies or infrastructure technologies in IT.

To accommodate the practices of technology transfer in Germany, Godt (2010) proposes a differentiated modular approach rather than licensing all results non-exclusively. The concept of equitable licensing is further developed, so that it does not necessarily imply to provide research results for free to anyone without providing some monopoly. In fact it is proposing a modular system of golden, silver and bronze versions of licenses (see Table 5.1), depending on the level of intellectual property rights at stake, while various options can be chosen.

Table 5.1 Equitable licensing concept

	Gold	Silver	Bronze
Distribution of rights	Non-exclusive licensing only	Differential licensing	Exclusive licensing possible, but obligatory realization plan
Conditions	Improve provision Technology building Education at target countries	Reasonable pricing	
Realization	Obligations milestones patronage [private person/NGO]	Licensees at target countries	
Controlling	Monitoring contractual penalty		
Grant-back	No charges to university for research and commercial use [share alike]	No charges to university for research but royalty charges for commercial use	Royalty free only for research use

Source: Based on Godt and Marschall (2010)

For all three versions, which are explained in detail at Godt and Marschall (2010) the patent rights should stay with the university or publicly funded research institute. With signing away the ownership on patents, the research institute cannot exert any control regarding the use of the patented technology. To secure the rights, patent application for joint research should not be solely in the name of industry partner, also not in designated countries nor second application.

The gold version allows only non-exclusive licenses. An unlimited number of companies can obtain rights to manufacture, sell and develop further the research results at stake. By this, it is assumed that the prices will be competitive. Moreover, licensees should commit to technology building and education in designated countries. Also, a realization plan with obligations, milestones, and responsibilities could be included. To secure that firms and academia really stick to these conditions, Godt (2011) advises the patronage of a private person or NGO operating in the public interests. Moreover, the licensing contract should include penalties and the implementation of agreements should be properly controlled and enforced. Further developments based on research results should fall under a share-alike agreement as known in copyright licensing, they should be granted back royalty free to the university, for research purposes and commercial use.

The silver license is a more differentiated licensing approach, which allows only non-exclusive licenses to low-income countries but exclusive licenses to high-income countries. The binding condition, however, should require the licensee to offer products at reasonable prices. Moreover, those using the licensed research results should be aiming to develop technologies targeting low-income countries in order to support technology building. As for grant-back clauses, further developments have to be grated back to university for research and commercial use, however, reasonable royalties could be charged.

An agreement involving two or more patent owners to license one or more patents to one another or to third parties including:

- agreed cross-license,
- agreed framework for out-licensing to third parties which do not contribute own patents (main difference to simple cross-licensing scheme), and
- agreed pricing and royalty sharing scheme.

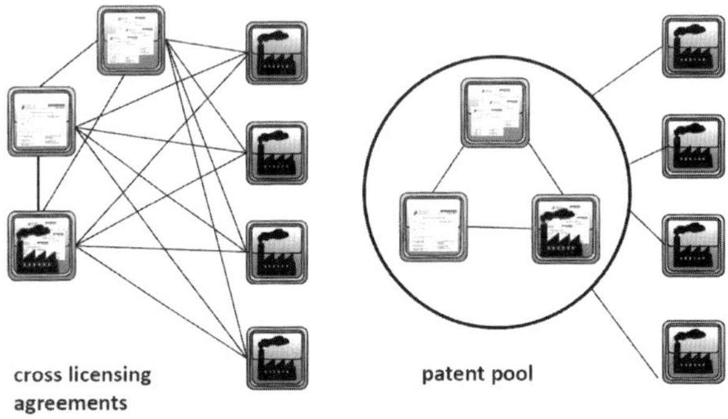

Fig. 5.2 Patent pool definition (Source: Based on McCarthy et al. 2004; Verbeuere et al. 2006)

Under the bronze license exclusive licenses are possible, but only in combination with an obligatory realization plan. The royalty free grant-back only applies to research, not to commercial use (Godt and Marschall 2010).

5.2.3 Patent Pools

When firms agree to share their patent rights, they usually decide on cross-licensing. But when more than two patent owners are involved, other forms such as patent pools can become more effective (see Fig. 5.2). A patent pool is a special co-operative patent arrangement that bundles patent rights for a specific technology in order to share this patent portfolio amongst the patent owners, and license it as a package to third parties. Instead of having to negotiate with each patent owner individually, interested parties can license a whole package with a single contract (Carlson 1999; McCarthy et al. 2004). Additionally, the licensor does not have to pay licensing fees to all relevant patent owners, which could sum up to high royalties. Consequently, patent pools can reduce transaction costs by shortening negotiation processes and counteracting royalty stacking (Merges 1999; Verbeuere et al. 2006).

The advantages of patent pools to foster innovation and overcome patent blocking and patent thickets were recognized in several studies (Clark et al. 2000; Shapiro 2000; Ziedonis 2004). Besides their potential to enable innovation when a multitude of patent owners agree on patent pools instead of blocking each other, they have also some negative effects that can hinder innovation and technology dissemination. Pools can enable agreements between firms resulting in market dominating positions (Gilbert 2004; Temple Lang 1994).

The majority of literature on patent pools focuses on suitable design from a legal and macro-economic perspective to avoid antitrust issues (e.g. Lerner and Tirole 2007), as the legal framework for patent pools evolved over time towards stricter guidelines. Due to several adjustments in antitrust law, nowadays patent pools have to be non-discriminating and should enable competition (Gilbert 2011). Interested parties are allowed to license patents on non-discriminatory terms, patents are limited to essential patents, and also grant-back clauses only apply to patents that found to be essential to the technology (EC 2004; Stumpf and Gross 2005; US Department of Justice and Federal Trade Commission 1995). Best-practice is to have an independent expert valuing the patents to determine essentiality and advice on fair and reasonable royalty rates (Verbeure et al. 2006; Van Overwalle 2009). On the other hand, the advantages and difficulties associated with setting up and executing patent pools for strategic innovation partnerships should not be underestimated. Especially universities underutilize patent pools for technology transfer as reports by technology transfer offices reveal (Astor et al. 2010). Consequently, the need for further research on patent pools exists.

Both concepts, equitable licensing and patent pools, despite their capability to disseminate technologies more broadly and although increasing known by researchers and technology transfer officers at publicly funded research institutions, are not yet widely utilised. Assessing technology transfer activities in Germany in 2010 showed that the transfer method of choice is still exclusive licensing (Astor et al. 2010).

5.3 Research Question and Research Design

With our analysis we aim to contribute to the responsibility debate of technology transfer from publicly funded research results to private industries. Therefore we identified and investigated subjective reasoning and objective obstacles of researchers, technology transfer officers and private industries and analyzed their different perspectives towards employing new technology transfer concepts to increase dissemination of research outcomes. While qualitatively analyzing our results, we suggest a quantitative method, based on actor-centered institutionalism, to establish evidence for the advantages of individual technology transfer strategies.

In order to investigate the potentials of equitable licensing and patent pools for technology transfer from publicly funded research institutes and universities to industry, we use the following guiding questions:

1. What are the constraints for publicly funded research institutes and universities to make use of equitable licensing and patent pools for technology transfer?
2. How can technology transfer offices and policy makers promote the use of these concepts in order to achieve higher dissemination of research outcomes?

We use actor-centered institutionalism as our analytical framework to analyze incentives and constraints, both internal and external, to publicly funded universities, publicly funded research institutes, and enterprises involved in technology transfer. The actor-centered institutionalism approach guides the analytical focus on actor's cognitive (knowledge, norms, believes) and motivational orientation (interests and goals, perceived opportunities and outcomes), their constellations (cooperative, competitive, profit maximizing or hostile) depending on their resources, their attitudes and engagements, and the institutional context (institutions in the sense of organizations, legal and political framework, and broader societal norms) (Mayntz and Scharpf 1995). Different to other streams of institutionalism, the decision on a certain course of action is therefore not only considered to depend on norms, interests, available resources and institutions, but also on actor constellations depending on the specific resources and perceived bargaining power (Scharpf 1997). This theoretical approach from governance studies is believed to hold advantages for analyzing the actors involved in innovation and technology dissemination (Schimank 2004). It is assumed that especially while negotiating and making strategic choices, where the outcome depends on all actors involved, actors cannot be sure that the other actors will adhere to agreements, even though they decide together on one course of action. We apply parts of this approach to analyze motives and preferences to priorities the technology transfer options and actor constellations during negotiation situations.

It is further on assumed that universities and other publicly funded research institutions have different interests and motives compared to private industry. Therefore, mutual agreements on technology transfer are expected to be somewhat difficult and that they can develop into long negotiations or even fail, if the interests are not compatible. Scharpf (1997) assesses the bargaining situations with models from game theory, informed by field observation. This approach of actor-centered institutionalism is the basis for the development of our model. The model is used to describe and analyze the bargaining situation in technology transfer, to translate the negotiation power and desired outcome and to calculate the pay-off functions. Accordingly, we analyzed strategic choices and payoffs for publicly funded research institutes including universities, and for private firms, while comparing the three technology transfer options: exclusive licensing, equitable licensing and patent pools. We assume that the patent ownership will stay with the university or publicly funded research organization as it is practiced in the U.S.A., hence the option of direct sale of patents is excluded.

Consequently, the concept of equitable licensing and patent pooling are critically compared and appraised from the perspective of universities and publicly funded research institutions, and of industry.

Relevant data and information underlying the analysis were gathered from mission and strategy statements of TTOs and firms, licensing agreements, press releases, and a standardized telephone interview was conducted with representatives of ten German TTOs and six pharmaceutical companies, willing and agreeing to speak to the researcher.

Critical comparison and appraisal are discussed in light of the various factors of Bozeman's technology transfer efficiency model. Focusing on the potential impact, incentives, advantages and disadvantages from an academia and from an industry perspective, the three different transfer options are discussed in the model. Based on results and discussion, recommendations for TTOs and policy makers for the implementation of equitable licensing and patent pools concepts into transfer technology strategies are made.

5.4 Research Results

Whether technology transfer of research results happens most efficiently from a societal perspective when research results are free accessible to everyone or when they are exclusively licensed to a single firm is still a matter of debate. The argument for exclusive licensing considers that monopolies are intrinsically interconnected with financial incentives to bring technology to the market. The argument favoring free access considers competition as a necessary attribute to level prices for consumers and force private companies to apply research results most efficiently. Moreover, the reasoning against access restriction of research results through patenting claims, that private companies take up valuable research results regardless whether they are protected because their competitors would do so as well. Additionally, the transfer mode is related to the type of research result. When we consider innovation as a cumulative process, basic research results which are protected by patents with a broad scope need to be distributed much wider than special applications covered by narrow patent scopes because they are applied in more products. Patents are an instrument to convert free knowledge goods into private property. They provide the patent owners with the legal right to control the use of the technology at stake. Consequently, the access to the particular technology and its dissemination is reduced while at the same time competition for it is introduced. However, the patent owner can decide to turn the private property into club goods or common goods when patents are licensed on a non-exclusive basis or royalty free and open to everyone. The following figure illustrates the three different technology transfer options exclusive licensing, equitable licensing and patent pools in their relation to knowledge goods, competition and dissemination of technology. It shows that equitable licensing and patent pools could increase the dissemination of technology in terms of providing access to more than one firm.

We argue that even technologies covered by patents with a narrow scope are transferred with higher economic and societal impact when licensed to more than one private company for two reasons. Firstly, it increases competition between these firms. When two or more firms compete with the same or similar products, they are forced to compete on quality, price, additional features and services which benefit consumers. Secondly, markets are often too large to be served by a single firm, even for multinationals. This often results in high income countries are served immediately while low income countries are neglected.

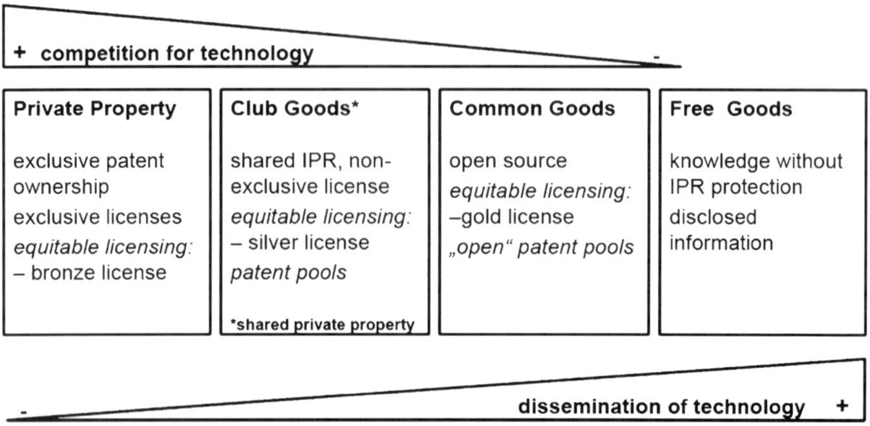

Fig. 5.3 Types of goods and technology transfer options (Source: Own consideration)

Table 5.2 Incentive to license patents under equitable licensing scheme or to patent pools

Dimension	Factor	Incentive to license under the equitable licensing scheme	Incentive to license to patent pools
Transfer agent	Mission	–	–
	Resources	–	–
	Organizational design	–	–
	Management style	–	–
	Political constraints	+/–	+/–
Transfer recipient	Business strategy	–	–
	Patent policies	–	+/–
	Marketing	+	+
Effectiveness	Political	+/–	+/–
	Economic development	+/–	+/–
	Out-the-door	–	–

+ incentive; – barrier; +/– neutral, neither incentive nor barrier

Although equitable licensing and patent pools appear to be interesting concepts to transfer technology broadly as illustrated in Fig. 5.3, they are hardly applied yet. In our interviews members of TTOs at universities and publicly funded research institutes admitted, that they are familiar with the concepts of equitable licensing and patent pools. However, they do not utilize them for a variety of reasons. Using Bozeman's technology transfer effectiveness model (Bozeman 2000), we categorized the reasons stated by research organizations and private firms, we analyzed the applicable factors for transfer agent (TTOs of universities and publicly funded research institutes), transfer recipient (private firms) and effectiveness measures (determined by funding organization and transfer agents), while various factors of transfer objects (patented technology) and the demand environment (e.g. treatment for neglected diseases) were not considered any further because they are expected to remain unchanged. In Table 5.2 we provide an overview of our findings on current preferences, incentives and barriers for transfer agent, transfer recipient, and effectiveness measures for the transfer concepts equitable licensing and patent pooling in comparison to exclusive licensing. In the following, we explain the factors of transfer agent, transfer recipient and effectiveness.

The transfer agent's *mission*, acting as a service organization to the universities and publicly funded research institutes, is to transfer research results into private industry under the premise of maximum commercial remuneration gains. The service of the transfer agent to the research institution involves filing for patent protection of research results, identifying the ideal recipient of the research results, negotiating deals and consequently transferring the technology as efficient as possible. Most technology transfer officers do neither explicitly nor inexplicitly refer to societal responsibility in connection to neither their technology transfer nor the research organizations. Accordingly, as most TTOs are still relatively young and in the process of building up contacts and linkages with private industry, their main focus on technology transfer to private industry is to accommodate private industry with licensing obligations.

The transfer recipients prefer exclusive licenses to obtain competitive advantages through monopolies. Currently, neither the transfer agent's *mission* nor the transfer recipient's *business strategy* provides incentives to license research results under equitable licenses or into patent pools.

On the contrary, the *resources*, *organizational designs* and *management styles* of TTOs do provide incentives for exclusive licensing. TTOs have rather restricted resources. They mostly employ scientists who recently finished a PhD in natural sciences and consequently have only limited industry contacts, negotiation and licensing experience. To compensate these resource constraints, they prefer short negotiations with a single recipient over lengthy negotiations with several potential licensees. Additionally, the funding of TTOs is often directly linked to the amount of royalties that they obtain through their technology transfer activities. Consequently, in order to expand their business capacity it is more advantages to TTOs if they do license out technology under most lucrative conditions. Since private companies argue that they are prepared to pay higher royalties in exchange for exclusivity, these types of contracts are still those most often negotiated. TTOs usually license patents under partly-exclusivity terms with restrictions to a particular field of use in order to license the patent also to other firms for different applications. Licensing to several firms for the same use rarely happens.

We could not identify any *political constraints* at the transfer agent. Within most of the public research funding schemes, institutes are obliged to transfer the results as effective as possible within their means. That does not imply that they cannot provide royalty free licenses to everyone who is interested. On the contrary, if it was considered to be the most effective transfer mode, it would be even incentivized.

One main argument of the transfer agent for not using equitable licensing or patent pools was that most firms would not agree to such conditions. If they agreed, obtainable royalties would be lower. The *management style* of TTOs with *out-the-door* licensing effectiveness measures and relying mainly on earned royalty reinforces the existing measures as success factors and by this licensing negotiations with rather one single private industrial partner and a single payment instead of numerous partners paying reduced royalty rates.

The transfer agent's political constraints and the effectiveness measures concerning *political effectiveness* and *economic development*, as determined by the funding organizations, are neutral towards the licensing type. Funding organizations require research institutes to disseminate results broadly. However, specific policies on how or requirements for further funding do not exist. If research institutes prefer that their research results IP are ideally transferred through exclusive licenses, they are entitled to do so. So far, research funding organizations, not even public funding organizations, request TTOs to try to achieve socially responsible licensing or to enhance competition by licensing to more than one private industry partner.

The transfer recipients' *business strategy* usually does not provide for equitable licensing or licensing to patent pools either. The dominant industry logic is to obtain exclusive rights and exert monopolies whenever possible. Private industries' *patent policies* as well usually outline the use of patents to secure competitive advantages by means of generating monopolies. In cases where they have to share patents and technologies, they prefer strategic alliances with a few selected partners only. Incentive to share patents in patent pools depends on the type of technology and the patent situation. While in telecommunication, patent pools are standard and foster efficient sharing of patents on many components using data compression and data transmission, we envision that for medical or environmental technology transfer patent pools could be more efficient than multiple cross-licensing agreements as discussed above in Sect. 5.2.3.

Moreover, equitable licensing and patent pools provide direct *marketing* benefits to private companies. Licensing under an equitable licensing scheme provide a competitive marketing advantage, as equitable licenses can be used to enhance the public reputation of all involved. When several firms use the same technology, private companies can benefit from the marketing activities of other patent pool participants. Additionally, licensing patents to pools for humanitarian targets can improve private companies' image as well.

In summary, although there are no legal or political constraints to make use of equitable licensing and patent pool concepts, organizational barriers such as resource constraints, organizational design, management style and mission impacts heavily on the choice of transfer method. More importantly, the current incentives for transfer agents and transfer recipients to choose equitable licensing or patent pools are considered low and inadequate compared to incentives gained through exclusive licensing.

On the contrary, exclusive licenses allow maximum financial remuneration only on basis of market monopolies and are a barrier to broad knowledge and technology dissemination. Since monopolies are hampering free market competition, it yet has to be established how equitable licensing and patent pools could not only improve wider knowledge and technology dissemination but also reduce high costs for society by allowing competition.

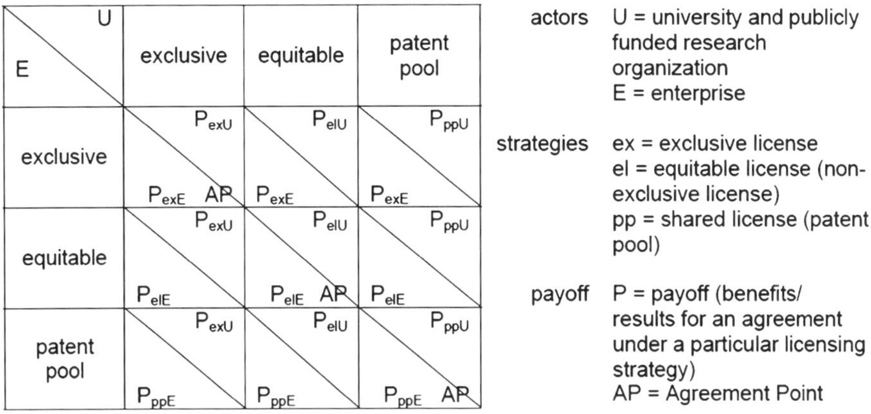

Fig. 5.4 Strategic choice analysis model

In order to discuss how incentives could be shifted from exclusive licensing towards licensing practices which enable broader dissemination of technology we developed a model under assumptions of the game theory.

Based on above findings on the strategic choices to license under a particular licensing scheme, our model includes two actors: university (U), as a placeholder for universities and publicly funded research institutes, and enterprise (E), for private industries, and three choices: exclusive licensing (ex), equitable licensing (el) and patent pools (pp) (Fig. 5.4).

First, the payoff for enterprises is developed: Patents provide firms with a significant competitive advantage above competitors. Consequently, they prefer securing a monopoly by obtaining an exclusive license. In cases where the patent is essential to improve the firm's position, it can be assumed that the firm would agree to get a non-exclusive (equitable) or a shared license (patent pool). However, these are less preferred options. Whether enterprises prefer to license from a patent pool or obtain a license under the equitable licensing scheme depends on the overall licensing conditions, market dynamics and prior experience with either of these methods, even when in both cases this reduces their monopoly.

Both, equitable licensing and patent pools are expected to foster market entry. Their positive impact on reputation of patent owners is expected to be higher for equitable licensing than for patent pools, but still higher for pools than for exclusive licenses. Additionally, patent pools could provide higher predictability for business planning. In particular, when a high number of manufacturers and consumers use the technology, the market size could be higher and more stable, hence predictable. Moreover, collaborating in patent pools could improve business networks.

Beside the benefit of lowering transaction costs, a patent pool is an administrative burden that consumes extra resources. However, it offers a competitive advantage when it bundles essential patents that would otherwise block companies to use and market a specific technology. In a situation where patents from other research organizations or firms are required, a license from a patent pool would be preferable to companies as it is expected to lower transaction costs. Consequently, it is expected that in the case that a patent pool for a specified technology is operating, a license from such pool is more preferable than a single license under the equitable licensing scheme:

$$P_{exE} \gg P_{ppE} > P_{elE} > 0,$$

As outlined above, policy incentives from funding organizations motivate universities and other research organizations to license patents as exclusive licenses to enterprises, because academia is restricted by the necessity to license the patent and by the resource constraints to contact all possible licensees. Funding organizations state, that they do prefer broad dissemination to foster economic growth. They do not, however, prescribe how to transfer research results.

According to the reasoning of the Bayh-Dole Act, exclusive licenses are a reasonable way to transfer technology because firms would not be interested to share technology with competitors. Also, TTOs have only limited resources to identify, contact and negotiate with potential transfer recipients. Because TTOs are eager to transfer technology, given it is their main objective, and enterprises are less dependent on receiving the technology from TTOs, enterprises have a stronger bargaining position in terms of determining licensing conditions. Consequently university's (U) higher payout is given when the enterprise (E) is satisfied, as it shortens costly negotiations.

Incentives to participate in patent pools are given by the advantage that they are a one-stop solution reducing negotiation costs. The equitable licensing scheme holds more advantages. Amongst them are the potential to contribute more to royalty streams, improve reputation and by this foster new funding, and attract better researchers and students. Since TTOs do not consider these complex benefits, given that their focus is on technology transfer rather than on reputation building and growth of their university or research organization, it is assumed that their preferable technology transfer strategy is exclusive licensing, compared to negotiating equitable licensing or participating in patent pools. Still, a license under the equitable licensing scheme is far more preferable than no license at all:

$$P_{exU} > P_{ppU} > P_{elU}, \quad \text{but } P_{elU} \gg 0$$

The payoff function of an exclusive license is for both positive and higher for E than for U. Because to U any license is preferred over a non-licensing option but for E the benefits of an exclusive license outweigh significantly:

$$P_{exE} > P_{exU} > 0$$

Whereas the payoff of a single license is to both U and E positive and would be slightly better to U than to E, because to U a single license means the option to further license the patent and to E it means higher competition:

Fig. 5.5 Strategic choice analysis

E \ U	exclusive	equitable	patent pool
exclusive	4 6 AP	2 3	3 3
equitable	1 -1	2 1 AP	1 -1
patent pool	2 -1	2 2	3 3 AP

$$P_{elU} > P_{elE} > 0$$

The payoff function of patent pools is to U more attractive than to E, but for both preferred over a non-agreement situation:

$$P_{ppU} > P_{ppE} > 0$$

When E now agrees on a non-exclusive license under the equitable licensing scheme or patent pool model, although U would have agreed on an exclusive one, the payout for E turns negative because it does not achieve the potential gain. To U this would be as good as agreeing on a single license:

$$P_{exU} > 0 > P_{elE} = P_{ppE},$$

When E enforces an exclusive license although U would have preferred to give only a single license under the equitable licensing or patent pool scheme, E's payout is lower than when U would have agreed without hesitations, as it strains the relationship. However, it is still higher than U's payout, as it obtains its desired outcome:

$$P_{exE} > P_{elU} = P_{ppU} > 0$$

We convert the above assumptions regarding interests, bargaining positions and payoff in the model as following (Fig. 5.5).

The most stable outcome is an exclusive license for both academia and enterprises. Patent pools provide moderate incentives; however, they are expected to be highly depending on the technology and industry-specific patent situation and the business models. Equitable licensing provides the least incentive to both academia

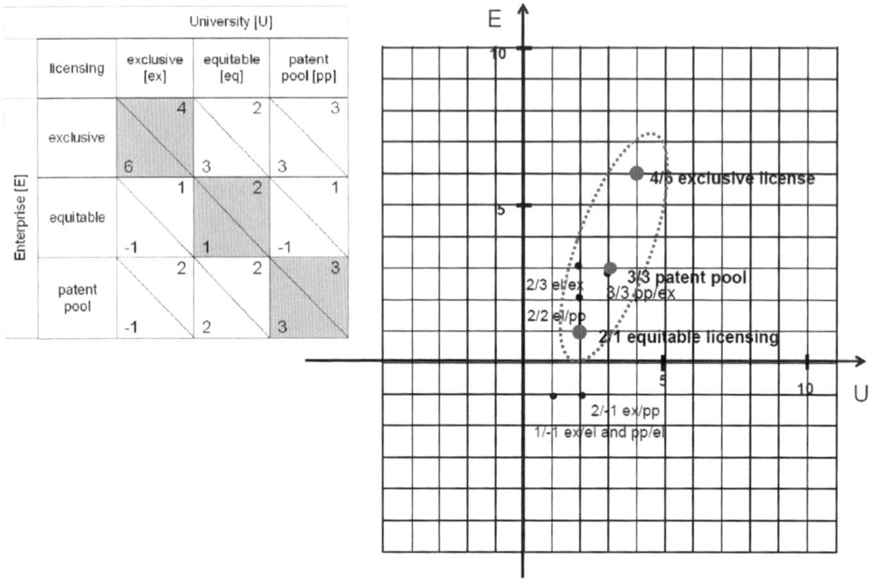

Fig. 5.6 Strategic choice payoff matrix

and industry. But equitable licensing is expected to be very interesting to both when for example an increase reputation is expected by any of the license contracting partners (Fig. 5.6).

When applying solidarity actor constellation + (1/1), as possible if future transactions enhances the decision to cooperate or incentives are provided to foster long-term cooperation, the decision space only shifts, but the outcome stays in the same relation.

Because non-exclusive technology transfer concepts such as equitable licensing and patent pools lack sufficient incentives at transfer agent, transfer recipient and effectiveness measures, it is suggested to adjust incentives if a more social responsible technology transfer of research results is the aim. The two options to provide incentives for more responsible technology transfer are:

1. increase benefits of non-exclusive licenses to companies, or
2. increase benefits of non-exclusive licenses to universities and publicly funded research institutes.

Both adjustments require incentives for technology transfer, which differ from those currently existing, e.g. measuring the performance in terms of number of licensing contracts and users, not licensing revenues. Increasing the benefits of non-exclusive licenses to firms appears only possible when the patented research results are essential to them.

In order to increase the number of non-exclusive licenses, the most effective options probably exist at the level of the funding organization. They could restrict the possible outcome to single licenses, by making it a condition of publicly funded research projects to transfer any IPR protected technology only as non-exclusive license, in order to foster competition between several manufacturers. This shift from exclusive transfer towards competition is expected to hold a high potential to increase market penetration of new technologies and lower their market prices.

Another option is to assess technology transfer effectiveness differently than proposed in Bozeman's model to provide for the public interest. Given that quality of life and social welfare is not simply related to economic growth, we propose that an additional effectiveness criterion should be added to include developments which target to accommodate technologies where market incentives fail. In addition to the factors (1) out-the-door, (2) market impact, (3) economic development, (4) political reward, (5) opportunity costs (see Sect. 5.2.1), it is important to add:

(7) soc*ietal impact*: the impact on quality of life, e.g. to solve societal issues, to increase overall welfare, societal and environmental sustainability.

Societal impact as an effectiveness measure could be determined by the difference research results used or technology transfer makes to quality of live, combining measures such as health care improvement, environmental sustainability, and improved access to clean water, nutrition's food, and education. This measure is especially important in areas such as drugs for neglected and rare diseases, and clean technologies. In order to achieve high societal impact, the technology at stake should be accessible at an adequate quality standard to as many people as possible while keeping prices low. However, how to measure such societal impact would depend on the research results or technology. For example, the number of suitable licensees is related to the industry architecture and the applications of the technology. In some cases, a high number is most preferred, in others only a few manufacturers can best achieve providing a wide public with low priced goods. Particularly, when this can best be achieved through economies of scale in manufacturing, a limited number of licensees might be more effective because this combines benefits of competition amongst them while providing them with the advantages of a limited oligopoly. In any case, it is important to consider the long term impact. Even when in the short term it might be efficient to license patents to selected private companies who operate already in the field of use, especially in low income countries new manufacturers have to be developed to achieve long term development effects and broader technology dissemination.

5.5 Conclusion and Outlook

Any IP resulting from publicly funded research should be transferred in a way that public interest is served. If higher dissemination of research results, and lower market pricing of products based on those are the aim of adjustments to the existing research and technology transfer system, we strongly suggest developing new guidelines for TTOs and advice for policy makers in order to implement technology transfer strategies with higher social impact.

Already some corporations include in their business objectives societal impact under the term corporate social responsibility. To what extend this is taken serious or only for marketing reasons is debatable. We noted that publicly funded research institutes and universities do not commit at large to social responsibility yet. Publicly funded institution may not consider an official commitment as important being a public institution however their main goal is intrinsically to serve the public good.

Two very interesting concepts that allow multiple licenses are equitable licensing and patent pools. The concepts of equitable licensing and patent pools could be used complementary, as both offer benefits to increase social impact of technology transfer.

In research areas with a high societal importance, e.g. medicines, interests of the funding public and consequently the responsibility to the tax payer, should be considered. Such social responsibilities could be protected by implementing legal frameworks, or through policies of research organizations and organizations funding research.

The research funding organizations role and their inherent public responsibility should be a matter for further exploration and open public debate. Our analysis is only a first assessment of incentives and barriers at German universities and publicly funded research institutes. As some academic institutions in the U.S. make already use of humanitarian licenses and patent pools, the studying of these examples could provide further valuable insights to better understand how the concepts can be applied more broadly.

Acknowledgements The authors would like to thank Prof. Dr. Christine Godt and Dr. Christian Wagner-Ahlfs for their support in the preparation of the manuscript. The research of Dr. Peter Tinnemann is founded by the VolkswagenStiftung.

References

Anderson, M. 2007. Making drugs available at affordable prices: How universities' technology transfer offices can help developing countries. *Journal of Intellectual Property Law Practice* 2(3): 145–152.

Astor, M., U. Glöckner, D. Riesenberg, and C. Czychowski. 2010, April. Abschlussbericht. Evaluierung des SIGNOFörderprogramms des BMWi in seiner ganzen Breite und Tiefe. Prognos. http://www.signo-deutschland.de/signo-unternehmen/content/e5072/e6287/SIGNO-EvaluationAbschlussberichtApril2010_ger.pdf/. Accessed 8 Oct 2010.

Baldini, N. 2009. Implementing Bayh-Dole-like laws: Faculty problems and their impact on university patenting activity. *Research Policy* 38(8): 1217–1224.

Bozeman, B. 2000. Technology transfer and public policy: A review of research and theory. *Research Policy* 29(4–5): 627–655.

Carlson, S.C. 1999. Patent pools and the antitrust dilemma. *Yale Journal on Regulation* 16: 359–399.

Clark, J., J. Piccolo, B. Stanton, and K. Tyson. 2000. Patent Pools: A Solution to the problem of access in biotechnology patents? United States Patent and Trademark Office. http://www.uspto.gov/web/offices/pac/dapp/opla/patentpool.pdf/. Accessed 30 Nov 2010.

European Commission. 2004, April 27. *Leitlinien zur Anwendung von Artikel 81 Absatz 3 EG-Vertrag*. Amtsblatt der Europäischen Union.

Gilbert, R.J. 2004. Antitrust for patent pools: A century of policy evolution. *Stanford Technology Law Review* (3).

Gilbert, R.J. 2011. Ties that bind: Policies to promote (good) patent pools. *Antitrust Law Journal* 77(1): 1–48.

Godt, C. 2010. Differential pricing of patent-protected pharmaceuticals for life-threatening infectious-diseases inside Europe: Can compulsory licences be employed? In *Differential pricing of pharmaceuticals inside Europe: Exploring compulsory licenses and exhaustion for access to patented essential medicines*, ed. C. Godt, 25–73. Baden-Baden: Nomos.

Godt, C. 2011. Equitable licenses in university-industry technology transfer. *GRUR International*: 377–385.

Godt, C., and T. Marschall. 2010. Equitable licensing: Lizenzpolitik & Vertragsbausteine. http://www.med4all.org/fileadmin/med/pdf/lizenz_med4all_final.pdf/. Accessed 30 Nov 2010.

Grassier, F., and M.A. Capria. 2003. Patent pooling: Uncorking a technology transfer bottleneck and creating value in the biomedical research field. *Journal of Commercial Biotechnology* 9(2): 111.

Lerner, J., and J. Tirole. 2007. Public policy toward patent pools. *Innovation Policy and the Economy* 8: 157–186.

Mayntz, R., and F. Scharpf. 1995. *Gesellschaftliche Selbstregelung und politische Steuerung*, Schriften des Max-Planck-Instituts für Gesellschaftsforschung. Frankfurt am Main: Campus-Verlag.

McCarthy, J., R. Schechter, and D. Frankly. 2004. *McCarthy's desk encyclopedia of intellectual property*, 3rd ed. Washington, DC: Bureau of National Affairs, Inc.

Merges, R.P. 1999. Institutions for intellectual property transactions: The case of patent pools. Working paper University of California at Berkeley. http://www.law.berkeley.edu/institutes/bclt/pubs/merges/pools.pdf/. Accessed 30 Nov 2010.

Mimura, C. 2010. Nuanced management of IP rights: Shaping industry-university relationships to promote social impact. In *Working within the boundaries of intellectual property*, ed. R. Dreyfuss, H. First, and D. Zimmerman. Oxford: Oxford University Press.

Mowery, D.C., R. Nelson, B. Sampat, and A. Ziedonis. 2001. The growth of patenting and licensing by U.S. universities: An assessment of the effects of the Bayh-Dole act of 1980. *Research Policy* 30(1): 99–119.

Murray, F., and S. Stern. 2006. When ideas are not free: The impact of patents on scientific research. *Innovation Policy and the Economy* 7: 33–69.

Sampat, B.N. 2009. Academic patents and access to medicines in developing countries. *American Journal of Public Health* 99(1): 9–17.

Scharpf, F.W. 1997. *Games real actors play: Actor-centered institutionalism in policy research*. Boulder: Westview Press.

Schimank, U. 2004. Der akteurzentrierte Institutionalismus. In *Paradigmen der akteurszentrierten Soziologie*, ed. M. Gabriel, 287–301. Wiesbaden: VS Verlag für Sozialwissenschaften.

Shapiro, C. 2000. Navigating the patent thicket: Cross licenses, patent pools, and standard setting. *Innovation Policy and the Economy* 1: 119–150.

Straus, J. 2008. Intellectual property versus academic freedom? A complex relationship within the innovation ecosystem. In *The university in the market*, ed. L. Engwal and D. Weaire, 53–65. London: Portland Press.

Stumpf, H., and M. Gross. 2005. *Der Lizenzvertrag*. Frankfurt a.M.: Verlag Recht und Wirtschaft.

't Hoen, E. 2011. The medicines patent pool: Stimulating innovation, improving access. Presentation at WIPO, Geneva, 18 May 2011.

Teece, D. 2008. *The transfer and licensing of know-how and intellectual property: Understanding the multinational enterprise in the modern world*. Hackensack: World Scientific.

Temple Lang, J. 1994. Defining legitimate competition: Companies' duties to supply competitors and access to essential facilities. *Fordham International Law Journal* 18(2): S.437.

Tinnemann, P., J. Özbay, V.A. Saint, and S.N. Willich. 2010. Patenting of university and non-university public research organizations in Germany: Evidence from patent applications for medical research results. *PLoS ONE* 5(11): e14059.

US Department of Justice and Federal Trade Commission. 1995, April 6. Antitruts guidelines for the licensing of intellectual property. http://www.justice.gov/atr/public/guidelines/0558.pdf/. Accessed 3 May 2010.

Van Overwalle, G. 2009. *Gene patents and collaborative licensing models: Patent pools, clearing-houses, open source models, and liability regimes*. Cambridge: Cambridge University Press.

Verbeure, B., E. van Zimmeren, G. Matthijs, and G. Van Overwalle. 2006. Patent pools and diagnostic testing. *Trends in Biotechnology* 24(3): 115–120.

Wagner-Ahlfs, C. 2009. Medizinische Forschung: Der Allgemeinheit verpflichtet. "Equitable licences" für Ergebnisse öffentlich geförderter medizinischer Forschung. Available from: http://www.bukopharma.de/uploads/file/Pharma-Brief/2009_01_spezial.pdf. Accessed 18 Oct 2010.

Wagner-Ahlfs, C. 2010. Oeffentliche Forschung für Entwicklungslaender: Zwischen Markt und sozialer Verantwortung. *Deutsches Aerzteblatt* 107(23). http://www.aerzteblatt.de/archiv/76421/. Accessed 3 Feb 2011.

WIPO. 1996. Agreement between the World Intellectual Property Organization and the World Trade Organization (1995), Agreement on Trade-Related Aspects of Intellectual Property Rights (TRIPS Agreement) (1994). *WIPO Publication* No. 223 (E).

Ziedonis, R.H. 2004. Don't Fence me in: Fragmented markets for technology and the patent acquisition strategies of firms. *Management Science* 50(6): 804–820.

Chapter 6
The Assumption of Scientific Responsibility by Ethical Codes – An European Dilemma of Fundamental Rights

Hans Christian Wilms

Abstract The latest efforts by research institutions and the European Union to steer scientists into the direction of scientific responsibility are subject to this article. Ethical codes as a mean to achieve this goal are interesting for legal sciences in two aspects. They both stress the concept of normativity and raise questions of fundamental rights. By disclaiming legal validity they could be classified as extra-legal or non-binding norms at first glance. But the non-binding character of these ethical codes put the concept of normativity in question as they are able to interfere with the legal guarantee of freedom of science. It will be shown that the sensitivity of the mechanisms of science demands a consideration of this fundamental right, even if the effects are rather indirect and caused by non-binding measures. The final resolution of ethical conflicts in science is thus not to be found in voluntary ethical codes or recommendations when these norms factually influence their addressees' behavior in a manner which is contrary to constitutional guarantees.

6.1 Introduction

Responsibility for the outcomes of research is a controversial discussed topic since science adopted a dominant role in society in the past centuries. Most notably since researchers who took part in the development of nuclear weapons called for

H.C. Wilms (✉)
Max Planck Research Group "Democratic Legitimacy of Ethical Decisions: Ethics and Law in the Areas of Biotechnology and Biomedicine", Max Planck Institute for Comparative Public Law and International Law, Im Neuenheimer Feld 535, 69120 Heidelberg, Germany
e-mail: hcwilms@yahoo.com

J. van den Hoven et al. (eds.), *Responsible Innovation 1: Innovative Solutions for Global Issues*, DOI 10.1007/978-94-017-8956-1_6,

the abolition of their products, the question of scientific responsibility has arisen in society and science itself. The following remarks will concern the European regulatory aspects in this matter from the legal angle.

6.2 The Regulatory Frame

The relationship between science and society was in the past often dominated by the conflict whether science could or should enjoy a legal sphere of protection and if so, which limitations such a sphere of freedom should have. Freedom of science is nowadays guaranteed as a fundamental right in many European constitutions and since 2001 also in Art. 13 of the European Charter of Fundamental Rights (hereinafter "The Charter") which is applicable law since December 2009. Each particular system of fundamental rights guarantees thereby in its own way the dimension of protection to be enjoyed by researchers and sets particular boundaries. The occurring differences are enormous and so is the task to harmonize them on the European level.

A corresponding second conflict is intimately connected with the first one and governs the protected scientific system itself, raising the analogical question whether researchers should bind themselves to ethical values, and if so, which ones. The question of professional ethics is prominent in many areas of modern society, most recently in the financial branch. The same is true for science, but opinions in regard to the specific field of scientific responsibility are twofold. Two scientists involved in the development of nuclear bombs gave two different answers in this matter: Carl Friedrich von Weizsäcker, a German researcher involved in the Nazi projects during the Second World War, explained that his experience in research forced him to accept the hindsight that science must be responsible for its outcomes. Edward Teller on the other hand, one of the developers of the American hydrogen bomb, denied this by stating that scientists only produce knowledge. Responsibility results from the application of that knowledge and must therefore be handled by its users or by society and politics.

The latest efforts to steer these conflicts into the direction of scientific responsibility will be subject to this article. Both the scientific community itself and the organs of the European Union refrain currently from choosing mandatory norms. The preferred way to steer the behavior of scientists nowadays seem to be non-binding instruments. Ethical codes or codes of conduct as a mean to achieve the goal of scientific responsibility are interesting for legal sciences in two aspects. They both stress the concept of normativity and challenge the scope of freedom of science. By disclaiming legal validity these codes only call for voluntary compliance, expressed by the deontic operator "should" and titles like "rules and recommendations" or "ethical codes". At first glance they can thus be classified as extra-legal or non-binding norms. However, I will try to demonstrate here that there are crucial differences between such ethical codes, depending on their drafters and whether they stem from inside or outside the scientific community.

6.3 The Legal Relevance of Ethical Codes

The topic of the value of ethical codes has been discussed for many years, especially since social and legal scholars discovered the benefit of interdisciplinary norm setting in various areas. These instruments combine several advantages vis-à-vis regular legislative measures. First of all they facilitate the setting of norms by horizontal cooperation instead of hierarchical regulation. Since legitimacy of norms is at global discussion horizontal norm setting became more and more an attractive alternative to steer the behavior of certain communities. Especially when sensitive issues like ethics are at stake the inclusion of the addressed actors can on the one hand augment acceptance of norms and on the other hand turn to account the variety of faculties of the addressees to increase the quality of the norms. In accordance with the aims of this book ethical codes are the perfect instrument to fructify multidisciplinarity and cooperation.

Abroad from the general discussion about the legitimacy and production of ethical codes new issues arise when these codes are either used to replace or amend legislative norms or when state actors are included in such code setting processes. But also specific issues are to be addressed when these instrument are created to influence sensitive areas or sub-systems like science.

There is a broad spectrum of ethical codes or codes of conduct in science, most of them concerning scientific misconduct and best practices for laboratory security. Particularly in the United States codes are used by scientific institutions and companies for their researchers. Several research institutions in Europe tried to imitate these efforts, to be able to compete with the American researchers on the international level. However, currently more and more ethical codes are developed concerning scientific responsibility towards third persons, society or environment. For instance one of the leading research institutions in Germany, the Max-Planck-Society, published in 2010 "Rules and Recommendations for Responsible Practices of Freedom and Risks of Science" and so did several other scientific institutions. Remarkably, in 2008 the European Commission likewise adopted such an instrument, a code of conduct for responsible nanosciences and nanotechnologies research, attached to a recommendation which is an instrument that shall have no binding force according to Art. 288 para. 5 of the Treaty on the Functioning of the European Union (TFEU).

The ethical questions considered in this code concern the grey area beyond those direct perils resulting out of scientific research which would actually demand for mandatory norms to protect the citizens. They rather stress the concept of causality and are therefore part of philosophical or ethical discussion asking which perils would be acceptable and how far researchers can go.

The non-binding character of this code of conduct put the concept of normativity in question and therefore this concept must be scrutinized, although for space reasons, in an admittedly rough fashion. The problem is well-known, especially in the field of public international law where "*soft law*" is a controversially discussed notion (Goldmann 2012). As the term *soft law* suggests, a non-binding character of an instrument doesn't mean that its norms cannot influence their addressee's behavior. Depending on the quality, functionality and the originator of such norms there can be factual pressure to comply with their provisions, resulting from the social conditions or the relative strength of the individual addressee (Senden 2004; Peters 2007; Friedrich 2011).

This particularly applies in the field of fundamental rights, where protection is guaranteed for certain spheres of life. Citizens shall be protected from any unjustified interference with their activities by state actors or, in case of the Charter, by European Organs. Accordingly, special diligence is necessary if such an organ, which is in principle able to enact binding legislative measures, tries to steer ones behavior. If originators lack this ability, the respective codes perhaps are to be handled by private or labor law, but this topic unfortunately is outside the scope of this article, since the validity of fundamental rights in these areas is at least in dispute. Although discussion of various forms of governance in the European Union involving non-state-actors is crucial at the moment, this contribution has to omit this argument to avoid the same mistake the European Commission made. The fundamental rights granted by the European Charter only bind state or Union actors according to its Art. 51 para 1. In contrast to non-state-regulation every institution and body of the Union is bound to the boundaries set by the Charter (Ehlers 2007). However, the European Commission seemed to overlook this difference by adopting the same kind of instrument as non-state actors do, potentially to disburden itself from these boundaries.

The problem with such a regulatory approach results from the effects of this code of conduct and the consequential threats for the newly established fundamental right of scientific freedom. Depending on the originator's motivation to choose this way of regulation, practical effects can come along even with a formally non-binding instrument.

6.4 The European Code of Conduct for Responsible Nanosciences and Nanotechnologies Research

When the issue of the risks of nanosciences came up in the European Union, the European Organs felt themselves under pressure to act. The Commission intended to anticipate regulatory measures by the member states to attain a harmonious regulatory level in Europe, but lacked an explicit competence to this end.

Accordingly, the European Commission chose the option only to adopt a non-binding recommendation to establish a code of conduct of ethical nanosciences. Such an instrument combines several advantages for the Commission: It maintains the Union's flexibility to react to new developments in nanosciences and at the same time did not require the same conditions to be met as formal, binding steering instruments would have. The non-binding character of the code of conduct could moreover widen the scope of regulatory possibilities. By avoiding definite, mandatory norms it could involve regulations which would be able to handle the uncertainty of risks resulting from nanosciences. These regulations would have been questionable, if implemented in formal acts like regulations and directives, especially in anticipation of the European Charter of Fundamental Rights.

At the same time, the scope of regulatory possibilities by the member states is narrowed as the Union already has acted, albeit through a non-binding instrument. Thus, there is a high factual pressure to comply with the provisions which is even increased by demands to the member states within the recommendation to report about the efforts to implement the provisions of the code. Considering the principle of sincere cooperation as it is set out in Art. 4 para 3 of the Treaty on the European Union the member states are requested to act in favor of the European idea and are thus prone to fulfill such demands, even if the formal character of such an instrument is non-binding (Peters 2007).

These factors produce a kind of pressure to comply that make it possible to talk of legal effects as the European Court of Justice recognized them even for non-binding instruments. The Court stated that for the scrutiny of legal effects the formal character of an instrument is irrelevant, only the content must be evaluated. The content of the recommendation demands a harmonization of the member states research funding, according with the ethical view provided by the nano-code, monitored by the Commission itself.

However, whenever legal effects result from such instruments the question arises if they are compatible with fundamental rights like freedom of science. Although neither the Charter of Fundamental Rights was in force when the code of conduct was adopted in 2008, nor has there been such a fundamental right on European level before, the implementation of the Charter was at least intended for the following years and the Commission committed itself to it under President Romano Prodi in 2001. Hence, it is at least arguable to scrutinize the code for interferences with the Charter. In addition, the code of conduct should be reviewed every 2 years. At least this review process should comply with the fundamental rights which are applicable since the Treaty of Lisbon came into force in 2009.

Two passages of the code need special considerations in this context: Firstly the Commission demands in the code that "researchers and research organizations should remain accountable for the social, environmental and human health impacts that their [...] research may impose on present and future generations." This provision is highly arguable considering freedom of science as a fundamental right. One of the main arguments for a specific fundamental right for scientists in Europe and elsewhere is the sensitivity of the scientific system that results from its very professional specifics (Luhmann 2002). Science needs a sphere of freedom for its pursuit of knowledge or cognition. This sphere was always threatened by the amalgamation of the production of knowledge and its application in the eyes of public authorities and society. However, particularly basic research is in urgent need for this sphere of freedom to guarantee the possibility of basic thinking processes without restraints.

A profession of science that would consider each impact of its research would be very welcome and from an ethical point of view it would be also highly recommendable. An obligation to remain accountable for all the impacts science may have, would on the other hand interfere with the actual essence of the freedom of science. It would express a preference for every other ethical or legal interest to those of science. Hence, scientific freedom wouldn't be valued as equivalent to other fundamental rights, an unacceptable result from the perspective of the Charter.

Thus, changes in this code of conduct seem to be necessary. Given that the provisions are proportional, it would be possible to demand mandatory considerations by researchers concerning direct perils for fundamental rights of third persons from their research. But if the quasi legal character of the code and the fundamental right to freedom of science are taken into account, a general demand for accountability is certainly too wide.

A second arguable point is the prohibition of "research in areas which could involve the violation of fundamental rights or fundamental ethical principles, at either the research or development stages." Needless to say, research generally should not violate fundamental rights, but what implies the violation of fundamental ethical principles?

At this point the Commission tries to bridge the gap between ethics and law by the usage of a non-binding instrument. It represents a development, which was titled by some authors the "ethicalization of law." The Commission certainly only follows here the tendency given by the European Council and the Parliament which already called generally for ethical research in their decisions that enacted the various Framework Programmes of the European Community. But the denial of funding by European organs due to ethical restrictions and the general prohibition of unethical research as it is set out by the code in an abstract and general manner concern two different sides of the freedom of science.

Of course, the Commission can decide freely which projects within the Framework Programmes would be funded and hence can impose certain ethical conditions on researchers which try to participate therein. A different question is a quasi legal steering instrument prohibiting research to all funding bodies in the member states. It is highly questionable both if there is a Union's competence to enact such a provision and if it could be in conformity with Art. 13 of the Charter of Fundamental Rights.

The main problem regarding the interference with the Charter is the intermixture of two different yardsticks. On the one hand we have the written catalogue of fundamental rights which represents those fundamental ethical principles the European Union already acknowledged as binding for itself. On the other hand another normative system shall be implemented hereby that consists not only of the fundamental rights in the Charter. According to the council decisions concerning the specific programmes within the Seventh Framework Programme, it is furthermore necessary to take into account relevant international conventions, guidelines and codes of conduct, such as the Helsinki Declaration and the Convention of the Council of Europe on Human Rights and Biomedicine. While the latter is a binding convention of public international law, the Declaration of Helsinki is a non-binding international guideline for and by the medical profession. This mixture demonstrates that no longer the legal quality of the sources is decisive, but the ethical content thereof. Hence, the Commission tried to establish the limitation of scientific freedom not solely through legal, but indefinite ethical criteria.

Another indicator for this intention is the reference to the Opinions of the European Group on Ethics in Science and New Technologies, which shall be likewise taken into account to discover the ethical boundaries of research in Europe. The group only advises the European Commission and is hence not able to adopt binding measures (Plomer 2008; Busby et al. 2008). Nevertheless its opinions shall be able to shape the European freedom of science.

Apart from the dubious formulation of the provision, which is too vague considering the general principle of clarity and definiteness, the Commission also implemented another yardstick into its policy, which is not in conformity with the actual possibilities to limit fundamental rights as they are set out in Art. 52 of the Charter. It tries to intermingle different standards to achieve acceptable results for the aspired European research area. This development has to be dismissed as the European Union is founded on normative provisions in its primary law, which have to be the sole standard for European research policy. The stability of the law and the rule of law are two of the main principles governing the European legal order. Undermining them could result in dangerous confusion.

6.5 Conclusion

In fact, the Commission tried to bridge the existing gap between pure ethical and legal discourse in science by using non-binding instruments like ethical codes. In my opinion this way shouldn't be continued. Despite the fact that ethical codes are a common instrument in many branches to achieve voluntary compliance and to avoid actual legislative measures, the Commission has to acknowledge that these ethical codes are usually generated by private bodies like companies or institutions. An executive organ of the European Union, committed to fundamental rights, cannot act the same way. Since the European Charter of Fundamental Rights came into force in 2009 it is first and foremost the challenge of the European Union to discover its ethical foundations and limits, not to generate parallel or diverging yardsticks of ethical evaluation for research projects.

To achieve the aim of ethical research in Europe a more co-operative approach should be chosen. By integrating the various actors of the specific scientific branches into a specific ethical discourse, a more definite and detailed regulatory instrument could be adopted, which would nevertheless had to comply with fundamental guarantees of the European Union like freedom of science and the principle of proportionality, if a harmonious European regulation is aspired.

If such an alternative European approach is not possible, and it is rather probable that there will be no European consensus about ethical issues in research, the European Organs have to realize that the ethical question of responsible innovation is an issue to be handled by the member states themselves, since their ethical and legal fundament should be more tightened. Perhaps the European Union should sometimes also be able to refrain from harmonization in favor of the sovereign member states, particularly with regard to ethical issues, instead of implementing questionable instruments.

References

Busby, Helen, Tamara Hervey, and Alison Mohr. 2008. Ethical EU law? The influence of the European Group on Ethics in Science and New Technologies. *European Law Review* 33(6): 803–842.

Ehlers, Dirk. 2007. *European fundamental rights and freedoms*, De Gruyter textbook. Berlin: De Gruyter.

Friedrich, Jürgen. 2011. Codes of conduct. In *Max Planck encyclopedia of public international law*, ed. Rüdiger Wolfrum, 264–276. Oxford: Oxford University Press.

Goldmann, Matthias. 2012. We need to cut off the head of the king: Past, present, and future approaches to international soft law. *Leiden Journal of International Law* 25(2): 335–368.

Luhmann, Niklas. 2002. *Die Wissenschaft der Gesellschaft*, 4th ed. Suhrkamp: Frankfurt am Main.

Peters, Anne. 2007. Typology, utility and legitimacy of European soft law. In *Die Herausforderung von Grenzen*, ed. Epiney Astrid, Haag Marcel, and Heinemann Andreas, 405–428. Baden-Baden: Nomos.

Plomer, Aurora. 2008. The European Group on Ethics – Law, politics and the limits of moral integration in Europe. *European Law Journal* 14(6): 839–859.

Senden, Linda. 2004. *Soft law in European Community Law*, 1st ed. Oxford: Hart.

Chapter 7
How (Not) to Reform Biomedical Research: A Review of Some Policy Proposals

Jan De Winter

Abstract In a recent article, Julian Reiss has identified some very important epistemic, moral and socio-economic failures in current biomedical research, and he argues that philosophers of science should reflect on how to (re)organize biomedical research in order to remedy these failures. In this chapter, several possible reforms of biomedical research are evaluated. I will reflect on how to tackle the epistemic failures by comparing the solution suggested by Julian Reiss to an alternative policy option. Most attention will, however, be paid to one of the moral failures: the fact that a disproportionately small part of the money devoted to health research goes to research into diseases that mainly affect third-world countries (the problem of neglected diseases). The most important advantages and disadvantages of some prominent proposals for a solution are disclosed – I will consider the proposals of Thomas Pogge, Joseph Stiglitz, Julian Reiss, and James Robert Brown – and I will also draw attention to an alternative policy proposal.

7.1 Introduction

Julian Reiss (2010) argues that philosophers of science should reflect on how to organize biomedical research, and he suits the action to the word by elaborating on his own reflections. As Reiss's account of the problems in biomedical research and potential solutions is, in my opinion, one of the most advanced contributions to the debate so far, I will start my contribution by briefly recapitulating his account in the next section. I will summarize Reiss's analysis of the failures in current biomedical research (Sect. 7.2.1), his main objections against the solutions proposed by Thomas Pogge and Joseph Stiglitz (Sect. 7.2.2), and his own policy proposal (Sect. 7.2.3).

J. De Winter (✉)
Centre for Logic and Philosophy of Science, Department of Philosophy and Moral Sciences, Ghent University, Ghent, Belgium
e-mail: Jan.DeWinter@UGent.be

J. van den Hoven et al. (eds.), *Responsible Innovation 1: Innovative Solutions for Global Issues*, DOI 10.1007/978-94-017-8956-1_7,

While I endorse Reiss's view on the existing problems in biomedical research and his objections to the proposals of Pogge and Stiglitz, I have some concerns with respect to his own proposal. These concerns, and some alternative strategies to deal with the failures identified by Reiss, are presented in Sects. 7.3 and 7.4.

Another philosopher of science who has paid attention to the organization of biomedical research, is James Robert Brown (2008a, b). Reiss raises a convincing objection to Brown's policy proposal, but even so it has some undeniable advantages over the other proposals discussed in this paper (including my own proposal), which are ignored by Reiss. Therefore, I consider the reform suggested by Brown worthy of further exploration. Such an exploration is offered in Sect. 7.5. It is argued that, although Brown's scheme is in itself probably not sufficient to achieve certain health goals, we should not be too quick to dismiss it entirely. My findings are summarized in Sect. 7.6.

Before I proceed, an important restriction of this paper should be mentioned. It assesses different policy proposals on the basis of their cost-effectiveness, i.e. how effectively do the different policies solve the relevant problems, and at what cost (for the public)? I will not consider arguments about the rights and duties of the different parties involved (governments, patients, pharmaceutical companies, medical researchers, etc.); the arguments discussed in the paper are therefore strictly utilitarian. This is not because I believe that such rights or duties do not exist, or are irrelevant, but because they are, in my opinion, outweighed by utilitarian concerns.

7.2 Julian Reiss's Contribution to the Debate

7.2.1 Failures

Reiss (2010) distinguishes three kinds of failures in current biomedical research: epistemic, moral, and socio-economic. Epistemic failures concern the inadequacy of the procedures and methods used in biomedical research for the generation of the knowledge required for health development. A first epistemic concern is that industry-sponsored research is more likely to draw pro-industry conclusions (e.g. a new industry product is superior to an alternative treatment) than research that is not sponsored by industry. This arouses the suspicion that the outcomes of industry-sponsored research are biased, which would be especially problematic given the fact that an increasing percentage of medical research is funded by industry.

A second epistemic concern is that commercialization has weakened the standards of biomedical research. Reiss worries that pharmaceutical companies use certain methods to generate research results favoring their products: they test their products on patients who are younger and healthier than the target population, resulting in an exaggeration of the product's effectiveness and an underestimation of its side effects, and they compare their products with products that are administered in insufficient doses or inadequate ways, which leads to an exaggeration of the relative effectiveness of the new product. Another way in which the standards are weakened according to Reiss, is by involving more and more community physicians who have no training in research, in the selection of patients.

Reiss's third epistemic concern is that research is reported inadequately. A number of clinical trials are not reported in sufficient detail to adequately assess their results, some are not reported at all, and there are long delays in the publication of research results. The latter is especially problematic when the research exposes certain side effects of products that have already been marketed.

Now consider the moral dimension. The question Reiss asks with respect to the moral dimension is: are the research priorities in current biomedical research adequate? A first problem in this context is the problem of neglected diseases. The idea is that a disproportionately small part of the money devoted to health research goes to research into diseases that mainly affect third-world countries. A second moral failure is that too much biomedical research aims at the development of drugs for conditions for which there are already very effective medical drugs or for which non-medical treatments such as exercise and diets are more effective. Thirdly, Reiss mentions the problem of disease mongering: pharmaceutical companies invent diseases in order to expand the market for their products.

Socio-economic failures, Reiss's third class of failures, concern the inefficiency of biomedical research. One kind of inefficiency is the fact that a lot of money is spent on the development of drugs that are no better than existing drugs. If that money was spent on new treatments with genuine medical value, biomedical research would be more efficient. Other costs to be avoided include lost welfare due to animal and human testing of new drugs that have no benefit over existing drugs. Reiss also points to the fact that health costs are exploding in the United States. Spending on prescription drugs and total health administration costs are much higher in the United States than in Canada.

7.2.2 *Proposals for a Solution*

Reiss (2010) discusses three proposals for a solution. The first was developed by Thomas Pogge. Reiss summarizes Pogge's proposal as follows:

> First, inventor firms should be rewarded with a 10-year monopoly on their inventions after market approval. Second, during this time they are rewarded, out of public funds, in proportion to the impact of their invention on the global disease burden. Third, the cost of this scheme is borne by the governments of advanced countries. (Reiss 2010, p. 438)

It should be noted that Pogge proposes this scheme as a supplement to, and not as a replacement of, the existing market system. So under Pogge's dual system, pharmaceutical companies can choose between two scenarios: they can claim payments from the newly created health impact fund, or they can make money in the traditional way, that is, by selling their patented products at prices far above marginal cost (Hollis and Pogge 2008; Pogge 2007, 2009a, b).

A similar solution is proposed by Joseph Stiglitz (2006a, b). According to Reiss, it consists of the following four elements:

(1) Introducing separate intellectual property regimes for different levels of development;
(2) The provision of drugs at cost to developing countries;
(3) Compelling innovating firms to provide licenses to (third-world) generic drug producers in the case of lifesaving drugs;
(4) Creating a Medical Prize Fund (from public and philanthropic money). (Reiss 2010, p. 439)

Reiss's main objection against the systems proposed by Pogge and Stiglitz is that they are socio-economically inefficient. The reason is that the prizes they propose will only stimulate pharmaceutical companies to invest in research and development (R&D) for medicines to treat third-world diseases if they make such R&D projects at least as profitable as the projects currently pursued, that is, R&D projects for products that sell in affluent countries. Since the profits from the latter projects are extremely high (Reiss states that, in 2008, the pharmaceutical industry had a profit margin of 19.1 %, which made it the second most profitable industry), this means that the prizes would have to be high as well, unnecessarily high in Reiss's view.

The third proposal that Reiss discusses was developed by James Robert Brown. As a separate section of this paper is devoted to Brown's policy proposal, I include Reiss's comment on it in that part of the paper (see Sect. 7.5), instead of discussing it here.

7.2.3 Reiss's Policy Proposal

After criticizing the proposals of Pogge, Stiglitz, and Brown, Reiss (2010) develops his own policy proposal. It consists of five recommendations[1]:

(1) Patent duration and/or breadth[2] should be reduced;
(2) Clinical trials should be run by an independent body committed to neutral hypothesis testing and overlooked by a board whose members represent different stakeholders;
(3) Drugs should only be approved if they are better than all existing therapies, including non-medical options;
(4) Research into neglected diseases should be stimulated by establishing Global Institutes of Health (in analogy with the U.S. National Institutes of Health but committed to global health issues), by advance purchase commitments (APCs), by awards for research into neglected diseases, and/or by tax breaks for such research;

[1] Reiss groups the third and the fourth recommendation in one section, under the heading "Aligning commercial and (global) patients' incentives" (Reiss 2010, p. 444).

[2] By breadth, Reiss means "the range of ideas that are considered worthy of patent protection" (Reiss 2010, p. 441). Patent breadth can be reduced by making things that are patentable under the existing regime (e.g. new uses of existing drugs, combinations of existing drugs) non-patentable.

(5) Socially harmful practices such as direct-to-customer advertising, industry sponsorship of continuing education events, advertising in medical journals, and payments from industry to doctors in the form of consulting fees, gifts, dinners, or finder's fees should be prohibited, and these prohibitions should be enforced.

This proposal leaves open some questions: Should patent duration be reduced, or patent breadth, or both? How much should patent duration and/or breadth be reduced? How should research into neglected diseases be stimulated: by establishing Global Institutes of Health, by APCs, by awards for neglected-disease research, by tax breaks, or by a combination of these? The reason why Reiss does not answer these questions, is that he thinks further empirical research is required to answer them adequately. More specifically, he thinks that if we want to know how much patent duration should be reduced exactly, we should gradually reduce patent duration, observe and assess the effects, and continue this process until a satisfactory outcome has been reached. Optimal patent breadth can be determined in the same way. If we want to know how to stimulate neglected-disease research, we should implement the different strategies and select the one that promotes global health most efficiently and effectively.

7.3 Dealing with the Epistemic Failures

A first element I would like to comment on, is Reiss's suggestion "to leave the running of clinical trials to an independent body committed to neutral hypothesis testing and overlooked by a board whose members represent different stakeholders" (Reiss 2010, p. 443). Reiss states that the costs could be covered with public money and/or by membership fees from companies that seek to test new drugs (Reiss 2010, p. 443). I do not reject this proposal, but I think additional arguments are needed before we can definitively accept it. An additional argument is offered in this section, and I specify the kind of empirical research that can help us further assess the proposal's desirability.

In a nutshell, Reiss's argument is that the adversarial system for drug approvals that Justin Biddle (2007) proposes (in which advocates of industry and advocates of the public argue before a panel of independent judges over questions such as whether a drug should be allowed on the market or not) is inadequate to deal with the epistemic failures, and that he sees no alternative option besides the one he proposes. The problem with Biddle's proposal is that if only pharmaceutical companies run clinical trials, the public's advocates could not provide independent reasons for or against allowing a drug on the market. Pharmaceutical companies could manipulate research in order to obtain desirable results, and the public's advocates would have to rely on these results. What is needed according to Reiss, are clinical trials that are conducted in a neutral manner, and of which the outcomes are reported in a neutral manner, and therefore he suggests the establishment of an independent body that is committed to neutral hypothesis testing (Reiss 2010, p. 443).

But there is also an alternative strategy to make sure that clinical trials are conducted and reported in a neutral manner, which is ignored by Reiss:

(1) Agencies responsible for the evaluation and approval of medicinal products, such as the U.S. Food and Drug Administration (FDA) and the European Medicines Agency (EMA), should not accept results of research that is designed to make a product look more effective or safer than it actually is;
(2) We should make sure that physicians know about the methods companies use to produce desired results (e.g. by including an overview of these methods in their education);
(3) Pharmaceutical companies should be obliged to register clinical trials and to make sufficiently detailed, genuine reports of these trials publicly available not long after their termination. To make sure that they fulfill these obligations, effective inspection by independent audits and severe punishment of offenders is needed.

(1) and (2) are meant to prevent that pharmaceutical companies would use certain methods to exaggerate a product's effectiveness and/or underestimate its side effects (e.g. test the product on patients who are younger and healthier than the target population). If (1) is implemented, pharmaceutical companies cannot get their products approved on the basis of the results of research in which such methods are used. If (2) is implemented, they cannot use such results to convince physicians to prescribe their products either. Physicians will see that these results make the product look more effective and/or safer than it may actually be. Rather than convincing physicians of the effectiveness and safety of the promoted product, it will make them suspicious. Why didn't the company use more honest methods to prove the product's effectiveness and safety? It seems, then, that if (1) and (2) are implemented, pharmaceutical companies are no longer stimulated to use methods to exaggerate a product's effectiveness and/or underestimate its side effects, on the contrary.

(3) is meant to avoid that research is reported inadequately (e.g. only reporting research that has favorable results, counterfeiting data). It should be noted that the basis for a compulsory registration system as proposed in (3) already exists in the United States.[3] Reiss acknowledges the existence of a registration system in the United States. He states that "the FDA now requires certain trials to be registered with clinicaltrials.gov, and medical journals will only publish results of registered trials, which makes suppression and delay of publication harder (albeit not impossible)" (Reiss 2010, p. 432n). Two remarks are in place here. Firstly, it is not because malpractices such as suppression and delay of publication, and counterfeiting of data are still possible under the current system in the United States, that this possibility cannot be excluded (or at least made highly unlikely) by a more strict and more demanding compulsory registration system that includes effective inspection and severe punishment of offenders. Secondly, even if suppression and delay of publication, and counterfeiting of data would not be entirely excluded by implementing (3), then this still does not mean that publications are more likely to

[3] See http://clinicaltrials.gov/. Accessed 17 August 2012.

be suppressed or delayed, or that data are more likely to be counterfeited under such a system than under the system Reiss proposes.

It seems that the threefold strategy I presented may be just as effective in eliminating the epistemic failures in biomedical research as implementing Reiss's proposal. But which strategy is most *cost*-effective? A disadvantage of the threefold strategy is that inspection of pharmaceutical companies by independent audits (see (3)) may be expensive (Steneck 2002, p. 11). These costs are avoided under Reiss's scheme. An advantage of the threefold strategy is that it leaves clinical trials to companies that have a strong financial incentive to run clinical trials efficiently. Pharmaceutical companies are stimulated to run clinical trials efficiently because the fewer resources they waste, the higher their profits. The institute Reiss proposes may, on the other hand, waste resources due to the lack of such a strong incentive to proceed efficiently. It should, however, be noted that incentives to proceed efficiently can also be introduced in the institute Reiss proposes. For instance, if the institute meets certain management agreements, governments could make money available for bonuses for the directors of the institute, its executives, and the employees who have done the best job. Empirical research should reveal which kind of incentives are necessary and sufficient to maximize the institute's efficiency. Empirical research should also test the speculative claim that the system Reiss proposes can be more cost-effective than a system in which pharmaceutical companies run clinical trials and are inspected by independent audits, due to the high costs of effectively inspecting pharmaceutical companies. Such an analysis should also take the costs of implementation into account, which are probably much lower for the latter system since the basis for such a system (industry research, a registration system, etc.) already exists.

7.4 Neglected Diseases

7.4.1 Costs of Pull Funding

Now, let us turn to Reiss's fourth recommendation. To tackle the problem of neglected diseases Reiss suggests, among others, APCs and awards for neglected-disease research as possible means to stimulate such research. Here we should recall Reiss's main objection against Pogge and Stiglitz: prizes are a bad idea because they will only trigger neglected-disease research if they make such research just as profitable as alternative projects, which means that the prizes must be very high. Does this objection apply to the APCs and awards that Reiss proposes as well?

At first sight, it seems that Reiss's system mitigates the problem. The reform he suggests reduces the profits from the biomedical research projects currently pursued. This means that prizes do not have to be as high as they have to be in the existing system to make neglected-disease research equally profitable as the projects currently pursued (Reiss and Kitcher 2009, p. 278). It seems, then, that in Reiss's system, prizes can mitigate the problem of neglected diseases at a lower cost for the public. Is this line of reasoning correct?

Reiss and Kitcher mention a few numbers concerning the profits in the pharmaceutical industry:

- In the 1990s, the top-ten pharmaceutical companies had a profit margin of about 25 % of sales, which was larger than that of any other U.S. industry.
- In 2002 (which, incidentally, was a recession year) the combined profits of the top-ten pharmaceutical companies in the Fortune 500 ($35.9 billion) were greater than those of all other 490 businesses combined ($33.7 billion)
- The median profits for other industries are about 3–5 % of sales, for commercial banking, the second most profitable industry, 13 % of sales. (Reiss and Kitcher 2009, p. 266)

These numbers raise the impression that the profit margin of the pharmaceutical industry could be substantially reduced and that it could still be the most profitable industry. The problem is that the numbers mentioned by Reiss and Kitcher (2009) are outdated. In 2008, the pharmaceutical industry had a profit margin of 19.3 % of revenues, and two other industries, networks and other communications equipments (20.4 %) and internet services and retailing (19.4 %), were even more profitable.[4] My question is then: will a substantial reduction of the profit margin of the pharmaceutical industry not chase away private investors to other, more profitable industries? In order to avoid that private investment in the pharmaceutical sector decreases, the average profit margin of this sector should not be reduced too much. If the average profits from pharmaceutical projects remain high, then prizes to stimulate neglected-disease research should be high as well, since pharmaceutical companies will only fund neglected-disease projects if such projects are expected to be at least as profitable as alternative projects. So even in Reiss's system, high prize money may be required to stimulate neglected-disease research.

7.4.2 *Push Funding Versus Pull Funding*

We saw that while the costs of pull funding are a problem for Pogge and Stiglitz, they may be a problem for Reiss just as well. But, contrary to Pogge and Stiglitz, Reiss leaves open the possibility that all neglected-disease research is supported by push funding (if push funding turns out to be most efficient and effective at promoting global health). He suggests the creation of Global Institutes of Health, in analogy with the U.S. National Institutes of Health (NIH) but committed to global health issues. What can we expect of such Global Institutes of Health (GIH)?

According to the website of the NIH, "[m]ore than 80 % of the NIH's funding is awarded through almost 50,000 competitive grants to more than 300,000 researchers

[4]See http://money.cnn.com/magazines/fortune/fortune500/2009/performers/industries/profits/. Accessed 17 August 2012. Note that Reiss (2010, p. 438n) refers to this source as well, although he mentions a profit margin of 19.1 % instead of 19.3 %.

at more than 2,500 universities, medical schools, and other research institutions in every state and around the world."[5] Since the GIH are analogous to the NIH, we can expect most of the GIH's funding to be awarded through competitive grants as well. As GIH grants will primarily be used for research that aims at the promotion of health in developing countries, the problem of neglected diseases is solved, or at least mitigated.

An important advantage of research grants is that it requires less money from the public than pull funding because the public does not have to pay for high profits for private investors (De Winter 2012a, p. 79). But just as prizes, research grants are not entirely trouble-free. Let me sum up some problems identified by Hollis and Pogge (2008). A first problem is that the financial incentives of employees of granting agencies to select the projects that are most likely to result in valuable innovation, are relatively weak. For a for-profit company, spending less money on unsuccessful projects leads to higher profits, and its employees will financially benefit from this. Such a financial incentive is absent in granting agencies: employees of such agencies do not profit from selecting the most successful projects. Personal research interests, familiarity with the applicants, and political factors are then more likely to influence decisions on which projects are funded, which could lead to resources not being allocated to the projects with the greatest health impact (Hollis and Pogge 2008, pp. 101–102).

Secondly, the financial incentives of innovators to finish their research and translate their findings into health outcomes (for medicines, this is done by conducting clinical trials, marketing the medicine to physicians, and distributing it to patients) are relatively weak. For a for-profit company, bringing a product to market is usually required to recover its investments and make a profit, and this incentive is sufficient to get the company to support expensive clinical trials, marketing activities, and distribution to patients. Such a strong financial incentive is usually absent for recipients of research grants (Hollis and Pogge 2008, p. 102).

A third problem is that research grants do not guarantee that the medicines developed through the research granted are accessible to the poor. The medicines developed through publicly funded research can still be sold at high monopoly prices, hindering access for the poor. Pull mechanisms such as AMCs and the HIF, on the other hand, offer incentives to make medicines accessible to as many people as possible (Hollis and Pogge 2008, pp. 102–103).

7.4.3 An Alternative Proposal

I think the disadvantages of prizes and research grants can be avoided by a policy I propose and defend in De Winter (2012a). My proposal is that governments should allocate more funding to non-profit organizations that aim at promoting public

[5] See http://www.nih.gov/about/budget.htm. Accessed 17 August 2012.

health in developing countries (e.g. the Drugs for Neglected Diseases Initiative, the Special Programme for Research and Training in Tropical Diseases, the Program for Appropriate Technology in Health). More specifically, more government funding should go to those organizations that promote public health in the Third World most efficiently, while funding of organizations that proceed relatively inefficiently could be reduced.

Future empirical research should reveal whether this policy requires less money from the public than prizes to stimulate neglected-disease research, and whether it has a greater health impact in the Third World than a system based on research grants allocated by a central granting agency. To know which proposal (prizes, research grants, or non-profit organizations) is to be preferred, each should be put into practice (initially at a small scale), and assessed on the basis of how well it solves the problem of neglected diseases, its health impact, and its cost for the public. As long as such an empirical evaluation is lacking, we can only turn to speculative arguments. Such arguments are offered in De Winter (2012), suggesting that a strategy based on non-profit organizations is the most promising way to promote research that is tailored to the health problems of the poor. De Winter (2012a) shows why we can expect this strategy to work and to be less costly to the public than prizes to stimulate neglected-disease research, and how this strategy avoids the aforementioned disadvantages of research grants allocated by a central granting agency.

7.5 James Robert Brown's Policy Proposal

As I mentioned in the introduction, there is also a policy proposal that has not been discussed in this paper yet, but that does, in my opinion, deserve further exploration: the policy proposal of James Robert Brown. Firstly, I will describe this proposal and some objections against it; and secondly, I will offer some possible responses to these objections, and in doing so, some important advantages of Brown's proposal are revealed. Because of these advantages, Brown's proposal should not be dismissed too quickly.

7.5.1 Brown's Proposal and Objections

Brown offers the following recommendations:

> Socialize research. Eliminate intellectual property rights in medicine. Make all funding public (including government and independent foundations and charities). (Brown 2008a, p. 762)

If all funding was made public, a lot of private funding for medical research would be lost. Therefore, public funding should be raised. According to Brown

(2008b, pp. 209–210), public funding should be adjusted to appropriate levels. He does not think that this means that current levels of funding (including both private and public funding) should be matched. He states that:

> Drug companies claim that it costs on average more than $800 million to bring a new drug to market. This, however, is a gross exaggeration. Something like $100 million is a more reasonable estimate, since marketing costs (which they include) are not part of genuine research. Moreover, many research projects are for "me too" drugs, which bring little or no benefit to the public. When we take these factors into account, it is clear that we can maintain a very high level of research for considerably less public money. (Brown 2008b, p. 210)

This passage suggests that the reform Brown proposes would substantially improve the socio-economic efficiency of biomedical research. There are, however, some problems. Firstly, Brown's estimate of $100 million seems far too optimistic. DiMasi et al. (2003) estimate that total R&D cost per new drug is $802 million, and these costs do *not* include marketing costs. Secondly, Brown's insinuation of eliminating research for "me too" drugs in order to reduce the costs for the public seems problematic. This is because it is hard to tell in advance which drugs have genuine medical value and which bring little or no benefit to the public. It is only after we have investigated a certain drug that we can tell whether or not the public can substantially benefit from it.

A third point is that, even if we can maintain a very high level of research for considerably less public money, biomedical research may still not be more socio-economically efficient. For biomedical research to be socio-economically efficient, it is required that the results are translated into health outcomes, and this may not be the case if intellectual property rights are eliminated in medicine. In this context, we can refer to the fact that before the United States enacted the 1980 Bayh-Dole act, which permits government-funded agencies such as universities to obtain intellectual property rights on products that are developed using federal grant money, the results of publicly funded research were not adequately translated into health outcomes (Reiss and Kitcher 2009, p. 280; Reiss 2010, p. 440).

7.5.2 Responding to the Objections

Brown has tried to deal with concerns about the efficiency of socialized medical research. Briefly put, his argument is that because socialized medicine[6] is more efficient than non-socialized medicine, we can expect socialized medical research to be efficient as well. But, as is shown in De Winter (2012b), this argument is not convincing for at least two reasons. The first is that it is not because socialized medicine is more efficient than non-socialized medicine, that it is not highly inefficient, since outperforming a very inefficient way of organizing health

---

[6]By socialized medicine, Brown seems to mean publicly funded medicine (see De Winter 2012b).

care is not very difficult. The second is that it is not because socialized *medicine* can be efficient, that the same holds true for socialized *medical research*, because there are several differences between medicine and medical research. Due to these differences, it is possible that a policy that works quite well for medicine, leads to major inefficiencies if applied to medical research.[7]

As Brown's response to concerns about the efficiency of socialized medical research is not convincing, let me try to deal with them in an alternative way. Firstly, I would like to note that although Brown's insinuations may be problematic that $100 million is sufficient to bring a new drug to market, and that we should eliminate research for "me too" drugs in order to reduce the costs for the public, this is not to say that his proposal does not allow for some major cost savings. What does the public currently pay for? Under the current regime, new medicines are usually sold at prices far above marginal cost of production (Pogge 2009b, p. 79). By selling medicines at such artificially high prices, pharmaceutical companies recoup their investments in R&D, marketing, etc. and they also make high profits (otherwise, they would not have made these investments). So payments from the public cover the expenditures of pharmaceutical companies as well as high profits. Under the policy proposed by Brown, the public does not have to pay for high profits for pharmaceutical companies, and the expenditures to be covered can be reduced. Huge amounts are currently spent on filing for patents in several national jurisdictions, monitoring these jurisdictions for possible infringements of patents, and lawsuits between patent holders and generic companies (Pogge 2009b, p. 82). Such costs would disappear if patents were eliminated in medicine. Furthermore, the costs associated with advertising and marketing can be reduced under the policy proposed by Brown. Although some marketing may still be required (physicians have to be informed about new medicines), we do not need all the advertising and marketing activities that pharmaceutical companies currently support.

It should be noted that some of the cost savings that can be accomplished under Brown's scheme are not achieved under the other policies discussed in this paper. None of the other policies eliminates the costs associated with patent administration. Furthermore, none of them prevents pharmaceutical companies from making profits on the medicines they sell to the *non-poor*. While the reforms suggested by Pogge, Stiglitz, Reiss, and me are all supposed to result in low-price medicines for *poor* people, none of them prevents a situation where *non-poor* people still have to pay so much for their medicines that these payments do not only cover the pharmaceutical companies' (excessive) expenditures, but also high profits for these companies.

Now consider the point that we can expect that under the policy proposed by Brown research results will not be adequately translated into health outcomes, because the results of publicly funded research were not adequately translated into health outcomes before the enactment of the Bayh-Dole act in 1980. It is, however, not because the process of transforming the results of publicly funded research into

[7]For a more extensive inquiry into Brown's argument for the efficiency of socialized medical research, see De Winter (2012b).

health outcomes was inadequate in the pre-1980 system, that it is impossible to think of a new system that is, just as the pre-1980 system, not based on intellectual property rights, but in which this process is nevertheless adequate. The selection of biomedical research projects could, for instance, be left to non-profit organizations that aim at bringing effective medicines to market, instead of to central granting agencies such as the NIH. The failure of the pre-1980 system is no reason to expect the failure of such a system.

The inadequacy of the pre-1980 system does, however, indicate that eliminating intellectual property rights in medicine and making all funding public is in itself not sufficient to ensure optimal health outcomes. It is important that Brown's scheme is supplemented by specific strategies that guarantee that the results of publicly funded research are adequately translated into health outcomes. Such strategies should be developed and assessed in further research. This can be done by experimenting with different ways of organizing public funding for biomedical research, and learning how the socio-economic efficiency of publicly funded biomedical research can be maximized. Note that my proposal to increase government funding of (the most efficient) non-profit organizations that aim at promoting public health in developing countries (see Sect. 7.4.3) is very useful in this context. Implementing this proposal will give us a clearer view on the potential of non-profit organizations.

7.6 Summary

Reiss (2010) identifies some very important epistemic, moral, and socio-economic failures in current biomedical research, and his proposal for a solution includes some recommendations that I might endorse. For instance, I think it is a good idea to prohibit socially harmful practices such as direct-to-customer advertising. I have also offered an additional argument in favor of his proposal to establish an independent body that runs clinical trials, while pointing to the need for empirical research to assess this proposal.

The part of Reiss's proposal that I have paid most attention to in this article, is his solution to the problem of neglected diseases. He suggests the establishment of Global Institutes of Health, APCs, awards for research into neglected diseases, and/or tax breaks for such research. As both push funding by Global Institutes of Health and pull mechanisms such as APCs and awards for neglected-disease research are not entirely unproblematic, I have offered an alternative proposal that is based on increased government funding of non-profit organizations that aim at promoting public health in developing countries. The purpose was not so much to offer a fully-developed policy proposal, nor to argue that the policy I propose outperforms the solutions proposed by Reiss, as more research is needed to settle these issues. Rather, my goal was to draw attention to an alternative to the strategies Reiss proposes that may avoid the disadvantages of these strategies.

Furthermore, after presenting some objections against Brown's policy proposal, I have tried to bring his proposal back in the game by pointing to the cost savings it makes possible, and by remarking that it does not imply socio-economic inefficiency just because the pre-1980 system of publicly funded research was socio-economically inefficient. In itself it is, however, not sufficient to ensure optimal health outcomes; it should be supplemented by specific strategies to ensure that the results of publicly funded research are adequately translated into health outcomes.

Acknowledgments Jan De Winter is a Ph.D. fellow of the Research Foundation (FWO) – Flanders. I am very grateful to Erik Weber, Jeroen Van Bouwel, Julian Reiss, and an anonymous reviewer for reviewing earlier versions of this paper.

References

Biddle, Justin. 2007. Lessons from the Vioxx debacle: What the privatization of science can teach us about social epistemology. *Social Epistemology* 21: 21–39.

Brown, James R. 2008a. Politics, method, and medical research. *Philosophy of Science* 75: 756–766.

Brown, J.R. 2008b. The community of science®. In *The challenge of the social and the pressure of practice: Science and values revisited*, ed. Martin Carrier, Don Howard, and Janet Kourany, 189–216. Pittsburgh: University of Pittsburgh Press.

De Winter, Jan. 2012a. How to make the research agenda in the health sciences less distorted. *Theoria* 27: 75–93.

De Winter, J. 2012b. The distorted research agenda in the health sciences and James Robert Brown's policy proposal. In *Logic, philosophy and history of science in Belgium II. Proceedings of the Young Researchers Days 2010*, ed. Bart Van Kerkhove, Thierry Libert, Geert Vanpaemel, and Pierre Marage, 123–130. Brussels: Koninklijke Vlaamse Academie van België voor Wetenschappen en Kunsten.

DiMasi, Joseph A., Ronald W. Hansen, and Henry G. Grabowski. 2003. The price of innovation: New estimates of drug development costs. *Journal of Health Economics* 22: 151–185.

Hollis, Aidan, and Thomas Pogge. 2008. *The health impact fund: Making new medicines accessible for all*. New Haven: Incentives for Global Health.

Pogge, Thomas. 2007. Medicines for the world: Boosting innovation without obstructing free access. *Sur – International Journal on Human Rights* 5: 117–140.

Pogge, Thomas. 2009a. Health care reform that works for the U.S. and for the world's poor. *Global Health Governance* 2: 1–16.

Pogge, Thomas. 2009b. The health impact fund: Boosting pharmaceutical innovation without obstructing free access. *Cambridge Quarterly of Healthcare Ethics* 18: 78–86.

Reiss, Julian. 2010. In favour of a Millian proposal to reform biomedical research. *Synthese* 177: 427–447.

Reiss, Julian, and Philip Kitcher. 2009. Biomedical research, neglected diseases, and well-ordered science. *Theoria* 24: 263–282.

Steneck, N.H. 2002. Assessing the integrity of publicly funded research. In Investigating research integrity: Proceedings of the First ORI Research Conference on Research Integrity, ed. Nicholas H. Steneck and Mary D. Scheetz, 1–16. http://ori.hhs.gov/documents/proceedings_rri.pdf. Accessed 27 Feb 2013.

Stiglitz, Joseph. 2006a. *Making globalization work*. New York: Norton.

Stiglitz, Joseph. 2006b. Scrooge and intellectual property rights: A medical prize fund could improve the financing of drug innovations. *British Medical Journal* 333: 1279–1280.

Part III
Values in a Globalizing World

Chapter 8
Responsible Design and Product Innovation from a Capability Perspective

Annemarie Mink, Vikram Singh Parmar, and Prabhu V. Kandachar

Abstract This chapter is about designing responsible technological product innovations for the multidimensional poor people in developing countries, to improve their livelihoods and make available to them better products. Attention for this so-called 'design for development' has already been raised in the 1970s. However, despite several design efforts for the poor, significant efforts are still required. To advance socially responsible design, we suggest the integration of Sen's capability approach into the product design process. This approach focuses on enhancing people's real opportunities, their capabilities. In this paper we take a capability perspective towards a technological product designed for and implemented in rural India, to explore the potential, the advantages and disadvantages of using a capability perspective when designing and innovating for the multidimensional poor. We conclude that the capability approach can offer designers a comprehensive and holistic view which aids them to better understand the context and to better predict the consequences of their product innovations. The approach therefore appears promising to support product designers in their efforts to influence the change that the multidimensional poor need in their societies and in their lives.

8.1 Introduction

Product innovations are being used in daily life and play a significant role in shaping and changing the world. As all existing product innovations have at one point been designed, design can be an agent of change. If specifically designed

A. Mink, MSc. (✉) • P.V. Kandachar
Faculty of Industrial Design Engineering, Delft University of Technology, Delft, The Netherlands
e-mail: a.mink@tudelft.nl

V.S. Parmar
Faculty of Industrial Design Engineering, Delft University of Technology, Delft, The Netherlands

Center for Innovative Business Design, Ahmedabad University, Gujarat, India

J. van den Hoven et al. (eds.), *Responsible Innovation 1: Innovative Solutions for Global Issues*, DOI 10.1007/978-94-017-8956-1_8,

for poor people in developing countries, product innovations have the potential to significantly support them in their daily life. Papanek (1984) already raised attention in the 1970s to Design for Development (DfD). He stated that from an ethical and moral point of view 'we are all citizens of one global village and we have an obligation to those in need.' Thomas (2006) stated that although the extremely poor have little money to spend on design and designed goods and therefore have limited choice, design can still 'improve their livelihoods by increasing income and making available to them better goods, products, and equipment'. Since then, designers and design scholars have been paying attention to DfD (Amir 2004; Donaldson 2002). Thomas (2006) stated that DfD has indeed been taken up by some designers, but that it has not received mainstream attention. Multiple authors agree that significant efforts are still required (Margolin and Margolin 2002; Donaldson 2002; Amir 2004).

Our research project proposes the use of Sen's Capability Approach (CA) to enhance socially responsible design and innovation for the poor. The CA focuses on capabilities; the real opportunities that people have to be who they want to be, and to do what they want to do. The approach focuses on expanding these real opportunities, rather than only focusing on income or commodities. Kleine et al. (2012); Johnstone (2007); Oosterlaken (2009) already indicated the relevance of technology and design to expand capabilities. However, Oosterlaken (2009) also noted that 'philosophers working on the capability approach so far do not seem to have sufficiently realized the relevance of technology, engineering, and design for capability expansion.' Until now, limited research has been conducted to investigate the link between technology, design and the CA.

This paper is a step towards further investigating this link. According to Johnstone (2007) 'justice dictates that we must look first to the needs of those whose capabilities are already low'. That is why will investigate the influence of design and product innovation on the capabilities of the poor. For them specifically it is important to enhance their valuable opportunities. Our research postulates that, by integrating the CA into the existing design processes, product innovations can become more effective and socially responsible. We expect the CA, as an influential and increasingly applied view on development, to add a new perspective to design and product innovation for the poor in developing countries.

This paper is a preliminary exploration in which we analyse a DfD case from a capability perspective in order to gain insight in the added value of the CA for DfD, and to investigate the advantages and disadvantages of using the CA as a theoretical framework for promoting responsible innovation for the most capability deprived. We will first introduce the general notions of design and product innovation, the design process, current DfD design activities, and some DfD guidelines. In Sect. 8.2 the CA and capabilities are explained in more detail and some parallels between discussions in CA literature and in product design literature are mentioned. In Sect. 8.3 we describe the case of a Tasar silk reeling machine, towards we will critically apply a capability perspective with hindsight in Sect. 8.4, where after we reflect on this case analysis in Sect. 8.5, and conclude this chapter in Sect. 8.6.

8.1.1 Product Innovation and Product Design

Innovation is a broad concept that has been defined in many different ways. Schumpeter (1983) defined innovation as 'the commercial or industrial application of something new.' Besides 'new', other keywords used in definitions on innovation are 'value', 'creation', and 'successful' (Amabile et al. 1996; Harvard Business Press 2003; Diehl 2010; Redelinghuys 2006). However, the core of innovation seems to be that it brings along significant positive change (Berkun 2010). We agree with Rogers (1995) that this change not only concerns its first use or discovery, but an idea is an innovation if it seems new to the individual. Schumpeter (1983) made a distinction between product and process innovations, while later on more types of innovation have been identified. The word design can be used as a verb, a noun or an adjective (Birkett 2010). In this chapter we use the word design as a verb, which, according to Birkett (2010), refers to the action or process of designing. Heskett (2005) defined design as 'the human capacity to shape and make our environment in ways without precedent in nature, to serve our needs and give meaning to our lives.' Besides 'needs' and 'shape', other keywords used to define design are 'creative', 'human/people', and 'change' (e.g., Simon 1996; Buchanan 2001; Donaldson 2002; International Council of Societies of Industrial Design 2011).

The focus of this paper is on the design of product innovations. The profession of product design is closely linked to product innovation, as all products have once been designed. This is recognized by multiple authors (e.g., Redelinghuys 2006; Skogstad and Leifer 2011; Veryzer 2004; Thomas 2006), and also in the OECD Oslo Manual[1] (OECD and Eurostat 2005). We specifically focus on technological design of product innovations, which we define now as 'the successful creation of tangible, technological products or services that induce change to a new context'. This change can be positive and/or negative. Designers, therefore, have a 'high social and moral responsibility' for the consequences of their innovations (Papanek 1984).

8.1.2 The Design Process

Drucker (1998) stated that there are innovations that come from a stroke of genius, without being preceded by 'a conscious, purposeful search for innovation opportunities'. Grassroots innovations for example are purely constraint driven and have hardly any systematic and scientific preparation. However, Drucker (1998) also stated that most innovations result from a conscious, and purposeful search for innovation opportunities. Likewise, Owen (1992) argued that breakthrough thinking

[1] This manual of the Organisation for Economic Co-operation and Development, provides guidelines by which comparable innovation indicators can be developed in OECD countries. Since 2005 non-technological innovation, and linkages between different innovation types are taken into account.

is almost always preceded by extensive preparation. Therefore, throughout the years the design and product innovation process has been structured. Many scholars have been developing design methodologies to guide and assist designers to create product innovations (Diehl 2010).

Cross (2000) described the basic structure of design methodology in three phases: the analysis phase, followed by a synthesis phase and an evaluation phase. Many additions to this basic structure have been made,[2] but the overall structure is the same. At the faculty of Industrial Design Engineering of Delft University of Technology the most frequently adopted design methodology is the 'basic design cycle' of Roozenburg and Eekels (1998). This process is not a linear process, but an iterative, spirally 'trial-and-error' process, during which the designer goes through reductive and deductive steps, and often needs to return to earlier phases to re-evaluate previous decisions (Roozenburg and Eekels 1998). Due to these iterations, the knowledge about the problem and about the design itself increases (Roozenburg and Eekels 1998). The different phases of this cycle are: (1) Analysis: the design problem is analysed and defined, resulting in design requirements; (2) Synthesis: a temporary design proposal is made, and ideas are formed. The best ideas are chosen and conceptualized. Then, the best concept is chosen and elaborated into a preliminary design; (3) Simulation: forming an idea of the behaviour and characteristics of the designed product by reasoning or by building a prototype; (4) Evaluation: determining the value or quality of the preliminary design by comparing the expected properties with the desired properties. The above process encompasses what Roozenburg and Eekels (1998) called the 'strict development'. This strict development is preceded by a product planning phase, and succeeded by a realization phase, as there can be only 'innovation' if the new activity is actually realized.

Several methods exist for each phase of the design process, which aid the designer in developing products and services. A method is a 'diachronous structure, which is consciously applied to the action' (Roozenburg and Eekels 1998). Design methods can be any procedures, techniques, aids or tools, as brainstorming, context mapping, use of checklists or process trees, among others (Cross 2000). The designer might use and combine them into the overall design process.

8.1.3 *The Multidimensional Poor*

Prahalad (2005) noted that our economic world appears like a pyramid, with the poor people at the 'Base of the Pyramid' (BoP). The BoP represent two thirds

[2]By e.g. Archer (1984), Pahl et al. (1984), March (1984), Wheelwright and Clark (1992), Verein Deutscher Ingenieure (VDI guideline 2221, 1993), Roozenburg and Eekels (1995), Buijs (2003), Ulrich and Eppinger (2004), Buijs and Valkenburg (2005), Unger and Eppinger (2010), Meinel and Leifer (2011) among others.

of the world's population of seven billion people living on incomes of less than $1.500 per year. BoP refers to economic deprivation alone. The Multidimensional Poverty Index (MPI) on the other hand, 'complements money-based measures by considering multiple deprivations and their overlap' (UNDP 2012). 'Although deeply constrained by data limitations, the MPI reveals a different pattern of poverty than income poverty, as it illuminates a different set of deprivations' (Alkire and Santos 2011). Therefore, we use the concept of the multidimensional poverty index (MPI) to specify our target users. This index is grounded in the CA and is used by the United Nations Development Program (UNDP). According to this index, worldwide an estimated population of 1.75 billion people experience multidimensional poverty (UNDP 2010).

8.1.4 Design for Development

Prahalad (2005) argued that we should start recognizing the poor as resilient and creative entrepreneurs and value-conscious consumers. He suggested an approach to 'achieve sustainable win-win scenarios where the poor are actively engaged, and, at the same time, the companies providing products and services to them are profitable'. Thomas (2006) observed that Prahalad did not identify the value of design in his book, and argued that design is truly relevant to poverty alleviation, which is also recognized by Kandachar and Halme (2008) and Oosterlaken (2009). At the Faculty of Industrial Design Engineering of Delft University of Technology (TUD), where the authors work, extensive work is being carried out to address this gap. This so-called 'Design for Development' is defined by Donaldson (2006) as 'product design aimed at disadvantaged or marginalized populations' to advance social, human, and economic development. Besides advancing development, developing countries also represent a very big consumer market, and designing products 'at affordable costs for the harshest of conditions with minimal resources can [offer insights that] benefit all markets' (Viswanathan et al. 2011). This is not only recognized at TUD, other universities[3] and companies[4] have also shown interest in western countries to design and innovate for the poor in developing countries. Processes, methods and toolkits have been developed to better address the needs of these people. Most of

[3]The Massachusetts Institute of Technology in the USA (D-Lab), Stanford university in the USA (partner in D-Rev), the Institute of Design from IIT Chicago in the USA, Aalto University in Finland (BoP Network), University of Colorado-Boulder (Engineering for Developing Communities (EDC) program), Ateneo School of Government, Philippines (Science and Technology Innovations for the Base of the Pyramid in Southeast Asia program), among others.

[4]According to Donaldson (2006), the most prominent Non-Governmental Organizations designing products for less industrialized economies are: Intermediate Technology Development Group, KickStart (formerly ApproTEC), International Development Enterprises, TechnoServe, and EnterpriseWorks Worldwide.

these design aids assist companies in developing business strategies,[5] others have been developed specifically for NGO's, social enterprises or community workers.[6] Thereby, Sklar and Madsen (2010) expressed that the design process from Western countries might often be transferable to the developing world, and that the design approaches which are used to learn about the world are also applicable across the world. Still there are 'little theoretical or practical guidelines for innovative product development' in developing countries (Viswanathan and Sridharan 2012). And the focus of new product development is still on high-income countries (Viswanathan et al. 2011; van den Waeyenberg and Hens 2008).

Coming from a Western context, it is not an easy task to identify the needs of the multidimensional poor. Birkett (2010) stated that a designer's experiences and relationships influence the decisions a designer makes during the design process. Balaram (2011) also indicated that design in the West is naturally geared to its own needs, and its own socio-cultural environment, values and economy. He stated that 'it is suicidal to transplant solutions onto a completely different ground'. Due to *dissimilar ideologies* of the designer and the target-user, it is difficult for the designer to identify the true needs of the target group and design accordingly. According to Thackara (2005), many of the troubling situations in our world are the result of design decisions. However, for users 'without financial safety net to take risks', unsuited or poorly engineered technology 'can only be detrimental' (Donaldson 2006). It is always important for product designers to be sensitive to context, to relationships, and to consequences (Thackara 2005). However, especially when designing for the multidimensional poor, the consequences of products and services must be predicted really well, which asks for a thorough analysis and continuous reflection by the designer during the design process.

8.2 A Capability Approach to Design

Because product design can induce positive and/or negative change, design decisions must be made well-informed. Thackara (2005) wrote his book 'In the Bubble' from a belief that ethics and responsibility can inform design decisions without

[5] Among others, the BoP Protocol 2nd Edition of Simanis and Hart (http://www.bop-protocol.org, accessed January 2011), the Market Creation Toolbox of the BoP Learning Lab (2011, http://www.boplearninglab.dk, accessed February 2012), and the Design for Sustainability (D4S) manual of the United Nations Environment Program in collaboration with TUDelft (http://www.d4s-de.org, accessed January 2011). The American Society of Mechanical Engineers (ASME) wrote a report on engineering solutions for the BoP which includes four critical business strategies, but also five design principles for engineers (https://www.engineeringforchange.org, accessed January 2011).

[6] Among others, Frog's 'Collective Action Toolkit' (CAT), which emerged from frog's collaboration with Nike Foundation/Girl Effect (http://www.frogdesign.com, accessed November 2012). And IDEO developed the 'Human Centered Design Toolkit' (HCD) in 2009, Developed after a request of the Bill and Melinda Gates Foundation (http://www.ideo.com, accessed January 2011).

constraining social and technical innovation. We think that the capability approach (CA) might provide product designers with a framework that helps them to identify the required information for making deliberate and responsible design decisions. To explore the advantages and disadvantages of using the CA as a theoretical framework for Design for Development (DfD), we apply a capability perspective to a specific DfD case. In this section, we will first describe the CA and the concept of capabilities, and the identified parallels between CA literature and product design literature, after which we discuss the establishment of a general list of beings and doings that will be used to analyse our case.

8.2.1 The Capability Approach

The CA has been introduced and developed by economist and philosopher Amartya Sen and by philosopher Martha Nussbaum. Within this approach, development is seen 'as the expansion of human capability to lead more worthwhile and more free lives' (Sen 1999). The approach focuses on human capabilities; a person's effectively available valued beings and doings. The CA evaluates justice, equality and development not by income, commodities or utility, but by the real opportunities that people enjoy. The CA makes a clear distinction between what people are free to do to improve their well-being ('capabilities') and what they actually choose to do ('functionings'). Examples of valuable capabilities are, among others: the capability to move freely anywhere you want, the capability to receive education, the capability to participate in public debates, and the capability to have sufficient nutritional intake.

The CA provides a more complete picture of poverty and deprivation, because it takes into account all dimensions of human well-being (Robeyns 2005; Chiappero Martinetti 2008). It is 'a flexible, and multi-purpose framework', due to its 'open-ended and underspecified nature' (Robeyns 2011). Many researchers from different disciplines have 'taken up, discussed and elaborated' the CA (Anand et al. 2009). However, translating such a rich, theoretical argumentation into practice is a difficult task (Chiappero Martinetti 2000). Sympathizers of the CA acknowledged that operationalizing the CA into practice is a major challenge lying ahead, 'either due to its emphasis on value judgments with high informational requirements or its multidimensional nature' (Comim 2001).

Until now, the CA has mainly been operationalized for evaluative and descriptive purposes (Alkire 2008b). The CA is then used to look with hindsight which capabilities have been influenced. Our overall research aim is to operationalize the CA for what Alkire (2008b) calls the 'prospective use' of the CA. Putting the CA into DfD practice in a prospective way denotes that at the start of a DfD project, we will try to look forward to identify those capabilities that are relevant for this specific project. So far, the CA has not explicitly specified a methodology for prospective analysis, and 'it seems that the methods will be plural and the questions will vary by discipline, level of analysis, policy audience, region and context' (Alkire 2008b).

However, according to Johnstone (2007), the CA offers ample material 'that can be used as the basis for developing approaches to action, policy and intervention.'

8.2.2 *Capabilities and Their Characteristics*

Capabilities are valuable beings and doings that a person can achieve. The focus is on those capabilities that are actually open to people; the *real* opportunities that people have (Alkire 2005). They are opportunities that a person can choose from, but not the achievements itself. They are therefore not directly observable and in practice not always identifiable (Sen 1995). The CA sees capabilities as the ends of well-being and development (Robeyns 2005). Within the CA it is however recognized that capabilities can also be a means to another end, and in this way, next to being of intrinsic importance, capabilities can also have an instrumental role (Robeyns 2005).

Some have argued that capabilities are incommensurable (Robeyns 2011). Thereby, capabilities differ per person, per group and change over time. Alkire (2005) explained that the CA provides an 'analytical map of important variables', but this map must be 'adapted, shaped, and fitted to many different institutional levels, time periods, groups, and so on'. This raises discussions about which capabilities matter in which context and how to prioritize between different capabilities. Sen (1999) stated that the prioritization of capabilities it is 'a "social choice" exercise', which 'requires public discussion and a democratic understanding and acceptance.' For every purpose thus, it is important to identify the *real* relevant opportunities to fit an individual or a specific context at a certain time.

8.2.3 *Parallels Between the Capability Approach and Product Design*

In CA and product design literature we identified five similar discussions. These parallels are presented in this section, and will be used to evaluate the DfD process of the case presented in Sect. 8.3.

8.2.3.1 The Use of Additional Theories

In many cases, the CA should not replace other, more established, approaches, but provide complementary insights to them (Robeyns 2006). Product design is a multidisciplinary profession which uses additional bodies of knowledge. Designers use for example different ethnographic methods (Friess 2010). Designers need to think of the consequences of their products and services thoroughly (Papanek 1984; Thackara 2005). Therefore, they could use the CA as a complementary theory, in order to provide additional insights about the target-users and their contexts.

8.2.3.2 Concern for Human Diversity

Robeyns (2005) noted that the CA takes human diversity into account in two ways. First, the CA focuses on the plurality of functionings and capabilities as the evaluative space (Robeyns 2005). This broad view causes a focus on 'things that really matter' and avoids 'the neglect of crucially important subjects' (Sen 1999). Second, the CA takes conversion factors into account. Conversion factors influence the 'transformation of resources into achieved functionings' (Frediani 2010). Robeyns (2005) mentioned three types of conversion factors; personal factors (e.g., physical condition, intelligence, sex), social factors (e.g., public policies, social norms, gender roles), and environmental factors (e.g., climate, geographical location).

Oosterlaken (2009) and Toboso (2011) already related the CA and human diversity to inclusive/universal design. Inclusive/universal design is the development of products and/or services that are accessible to, and usable by, as many people as reasonably possible (Keates and Clarkson 2004). The design process therefore starts with 'a rigorous and exhaustive analysis of user requirements and other basic features of the problem' (Roozenburg and Cross 1991). By doing so, designers do not only look at the technical function, but also at the psychological, social, economic and cultural functions that a product has to fulfil (Roozenburg and Eekels 1998). However, Nieusma (2004) argued that universal design 'implies embracing ever-greater diversity in design', while knowing that we can never develop one system that meets everyone's needs. Roozenburg and Eekels (1998) also acknowledged that a design outcome can never fulfil all requirements. They state that every solution for a design problem means compromising between contradictory criteria. Sklar and Madsen (2010) stated that appropriate choices need to be made to satisfy the priorities of the target-group, and those of the involved stakeholders as well. The CA might aid designers to consider multiple dimensions and to take conversion factors into account. In this way, designers might be able to make more deliberate trade-offs, and to minimize exclusion of users.

8.2.3.3 Involvement of the People Concerned

Sen (1999) argued that the involvement of the people concerned is a requirement when enhancing capabilities. He also stated that capability selection is not a task for outsiders, but it needs to be a participatory, democratic process. Oosterlaken (2009) therefore connects the CA to participatory design, which is also propagated in IDEO's, and Frog's Toolkits. For DfD, Donaldson (2002) and Viswanathan and Sridharan (2012) stressed the importance of truly addressing peoples' needs. Sklar and Madsen (2010) emphasized that to be able to do so, designers should see the world from the point of view of their target-users, and should understand their motivations and aspirations. User participation in a design process has already been developed in the 1970s (Bødker and Pekkola 2010). It has served the discipline of design very well, as it gives design 'a purpose, a structure, and [...] a story to tell'

(Friess 2010). Donaldson (2002) stressed that continuous interaction with potential users is important throughout the design project, and also after its completion. She also referred to feedback loops as being vital. Hanington (2010) mentioned design ethnography, participatory design and design testing as good methods to be used in the different design phases. Although participation can make the design process long lasting and does not ensure agreement due to different preferences and opinions, we do acknowledge the use of research and participation in the local context, to make the designer understand the context and the values of the target-users, to stimulate deliberate decision-making.

8.2.3.4 Concern for the Individual and for Communities

Within the CA, there is an on-going debate about a focus on individual and/or collective capabilities (Robeyns 2006). Frediani (2010) referred to Gore, who observed that the CA measures well-being in terms of an individual ability, while some capabilities belong more to societies or groups than to individuals. Therefore, it might be useful to take both into account. Designers need to consider more than the individual, they need to balance individual and community needs (Sklar and Madsen 2010). A focus on collective capabilities, however, can complicate the process of agreeing on a capability set (Kleine 2010). Clark (2009) noted that 'even in cases where deliberative forms of democracy function well and everyone's voice is heard, there may be grounds for concern insofar as majority rule is allowed to trump individual values.' Another difficulty is the question how far preferences must be respected and can be justified (Robeyns 2006). The consequences for designers are that they have to identify individual and communal capabilities and need to balance these capabilities, while making deliberate choices about which capabilities will be addressed.

8.2.3.5 Focus on People's Personal Choice

The CA focuses particularly on people's capability to choose the lives they have reason to value (Sen 1999). However, not all choices are relevant, only the choices between valued opportunities are (Johnstone 2007). Kleine (2011) stated that choice does not only has an instrumental role, but also has intrinsic value, as 'being able to pursue one's own choices is part of being fully human.' She mentioned four dimensions of choice; the existence, the sense, the use, and the achievement. However, people's use of choice might be influenced by a phenomenon which is called 'adaptive preferences'. Sen (1999) described this phenomenon as 'the adjustment of people's desires and expectations to what they unambitiously see as feasible due to their deprivation.' According to Clark (2009) adaptive preferences come into existence due to several reasons: (1) the malleability of people's aspirations and desires to the circumstances in which they live; (2) the social conditioning or cultural and religious indoctrination; and (3) the more general form of people's own limitations to make informed judgments and rational choices. Clark (2009) argued that democratic and participatory techniques might not be sufficient to identify adaptive preferences. He reasoned that it might be the best strategy to engage directly with the experiences and views of the poor. Users' personal choice is also relevant for designers. A designer has to make informed design decisions in order to offer the user appropriate design solutions that provides the user with sufficient choice to achieve their desired outcomes. It is therefore important for the designer to identify adapted user preferences. Then, designers might choose to persuade the user to behave in a certain manner. Parmar (2009) argued that persuasive technology is known in the western world as a strategy for changing people's social and health-related attitudes, but is rarely used in the rural context and at a community level without fully taking away the user's freedom to act differently. However, when the designer develops a product which is considered the best option for the potential user and the user is not given the freedom to act differently, this is called paternalism. Suber (1999) described paternalism as 'to act for the good of another person without that person's consent.' Considering the influence of product designs on a user's choice, it becomes clear that a designer has to identify people's true valued beings and doings, and has to be aware of the existence of adaptive preferences, to be able to make deliberate design choices. The designer might consider persuading the target-user to behave in a certain manner, but must be careful not to become paternalistic.

8.2.4 A List of General Beings and Doings

In this paper we analyse a DfD case from a capability perspective. We explore how a product, specifically designed for the rural poor in India, influenced their capabilities. We will base our case on, what Alkire (2008a) recognized as, informed guesses from the researchers. In order to make these guesses we first developed a list of beings and doings. The use of a list is an issue which is highly debated within the CA. Nussbaum (2000) created a list of ten capabilities which every human being should be entitled to. According to Nussbaum, her list is formulated at an abstract level, and the translation to implementation and policies should be done at a local level, taking into account local differences (Anand et al. 2009). Sen has explicitly refrained from defending a well-defined list of capabilities (Robeyns 2006). Sen argued that important capabilities and their weight should be selected in the light of the purpose of the study and the values of the referent populations (Alkire 2008a). Alkire (2008a) also refrained from using a single list of poverty dimensions, but she did, however, identify 37 lists (including Nussbaum's list) that contain poverty dimensions. She also mentioned that it can be useful to make such a list for certain exercises, but that the 'the same list would not be helpful in diverse analyses.' While we do agree that capabilities are context-specific and that we cannot simply prescribe a specific set of capabilities that can be used for every product innovation process, we did develop a list of beings and doings, considering all lists that Alkire (2008a) identified. This list contains *general* beings and doings, for example being literate. These general terms are according to Robeyns (2011) the focus of the CA. However, if 'a particular person then decides to translate these general capabilities in the more specific capabilities A, B or C (e.g., reading street signs [...]) is up to them.' We will therefore analyse our case by going through this list of general beings and doings to be able to extract more specific capabilities.

We listed all the dimensions of the lists and classified the dimensions according to the seven aspects of well-being as identified by Williamson and Robinson (2006). They classified well-being by biological, mental, emotional, material, social, cultural, and spiritual aspects. We provided descriptions for every doing and being on the list. The full list, containing all dimensions and their descriptions, is given in Appendix A.

8.3 The Case of the Anna Tasar Reeling Machine

To explore a capability perspective towards Design for Development (DfD), we will analyse an innovative product that has been implemented in the field: the Anna Tasar Reeling Machine (ATRM), a reeling machine that processes Tasar silk cocoons into yarn. In this section we provide some background information about the development of the ATRM and its implementation. The development of this reeling machine for Tasar silk has been the graduation project of the first author and is implemented in rural eastern India, and was part of a larger project of the Indian non-governmental organization (NGO) PRADAN. This NGO organizes poor rural village women in so-called Self-Help-Groups (SHGs) and engages them in independent livelihood activities. One of these activities in the states of Bihar, Jharkhand, and Chhattisgarh is Tasar silk reeling. Information about this project comes from PRADAN and from the first author. It must be noted that the capability approach (CA) was never considered during the development of this machine. The CA perspective is only used to analyse this case after its development and implementation, which will be done in Sect. 8.4.

8.3.1 The Tasar Silk Reeling Project

The Tasar silk reeling activity has traditionally been a low-paying activity in the states of Jharkhand, Bihar and Chhattisgarh, carried out by poor rural women in their spare time. This is done mainly by women in weaver families (without any remuneration), or as an uncertain, low paying type of wage labour. PRADAN separated the yarn production from the weaving activity and promoted it as an independent, separate, and viable enterprise. They introduced existing machinery (a reeling and a re-reeling machine, see Figs. 8.1 and 8.2) to replace the primitive and rudimentary technology of palm or thigh reeling (see Fig. 8.3). They organized women from different SHG's into reeling groups who work together in a

Fig. 8.1 The Tasar silk reeling machine as introduced by PRADAN (Picture by first author)

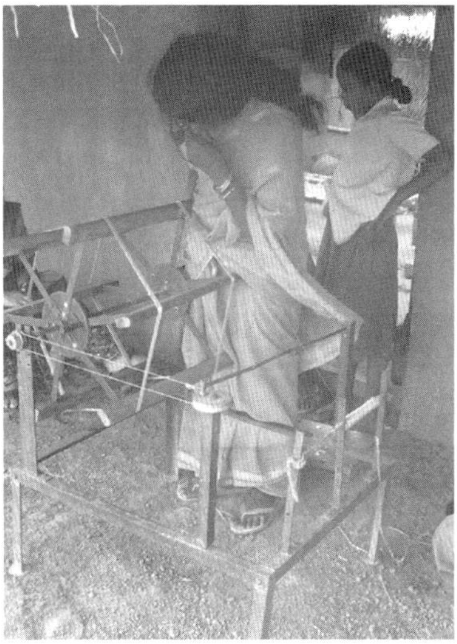

Fig. 8.2 The Tasar silk re-reeling machine as introduced by PRADAN (Picture by first author)

Fig. 8.3 Thigh reeling, a traditional method of Tasar reeling (Picture by PRADAN)

reeling centre. This reeling centre is specially built for this activity in the center of several villages, to allow women from multiple villages to join the activity. The women who engage in reeling generate income which enables them to better fulfil their basic needs. It also reduces the urge for the husband to migrate to the city for work. Then, the women gain more confidence, generating income in a dignified way and becoming more self-sufficient and independent.

PRADAN assist the women in obtaining the machine with government subsidy (the women are too poor to buy a machine by themselves), they provide reeling and entrepreneurial training, they help out in cocoon buying and storage and in the sales of the yarn. They also employ technicians for repairs of the machines. Moreover, they opened up new markets for Tasar silk. Because the activity flourished well, PRADAN organized the women in their own producers' company called MASUTA.

8.3.2 The Development of the Anna Tasar Reeling Machine

The reeling machine that PRADAN introduced did improve the working circumstances of the women highly, but the machinery suffered from several problems (e.g., energy-loss, failing materials, safety issues, physical problems due to running the machine by pedalling, and yarn quality problems). With help from one of their subsidiaries (ICCO, a Dutch NGO) the first author was appointed to re-design this machine as part of a Base of the Pyramid graduation project at the faculty of Industrial Design Engineering of Delft University of Technology (Mink 2006). This effort led to a vastly improved machine and was named 'Anna Tasar Reeling Machine'. Up scaling started leading to large scale utilization. Currently, November 2012, 219 machines are running in several villages.

The re-design of the ATRM was executed following the methodology described by Roozenburg and Eekels (1998). During the analysis phase all stakeholders were interviewed to identify the design requirements. Reelers, PRADAN staff (field workers, technicians, yarn graders, and team leaders), and the managing director of MASUTA were all interviewed about the use of the existing machine, the quality and characteristics of the reeled yarn, and about their preferences for a new machine. The full process from cocoon rearing up to yarn making, weaving, and fabric marketing was analysed to obtain a good view of this process and the requirements this process brings along. The reelers were also observed during their work on the machine, during SHG meetings, and during daily activities. Some of them were also interviewed about their lives, and because no anthropometric data was available of rural North-Indian women, measurements were taken of 24 women. From a technical point of view, the existing machine was fully analysed, as well as other silk reeling machines which are in use in India, and the production possibilities in India were explored. During this analysis phase, a lot of requirements were identified, mainly concerning the technical and economic function of the machine, and concerning the user comfort during the reeling activity.

Fig. 8.4 (**a–h**) Several prototypes of the Anna Tasar Reeling Machine (Pictures a, b by first author, pictures c-h by MASUTA)

Due to the high technological character of the machine development, the reelers were not involved in the synthesis phase of the machine development. They were not involved in idea generation, neither in choosing between several ideas and concepts. For this phase mainly technical knowledge of the reeling process and of machines was required, therefore only PRADAN and MASUTA staff were involved here. The preliminary design was manufactured in 2006 in Nagpur, and thereafter, the machine was extensively tested by reelers, in which they could suggest changes. During further adjustments, reelers were continuously involved by testing the machines (see Fig. 8.4a–h). Their feedback together with technical optimizations lead to the final machine design, which was ready for up scaling in 2010. Still now, the machine is continuously being optimized, with help from MASUTA's own technicians, the reelers, and the Central Silk Board of India.

8.3.3 Results After Implementation

The impact of the ATRM was evaluated after implementation, and it turned out to be that this machine further improved the reeling activity, ensuring a higher yield and a higher quality yarn, while the cost of the reeling machine is approximately the same as of the old machine (around INR 25,000). The reelers are able to extract more yarn from one cocoon, which is according to MASUTA's managing director probably because the reelers have more time to concentrate on extracting the yarn, and because there is less yarn breakage. Therefore, the reeler's income went up, compared to the income they earned with the old machine. This is shown by statistics from Danidih village in Jharkhand (Table 8.1).[7] This additional

[7]For each reeler, data are kept to capture the performance of each reeler and to be able to calculate the reeler's payment. These data concern the amount of days the reeler works, how many spindles

Table 8.1 Improvements for reelers due to the re-design of the reeling machine

Reeling statistics from the reeling centre in Danidih (Godda District, Jharkhand)		
	Old machine	Anna Tasar Reeling Machine
Yarn quality ratio (grade A:B:C)[a]	64:36:0	92:8:0
Reeled yarn per working day (gram)	127	171
Profit earned per working day (Indian Rupees)[b]	30	56

[a]A-grade yarn is the best quality, C-grade yarn is unsuitable for selling

[b]For heavy physical labour women are paid 10–20 rupees per full working day (8 h), this income is earned during the 4–6 h that a reeler spends on average in yarn production per day

income enables the women to better fulfil their basic needs, and to gain even more confidence, by becoming more self-sufficient and independent.

The ATRM is also more comfortable and easy to use, is more energy efficient and the safety of usage has improved. The ATRM thereby introduces the possibility of producing a new type of yarn, which ensures better sales. This type of yarn is called *untwisted* yarn,[8] and was, until now, produced by women in traditional weaving pockets by using the traditional methods. Solar panels are used to supply energy to the machines, and therefore pedalling is no longer required.

Some aspects concerning the machine were adjusted during prototype testing. Initially the machine was placed on the floor to work (as can be seen in Fig. 8.4d). Sitting on the floor is according to Indian culture, however, it worsened the working position of the women instead of improving it. Therefore, the reeling machines are now placed on a platform (Fig. 8.4e, f). Second, the machine was designed to have four spindles, but due to the increased speed of the spindles, the women were only able to use one or two of those spindles. Therefore, the ATRM has been downsized, and currently contains only two spindles.

The ATRM also had some undesirable effects; it is more difficult to mend the ends of the yarn after breakage, because the yarn entangles more on the bobbin. This is a challenge that still needs to be overcome during further optimization. Second, the covering of rotating parts makes the machine safer to use, but also makes maintenance a more difficult job. This has, however, been a deliberate choice. Third, the ATRM is easier to use, which is beneficial for the reeler, but might also encourage child labour. PRADAN keeps a close eye on keeping children from working fulltime in yarn production. The children do sometimes help their mothers during reeling, but mostly they do not reel themselves, as yarn reeled by occasional reelers is of low quality with low recovery. In some villages, grown-up girls who stopped going to school (due to the distance to high school girls are not always send there) start reeling as a full time business before their marriage. Lastly, PRADAN

she uses, the amount of cocoons she uses per day, the amount of yarn she reels, and the quality of the reeled yarn. These data are entered in a computer programme called Softyarn.

[8]The warp of a fabric requires twisted yarn for its strength, but untwisted yarn can be used for the weft of a fabric to give it a softer feel.

asked for a small and light-weight machine that can be taken home to be used there. For several reasons, PRADAN now started promoting only individual home-based reeling for new reeling villages.[9] Reeling at home does not require an unreliable reeling centre manager, and the reeling activity does not suffer any longer from closure of the reeling centre – considered common property by the community – due to community disputes. It therefore turned out that, where the ATRM was designed to give the reelers the choice to work in a reeling centre at home, new reelers do not actually have this choice.

8.4 The Anna Tasar Reeling Machine from a Capability Perspective

We applied a capability perspective to the case of the Anna Tasar Reeling Machine (ATRM) with hindsight, in order to capture a broader view of the impact of the ATRM on the lives of its users. The analysis of the ATRM from a capability perspective is done by using the established list of general beings and doings (Sect. 8.2.4). For every being and doing on the list we evaluated it to be meaningful for this case or not. The eventual identified capability parameters were validated by consulting the Producer's Company MASUTA[10] in Jharkhand, India. All the capabilities that are relevant to this project are summarized in Table 8.2. After this exercise, we looked at the development process of the ATRM by using the identified parallels between the CA and product design literature (Sect. 8.2.3), and use this case to reflect on capabilities and their characteristics (Sect. 8.2.2). As we have not determined how to measure the identified capabilities, we cannot provide any quantitative statements. However, as Sen (1995) stated 'having more of each relevant functioning or capability is a clear improvement, and this is decidable without waiting to get agreement on the relative weights to be attached to the different functionings and capabilities.' Therefore, our investigation is aimed at an increase or decrease of capabilities and functionings.

8.4.1 Evaluation of Desired Outcomes: Beings and Doings

By using the list of general beings and doings we identified several specific enhanced and decreased capabilities as a consequence of the usage of the ATRM. These beings and doings are discussed below and an overview can be found in Table 8.2.

---

[9] *Obtained from email-contact with Mr. M. Ray, Director of MASUTA Producer's Company Ltd.*

[10] *The information is gathered through email-contact with Mr. M. Ray, MASUTA's director who is also in close contact with the implementing non-governmental organization PRADAN.*

Table 8.2 Overview of enhanced, decreased and unchanged capabilities, extracted from the case of the ATRM

Biological aspects of well-being	**Influence on capabilities**
Health	*Enhanced*
Being able to have good bodily health	Due to covering of the machine, the safety for the reelers and their children improved
	When the machine is placed on a table, the ergonomic posture of the reeler improved
	Decreased
	When the machine is placed on the floor, the ergonomic posture of the reeler decreased
Mental aspects of well-being	**Influence on capabilities**
Education	*Enhanced*
Being able to receive education	Due to promotion of home based reeling, the solar panel attached to the roof enables children to study in the evening
Freedom of movement	*Decreased*
Being able to move freely from place to place	Due to promotion of home based reeling, the reeler is restricted in moving around freely.
Meaningful work	*Unchanged*
Being able to choose one's work, and to work as a human, to exercise practical reason, and to enter into meaningful relationships of mutual recognition with other workers	The machine still enables the women to work as a human
	Decreased
	Due to promotion of home based reeling, the reeler has less possibility to enter into meaningful relationships with other workers
Emotional aspects of well-being	**Influence on capabilities**
Happiness	*Enhanced*
Being able to lead a happy, enjoyable life	Additional income and/or time improve the reeler's ability to lead a happier, more enjoyable life
	A better ergonomic posture increases the reelers health, which enhances their happiness
Love	*Enhanced*
Being able to experience love, longing and grieve, and being able to give love and affection	Affection towards daughters might be enhanced when daughters run the machines for their mothers to ensure economic security
Worry-free	*Enhanced*
Having a prosperous life, without worries and with confidence in the future	Additional income and/or time improve the reeler's ability to lead a more prosperous life
Self-respect	*Enhanced*
Having the social bases of self-respect and non-humiliation	Additional income increases self-respect
	Unchanged
	Self-respect due to owning and using the machine by themselves

(continued)

Table 8.2 (continued)

	Decreased Due to covering the rotating parts, the women themselves have more difficulty to maintain the machine. This decreases their confidence and self-respect
Achievement Being able to accomplish one's aspirations, to demonstrate competence and making a lasting contribution	*Enhanced* The additional income gives the reeler more sense of achievement
Equality Being able to be treated as a dignified being whose worth is equal to that of others	*Unchanged* Reeling enables the women to be treated as a dignified being who is equal to others
Recognition Being recognized and having status	*Enhanced* The additional income increases the recognition and status of the reeler
Having power Having social status and prestige, and having control or a dominant position within the household and the more general social system (includes decision-responsibility)	*Enhanced* The additional income increases the dominant position of the reeler within the household
Material aspects of well-being	**Influence on capabilities**
Goods Being able to hold property/to have sufficient assets, control over material environment	*Enhanced* The machine is easier to use which gives the reeler more control over their material environment
	Unchanged Control over their material environment due to local reparation possibilities
	Decreased The machine is more difficult to maintain and therefore decreases the reeler's control over her environment
Economic security Being economically secure at present and in the future	*Enhanced* Additional income gives economic security
	Children can run the machine if the reeler herself is not able to, which increases the economic security of the family
	Unclear Does the reeler earn more income when she works at home (reel whenever she has time and use of light), or when she works in a reeling centre (away from household chores and children)?
Settings of interaction Having places to meet others	*Decreased* Due to promotion of home based reeling, the reeling centre is no longer a setting of social interaction

(continued)

Table 8.2 (continued)

Social aspects of well-being	**Influence on capabilities**
Significant relationships	*Decreased*
Being able to have attachments to people and things outside ourselves, to recognize and show concern for other humans, to engage in various forms of social interaction; to be able to imagine the situation of another	Due to promotion of home based reeling, attachments to friends decreased Due to promotion of home based reeling, engaged in various forms of social interaction decreased
	Unclear
	Not much attention was paid to the attachment of the reelers to the machine (shape, size, color), unclear if the machine characteristics will influence this
Family	*Enhanced*
Being able to care for, bring up, marry & settle children	Due to promotion of home based reeling, the additional time and the availability of light in the house increases the time to care for family
Friends	*Decreased*
Being able to form friendships and to enjoy companionship	Due to promotion of home based reeling, the possibility to form friendships decreased Due to promotion of home based reeling, enjoyment of companionship decreased
Community	*Decreased*
Being able to live in and participate in a community	Due to promotion of home based reeling, participation in the community decreased
Cultural aspects of well-being	**Influence on capabilities**
Cultural identity	*Enhanced*
Having respect, commitment, and acceptance of the customs and ideas that one's culture or religion impose on the individual, and being able to live according to culture	When the machine is placed on the floor: Working according to culture is enhanced by sitting on the floor Due to promotion of home based reeling, living according to culture increased
	Unchanged
	Reeling is a job which is more according to culture than heavy physical labour
	Decreased
	When the machine is placed on a table: Working according to culture is decreased
Instrumental role of capabilities and resources	**Influence on capabilities**
Multiple aspects of well-being can be enhanced/decreased by using capabilities, income or time by own choice	*Enhanced*
	Income is instrumentally important, the reeler family can choose which opportunities they want to enhance. For instance bodily health, or education

(continued)

Table 8.2 (continued)

	Time is instrumentally important, the reeler can choose how to use this extra time. For instance to enjoy leisure, time with her family, or additional time for her religion.
	Capabilities itself can also be instrumentally important, for example: good bodily health due to a good ergonomic posture might enhance income, and increase a reeler's happiness with her work, and control over her environment
	Decreased
	Time is instrumentally important. If a child has to run her mother's machine, she has less time for homework, or to play

8.4.1.1 Using the Machine

The ATRM is owned and used by the reelers, and can be repaired by local technicians, just like the old machine. Owning and using the machine gives the reeler self-respect, and the possibility for reparation at a local level gives the reeler control over her own environment. In this sense, not much has changed for the reeler. What has changed is that the ATRM is covered to shield the rotating parts, which makes maintenance for the reeler herself more difficult, and therefore slightly decreases the reeler's ability to have control over her own material environment.

Because the use of the machine is lightened and made easier, the machine ensures a good ergonomic posture (when placed on a platform), and the safety of the women and their children is improved by shielding the rotating parts, their capability to have good bodily health improved. Although placing the machine on the floor is more in accordance with culture, the reelers themselves prefer to place the machine on a platform.

The reelers' daughters do sometimes work in the reeling centre to help their mothers, but mainly after school. When their mother is not able to use the machine for some time, due to pregnancy, illness, or other causes, the family income is going down. By letting their daughter reel during these periods of time, a reeler family can secure their income. It is not unusual in these areas that children contribute to the household in some way, which adds to the basic survival capabilities of their families. And by helping their mothers, or by working on the reeling machine themselves, this might be a better working opportunity for these girls, than working in heavy physical labour. For the daughters themselves, they might like to reel on the machine out of affection for their family, and this might also enhance the affection of their family for them. However, a decrease in the capabilities of the daughters also comes into existence, as the daughters have less time to pursue other goals, as study or leisure. In this case, it is not clear what the daughters themselves see as their most valuable capability: the ability to perform meaningful work, or the ability to study, play or otherwise spend their time. However, all these considerations illustrate that 'child labour' comprises much more than is suspected at a first glance.

8.4.1.2 Working at Home or in the Reeling Centre?

The ATRM was meant to give the reelers the choice to work in a reeling centre or to work at home. Both workplaces have certain advantages and disadvantages, which all became clear due to this analysing exercise. The advantages of working in a reeling centre are that this allows the reeler to move around more freely, and to better focus on her work (as she is away from her household chores). It also allows her to socially interact with other reelers and have attachments to them, to form friendships, to enjoy companionship, to engage in various forms of social interaction, to participate in her community, and to enter into meaningful relationships with other workers. Thereby, from a community point of view, the reeling centre itself can be viewed as a setting of social interaction. Working at home, on the other side, is more in accordance with culture, and gives the reeler the opportunity to combine household chores with reeling work, due to which the reeler can use every spare moment to earn additional income. Not having to walk to the reeling centre also saves the reeler time which she can spend otherwise. Thereby, individual, home based reeling brings the advantage of installation of a solar panel to the roof of the reeler's houses. This solar panel provides sufficient energy to bring light to the house. This enables the reeler to work at night, but also to gather with the family, and to enable children to study in the evening. From this exercise, it remains however unclear which working environment gives the women most time to reel yarn of good quality, and thus earn most.

As can be concluded from above, both working environments enhance certain capabilities. Developing a machine that can be used at home, as well as in the reeling centre, did not lead to a choice for new reelers where to work, because PRADAN started introducing only home based reeling for new reeling villages. During validation MASUTA's director indicated that the reelers themselves prefer to work in a reeling centre (being away from the household chores is a relief for them), where the family wants the woman to work at home (this is more in accordance with culture). Therefore, if the reeler would have been given a choice by PRADAN, her personal preference could still be restricted by her family.

8.4.1.3 A Dignified Way of Generating Income

The reeling machine enables the women to have a job that is more in accordance with culture, as they now do not have to engage anymore in types of heavy physical labour with low status, which is looked down upon. They are able to work as a human, and are treated as a dignified being, equal to others. However, these capabilities already improved due to the introduction of the old machine. The ATRM only enhances the reeler's opportunity to live according to culture, as it enables the reeler to work at home. This last opportunity can however be debated, as it became clear that the reelers prefer to work in a reeling centre, but are only given the option to work at home. According to Sen (1999) the prohibition of outside employment is a serious violation of women's liberty and gender equity.

8.4.1.4 Appearance of the Machine

The appearance of the ATRM was not given much consideration during its development. According to PRADAN's field staff,[11] new reelers are not used to machinery; they are often scared to use machines. By involving the users in giving a product the right shape, size, and colour, this can improve the attachment of the users to the product. However, in this design process, the users were not involved in decisions concerning the appearance of the machine. The form giving of the machine was mainly based on covering all the machine parts, and making the machine as small as possible. The machine did go through a change of colour (from green to brown to blue to brown), however, MASUTA's director indicated that the change of colour was not for the purpose of enhancing the reeler's attachment to the machine. The machine was painted blue because for the manufacturer's convenience, and was changed to brown on request of PRADAN to enhance the contrast of the yarn colour with its background. During field trials, the reelers could have indicated their preference to the machine in this respect, but they were never specifically asked about it.

8.4.1.5 Additional Income and/or Time

Earning an income increases the reeler's self-respect, and gives her recognition and status, as well as a more dominant position within the household. Moreover, the additional income gives the reeler more sense of achievement. The additional income also gives the reeler's family economic security, and the possibility to lead a happier, more enjoyable, and more prosperous life. However, most of these capabilities already improved due to the implementation of the old machine. The ATRM just slightly further enhances these capabilities, because the reeler's productivity per working hour has gone up. Thereby, theoretically, home-based reeling provides the reeler the opportunity to reel in the evenings and during every moment of free time, and these reelers do not have to spend time walking to the reeling centre anymore. Therefore, theoretically, the total amount of working hours increases. It is, however, not yet clear if the women actually use this extra time for reeling. The reelers can also choose to reel less (because in less hours they can earn the same income as before), which gives them more time to pursue other goals, such

[11] *This information is obtained from field staff of PRADAN in 2006, in Deoghar District, Jharkhand State, India.*

as spending time with their family, working on the fields, or taking rest. However, the ATRM might contrarily decrease the time that daughters can spend on doings valued by them. If their daughters help their mothers during reeling, or work on the machine themselves, they have less time to spend on, for example, study, play, or performing other work.

8.4.1.6 Instrumental Role of Capabilities and Resources

As stated before, capabilities can have instrumental importance. In this case this instrumental role was also detected. For example, due to good bodily health because of a good ergonomic posture, a reeler can better concentrate on her work and can continue for a longer time. This gives her the ability to enhance her economic security. The good ergonomic posture also enhances the reeler's happiness with her job. The additional income that the reeler generates by using the ATRM is also instrumentally important. This resource can be used to achieve several opportunities; it can be spend for instance to improve bodily health, or the educational level of the children. Lastly, time is also an instrumentally important resource. The reeler can choose how to spend her extra time. For example, she can enjoy leisure, spent more time with her family, or spent more time on religion or cultural practices. As stated above, the reelers have more money and/or more time to spend, and can therefore increase several opportunities.

8.4.2 Evaluation of the ATRM Design Process

In this section we discuss the design process of the ATRM according to the five identified parallels between product design literature and CA literature (as mentioned in Sect. 8.2.3). These parallels are used to judge how process requirements have been used during the development of the ATRM which consequences this had on the eventual outcome.

8.4.2.1 The Use of Additional Theories

During the design process of the ATRM, no additional theories have been used. This analysis made clear that not all aspirations and motivations of the reelers and their families have been brought up during the design process. If additional theories were used, such as the CA or design ethnography, a broader view could have been captured.

8.4.2.2 Concern for Human Diversity

We can conclude that during the design process the view taken was not as broad as could have been taken. The focus of the project was mainly on the usage and the technical and economic function of the machine, and less on its psychological, social, and cultural functions. Several conversion factors were detected which are relevant to this case. First, several personal conversion factors can be identified. PRADAN only provides the machine to women, and only to women who have sufficient skills to reel yarn. The implementing NGO thus excludes men from reeling, and personal skills might prohibit a woman to use the machine.[12] Thereby, a reeler with a better physical condition, intelligence and skills is more likely to enhance her opportunities than a reeler with less skills (e.g., self-confidence, economic security, friendship, and status). Second, the social norm for women is to work at home, and to perform household work. If they are involved in income generation, this job should be a dignified job (in the eyes of the community). The reeling activity can thus be available, but if the household work is too demanding, or if the community rejects the reeling activity, a woman will still not be able to reel yarn. Third, environmental conversion factors can also be identified. The climate in Bihar, Jharkhand and Chhattisgarh is suitable for the tree on which the Tasar silk worm lives. Therefore, this area is suited for promoting the livelihood activity of Tasar silk reeling. If a poor rural woman lives in another area, and is able to purchase a reeling machine, it might still be difficult for her to obtain cocoons which she can reel. The walking distance to a reeling centre is another factor that might prohibit a reeler from working in reeling. The ATRM makes it easier for women to join the reeling activity, as this machine can be used at home. During the design process, not all valued opportunities and not all conversion factors were identified. Also, the in- or exclusion of children was not considered during the development of the ATRM. A broader view could have captured most of these opportunities and could have changed the design decisions taken.

8.4.2.3 Involvement of the People Concerned

When evaluating the used participatory methods, we can state that the reelers and other stakeholders were involved in the development of the machine, but mainly in the analysis, the simulation and evaluation phase. In the analysis phase, the participatory methods used were not as elaborate as used in design ethnography or the methods propagated by the toolkits of IDEO or Frog. In the synthesis phase participatory design was not practiced, mainly due to the high technological character of the design. If they had been involved, this could have caused a higher personal attachment to the ATRM. The reelers did, however, look forward to the

[12]It must be noted that, if a woman does not have sufficient skills for the reeling activity, PRADAN will engage her in another livelihood activity.

new machine, but mainly because PRADAN was involved in the development and they fully trust PRADAN. Ethnographic methods could have resulted in a deeper insight in the culture, as for example the social norm of women staying at home, and the importance of machine appearance.

8.4.2.4 Concern for the Individual and for Communities

During the analysis phase only the reelers themselves were interviewed, not their families. Therefore, mainly individual needs were identified, not the family and/or community needs. The reeling activity does not only change the reeler's life, it affects her family and the community as well. Making the machine suitable for everyone to use, and making the machine suitable for home-based reeling were more delicate issues than was anticipated upon by the designer. This exercise pointed out that capabilities of the individual and of the community are all relevant to consider in a DfD project, however, it turned out that the capabilities of the family are another type of capabilities that need consideration. By doing so, these capabilities can be properly weighed, before making a design decision.

8.4.2.5 A Focus on Choice

In this case, poor rural women in Bihar, Jharkhand and Chhattisgarh have the choice of working in the activity of Tasar silk reeling, because of the existence of the ATRM, the presence of PRADAN and MASUTA, and the presence of the Tasar silk worm in these areas. The sense of choice is generated by PRADAN who makes the poorest families in communities aware of the opportunity to participate in the activity of Tasar silk reeling. This sense is improved when the women gain confidence of being allowed and able to use the ATRM. If a woman will actually use this choice depends on her preference (she can also choose to engage in another livelihood activity), and on her husband (he has to allow his wife to work as a reeler, in a reeling centre or at home). The effectiveness of the choice depends on how well the use of the ATRM helps the women to achieve their desired outcomes. In this case, we identified one adaptive preference. If PRADAN would give new reelers the choice to work in a reeling centre or at home, reelers will not be able to use and achieve their choice to work in a reeling centre. Due to social conditioning or cultural indoctrination, the new reelers will work at home. This adapted preference was not detected by the designer. However, according to Sen (1999) people should be free to choose which traditions to follow. Social change already started in most reeler families, as the women gain more confidence and respect, and therefore are more involved in decision-making, but cultural aspects are not easy to influence or change without being paternalistic. If the machine was made in a way that it can only be used in a reeling centre, or in a way which excludes children from using it, this design decision goes beyond persuasion (as the choice of working at home will be ruled out), and can be seen as paternalistic.

8.4.3 Reflection on Capabilities and Their Characteristics

In Sect. 8.2.2 we discussed capabilities and their characteristics. In this section we will discuss how they played a role within this case, and we will also reflect on the consequences of these characteristics for a designer of a DfD project. If we look back, we can state that we mainly detected people's achieved beings and doings, not their capabilities. We did identify the reeler's valued opportunity of working in a reeling centre, but we are not certain if we identified all capabilities that are valued by the reeler by this analysis. Then, considering people's *real* opportunities turned out to be very relevant in this case. Cultural practices, choices of others, the absence of specific capabilities, or resources can all prohibit the reeler to actually fulfil the opportunities that she values. Conversion factors and the instrumental role of capabilities play a major role here.

With this exercise we tried to map how the valued capabilities of the reelers and their families changed because of the ATRM. What we did not identify is to what extent reelers value the different opportunities they obtained due to the ATRM, and which opportunities they value most. For example, how satisfied the reeler is with her job and how much she experiences a sense of achievement by using the ATRM is not known. Thereby, this analysis gave an overall view of the valued opportunities as perceived by the author and by MASUTA's managing director, and it might be possible that individual reelers have a different preference than these identified preferences. It is for example possible that some reelers prefer to work at home instead of working at a reeling centre. Of some of the consequences of the ATRM we could not identify what their impact on certain opportunities is. For example, in which workplace does the reeler earn most? Does the appearance of the machine enhances or decreases the reeler's attachment to it? And what is the effect on the opportunities of reeler families if children are involved in reeling? Although capabilities are not interchangeable, in this DfD case, trade-offs had to be made, because not all capabilities could be obtained at the same time. The machine is designed to be suitable for home based reeling, and to be easy to use. These choices enhance some of the reeler's capabilities, but unfortunately keeps other capabilities out of reach.

What can be concluded is that a designers have the power to influence which incommensurable capabilities their target-users will be entitled to, and which ones not. Therefore, it is important that the designer considers real opportunities, conversion factors and the instrumental role of capabilities during the design process. The list of beings and doings aids the designer to consider them. However, to be able to properly identify what the different reelers themselves perceive as their most valuable opportunities, how they perceive their change in capabilities, and to what extent they experience this change, the reelers themselves must be consulted.

8.5 Reflection on Case Analysis

The parallels between the Capability Approach (CA) and the product design literature suggest that product designers already include a lot of relevant perspectives in their design process. Still, the CA added new insights to the case of the Anna Tasar Reeling Machine (ATRM), by identifying aspects that were overlooked before. The list of beings and doings made us carefully rethink the impact of the ATRM on the lives of its users, their families and their communities. When these insights would have been detected during the design process, they could have influenced the decision making process. We can however not verify if these insights indeed would have led to different design decisions.

From the reflection on the design process, we can conclude that the design process had some shortcomings. Because the followed design process influences the outcome, these shortcomings might be an explanation for not identifying all possible capabilities or product consequences. However, we can also not verify if all these consequences would have been brought about when the followed design process had been without shortcomings.

The list of beings and doings was very useful to identify possible opportunities, but to find out if these opportunities comprise all the valued capabilities of the users, and if they are *real* and are actually achieved by the users, the users themselves should be consulted. Next to the list of general beings and doings, the characteristics of capabilities and the parallels between CA and DfD literature were helpful in providing deeper insight in the case and in the usefulness of applying a CA perspective to this case. Particularly the conversion factors, the instrumental role of capabilities, and adaptive preferences were detected to be very relevant aspects for this particular case. They played an important role in identifying people's real opportunities. However, the list, the characteristics, and the parallels might not be limited to the ones presented in this paper. They do, however, form a start for developing a capability inspired design framework for DfD.

8.6 Conclusion

Reverting back to innovation as significant positive change; product designers have the possibility to influence the change that the multidimensional poor need in their societies and in their lives to uplift themselves socially and economically. By taking into account the theoretical aspects of the capability approach (CA) in the design process from the early start up to the implementation of a product, a holistic and comprehensive view of the predictable consequences of the design can be drawn.

From this exercise it became clear that the CA does not inform designers which design decisions need to be made, but the approach informs the designer to make deliberate and responsible decisions during the design process, which diminishes unintended consequences of product innovations and enhances the innovative value of their product or service for the target-user. When designing for a totally different context, it is particularly important to gain this broader and deeper insight in that specific context. Therefore, we think that the CA can add a new body of knowledge to Design for Development (DfD). The CA is not the only approach that offers this body of knowledge, but it appears to be particularly useful to offer designers the insights they require to advance socially responsible design.

The case analysis, however, also pointed out some challenges of using a capability perspective. The CA does suggest to use participatory methods in order to involve people in decisions concerning their own lives, but the approach does not provide a specific methodology or methods on how to identify capabilities or adaptive preferences, how to select, weigh, or aggregate a set of incommensurable capabilities, or how to make appropriate trade-offs. The CA furthermore does not specify when a certain capability has been fully achieved, how the needs of the individual can be balanced with those of the community, or which preferences can be justified in a certain context. Several CA researchers and practitioners are working on these operationalization issues of the CA, and we will draw upon their work. However, other bodies of knowledge, as for example design ethnography, might also be useful to consider.

Those challenges indicate that, although the CA has the potential to offer a framework and a set of tools to designers, operationalizing the CA for DfD is a big task lying ahead. This exploration is just the start of a process in which we will continue to explore how the CA can best add value to DfD. We will try to integrate the CA into the design process in a prospective way. Because it is not feasible to teach designers all underlying philosophical foundations of the CA, we will try to take into account the philosophical foundations of the CA, but in a for designers understandable and useful way. Using the CA as Alkire (2008b) argued – 'to identify which concrete actions are likely to generate a greater stream of expanded capabilities' – we do not ensure responsible innovation, but a significant contribution can be made to let product innovations become more responsible and successful, as an effort to respond to Papanek's call in the 1970s to better address the true needs of the poor.

Acknowledgements This research has been made possible by a grant from NWO (the Netherlands Organisation for Scientific Research). We would also like to thank Ilse Oosterlaken (who is working on the same research project) for critically reflecting on the content of this chapter and for exchanging thoughts on several issues. This chapter has greatly benefited from the many discussions with her.

Appendix A: Table Containing Needs Derived from Alkire's Lists

Classified according to the seven aspects of well-being by Williamson and Robinson (2006)

Aspect of well-being	Being or doing
Biological	*Physical survival*
	Being able to live to the end of a human life of normal length
	Nutrition
	To be adequately nourished
	Health
	Being able to have good bodily and mental health
	Reproduction
	Being able to have good reproductive health
	Healthcare
	Being able to receive good healthcare
	Shelter
	Having adequate shelter
	Sanitation
	Having adequate water, sanitation and hygiene
	Rest and exercise
	Having adequate periodic rest, and adequate physical activity
	Physical security
	To be secure against harassment, pain, anxiety and violent assault, and being able to have pleasurable experiences, safety, harmony and stability
Mental	*Education*
	Being able to receive education, to experience and appreciate beauty, and to develop curiosity, learning, and understanding
	Practical reason
	Being able to form a conception of the good and to engage in critical reflection about the planning of one's life
	Identity and individuality
	Having a sense of the aspects that makes one unique
	Morality
	A sense of goodness, righteousness, duty, and obligation
	Freedom of sexual activity
	Having the opportunities for sexual satisfaction and for choice in matters of reproduction
	Freedom of movement and residence
	Being able to move freely from place to place, and to reside where one wants

(continued)

(continued)

Aspect of well-being	Being or doing
	Meaningful work
	Being able to choose one's work, and to work as a human, to exercise practical reason, and to enter into meaningful relationships of mutual recognition with other workers
	Leisure
	Being able to laugh, to play, to enjoy recreational activities
	Political liberty
	Having the right of political participation, protections of free speech and association
Emotional	*Freedom of mind*
	Having the freedom of thought, imagination, opinion
	Freedom of experiencing and expressing emotions
	Having the freedom to experience emotions and express oneself, not having one's emotional development blighted by fear and anxiety
	Happiness
	Being able to lead a happy, enjoyable life
	Love, longing, and grieve
	Being able to experience love, longing and grieve, and being able to give love and affection
	Worry-free
	Being able to live a prosperous life, without worries and with confidence in the future
	Self-respect
	Being able to have the social bases of self-respect and non-humiliation
	Aspirations and self-actualization
	Being able to express and activate all one's aspirations and capacities
	Achievement
	Being able to accomplish one's aspirations, to demonstrate competence and making a lasting contribution
	Equality
	Being able to be treated as a dignified being whose worth is equal to that of others
	Recognition
	Being recognized and having status
	Having power
	Having social status and prestige, and having control or a dominant position within the household and the more general social system (includes decision-responsibility)
	Acceptance and self-adjustment
	Being able to adjust to circumstances
	Self-acceptance
	Being able to accept oneself and one's circumstances
	Being able to hold property/to have sufficient assets, control over material environment

(continued)

(continued)

Aspect of well-being	Being or doing
	Services
	Having access to services concerning i.e. mobility and media services
	Housing
	Being able to own a house
	Economic security
	Being economically secure at present and in the future
	Settings of interaction
	Having places to meet others for educational, spiritual or creative purposes
Material	*Goods*
Social	*Significant relationships*
	Being able to have attachments to people and things outside ourselves, to recognize and show concern for other humans, to engage in various forms of social interaction; to be able to imagine the situation of another
	Family
	Being able to care for, bring up, marry & settle children
	Friends
	Being able to form friendships and to enjoy companionship
	Community
	Being able to live in and participate in a community
	Other species
	Being able to live with concern for and in relation to animals, plants, and the world of nature
	Social security
	Living in an open, just, and secure environment
	Privacy
	Being able to seclude oneself or information about oneself
Cultural	*Cultural identity*
	Having respect, commitment, and acceptance of the customs and ideas that one's culture or religion impose on the individual, and being able to live according to culture
Spiritual	*Peace of mind*
	Being able to find meaning, inner harmony and inner peace
	A spiritual life
	Being able to find meaning and value, and being free to believe in a greater than human source

References

Alkire, Sabina. 2005. Why the capability approach? *Journal of Human Development* 6(1): 115–135. doi:10.1080/146498805200034275.

Alkire, Sabina. 2008a. Choosing dimensions: The capability approach and multidimensional poverty. In *The many dimensions of poverty*, ed. Kakwani Nanak and Silber Jacques. New York: MacMillan.

Alkire, Sabina. 2008b. Using the capability approach: Prospective and evaluative analyses. In *The capability approach: Concepts, measures and applications*, 26–50. New York: Cambridge University Press.

Alkire, Sabina, and Maria Emma Santos. 2011. *Acute multidimensional poverty: A new index for developing countries*. New York: Human Development Report Office (HDRO), United Nations Development Programme (UNDP).

Amabile, Teresa M., Regina Conti, Heather Coon, Jeffrey Lazenby, and Michael Herron. 1996. Assessing the work environment for creativity. *The Academy of Management Journal* 39(5): 1154–1184.

Amir, Sulfikar. 2004. Rethinking design policy in the third world. *Design Issues* 20(4): 68–75.

Anand, Paul, Graham Hunter, Ian Carter, Keith Dowding, Francesco Guala, and Martin Van Hees. 2009. The development of capability indicators. *Journal of Human Development and Capabilities* 10(1): 125–152.

Archer, Leonard Bruce. 1984. Systematic method for designers. In *Developments in design methodology*, ed. Cross Nigel. Chichester: John Wiley & Sons, Ltd.

Balaram, Singanapalli. 2011. *Thinking design*. Kundli: Sage.

Berkun, Scott. 2010. *The myths of innovation*. Sebastopol: O'Reilly Media, Inc.

Birkett, Stacey. 2010. The development of responsibility in product designers. Ph.D. thesis, The Open University, Milton Keynes, UK.

Bødker, Susanne, and Samuli Pekkola. 2010. Introduction the debate section: A short review to the past and present of participatory design. *Scandinavian Journal of Information Systems* 22(1): 45–48.

Buchanan, Richard. 2001. Design research and the new learning. *Design Issues* 17(4): 3–23.

Buijs, Jan Arie. 2003. Modelling product innovation processes, from linear logic to circular chaos. *Creativity and Innovation Management* 12(2): 76–93.

Buijs, Jan Arie, and Adriana Cornelia Valkenburg. 2005. *Integrale productontwikkeling [integral product development]*, 3rd ed. The Hague: Boom Lemma Publishers.

Chiappero Martinetti, Enrica. 2000. A multidimensional assessment of well-being based on Sen's functioning approach. *Rivista Internazionale di Scienze Sociali* 2: 207–239.

Chiappero Martinetti, Enrica. 2008. Complexity and vagueness in the capability approach: Strengths or weaknesses? In *The capability approach: Concepts, measures and applications*, ed. Flavio Comim, Mozaffar Qizilbash, and Sabina Alkire. New York: Cambridge University Press.

Clark, David A. 2009. Adaptation, poverty and well-being: Some issues and observations with special reference to the capability approach and development studies. *Journal of Human Development and Capabilities* 10(1): 21–42.

Comim, Flavio. 2001. Operationalizing Sen's capability approach. *Paper presented at the conference justice and poverty: Examining Sen's capability approach*, Cambridge, 5–7 June.

Cross, Nigel. 2000. *Engineering design methods: Strategies for product design*. Chichester: Wiley.

Diehl, Jan Carel. 2010. Product innovation knowledge transfer for developing countries. Towards a systematic transfer approach. Ph.D. thesis, Delft University of Technology, Delft, The Netherlands.

Donaldson, Krista M. 2002. Recommendations for improved development by design. In *Development by design (dyd02) conference*, Bangalore, India.

Donaldson, Krista M. 2006. Product design in less industrialized economies: Constraints and opportunities in Kenya. *Research in Engineering Design* 17(3): 135–155. doi:10.1007/s00163-006-0017-3.

Drucker, Peter F. 1998. The discipline of innovation. *Harvard Business Review* 76(6): 149–157.

Frediani, Alexandre Apsan. 2010. Sen's capability approach as a framework to the practice of development. *Development in Practice* 20(2): 173–187. doi:10.1080/09614520903564181.

Friess, Erin. 2010. The sword of data: Does human-centered design fulfill its rhetorical responsibility? *Design Issues* 26(3): 40–50.

Hanington, Bruce M. 2010. Relevant and rigorous: Human-centered research and design education. *Design Issues* 26(3): 18–26.

Harvard Business Press. 2003. Managing creativity and innovation. In *Harvard business essentials series*, ed. Ralph Katz. Boston: Harvard Business School Publishing Corporation.

Heskett, John. 2005. *Design: A very short introduction*. New York: Oxford University Press.

International Council of Societies of Industrial Design, ICSID. 2011. *Definition of design*. http://www.icsid.org/about/about/articles31.htm. Accessed May 2011.

Johnstone, Justine. 2007. Technology as empowerment: A capability approach to computer ethics. *Ethics and Information Technology* 9(1): 73–87. doi:10.1007/s10676-006-9127-x.

Kandachar, Prabhu, and Minna Halme. 2008. Introduction. Farewell to pyramids: How can business and technology help to eradicate poverty? In *Sustainability challenges and solutions at the base of the pyramid: Business, technology and the poor*, 1–27. Sheffield: Greenleaf Publishing.

Keates, Simon, and John P. Clarkson. 2004. *Countering design exclusion: An introduction to inclusive design*. London: Springer.

Kleine, Dorothea. 2010. ICT4what? Using the choice framework to operationalise the capability approach to development. *Journal of International Development* 22(5): 674–692. doi:10.1002/jid.1719.

Kleine, Dorothea. 2011. The capability approach and the ‘medium of choice’: Steps towards conceptualising information and communication technologies for development. *Ethics and Information Technology* 13(2): 119–130. doi:10.1007/s10676-010-9251-5.

Kleine, Dorothea, Ann Light, and Maria-José Montero. 2012. Signifiers of the life we value? – Considering human development, technologies and fair trade from the perspective of the capabilities approach. *Information Technology for Development* 18(1): 42–60. doi:10.1080/02681102.2011.643208.

March, Lionel J. 1984. The logic of design. In *Developments in design methodology*, ed. Cross Nigel. Chichester: John Wiley & Sons, Ltd.

Margolin, Victor, and Sylvia Margolin. 2002. A “social model” of design: Issues of practice and research. *Design Issues* 18(4): 24–30. doi:10.1162/074793602320827406.

Meinel, Christoph, Larry Leifer, and Hasso Plattner. 2011. *Design thinking: Understand – improve – apply*. Berlin/Heidelberg: Springer-Verlag.

Mink, Annemarie. 2006. Reeling machine for women self help groups in Eastern India. Graduation thesis, Delft University of Technology, Delft, The Netherlands.

Nieusma, Dean. 2004. Alternative design scholarship: Working toward appropriate design. *Design Issues* 20(3): 13–24. doi:10.1162/0747936041423280.

Nussbaum, Martha C. 2000. *Women and human development: The capabilities approach*. Cambridge: Cambridge University Press.

OECD (Organisation for Economic Co-operation and Development), and Eurostat. 2005. *Oslo manual: Guidelines for collecting and interpreting innovation data*, 3rd ed. Aufl. Paris: OECD Publishing.

Oosterlaken, Ilse. 2009. Design for development: A capability approach. *Design Issues* 25(4): 91–102. doi:10.1162/desi.2009.25.4.91.

Owen, Charles L. 1992. Context for creativity. *Design Studies* 13(3): 216–228.

Pahl, Gerhard, Wolfgang Beitz, and Ken Wallace. 1984. *Engineering design*. London: Design Council.

Papanek, Victor J. 1984. *Design for the real world*, 2nd ed. Aufl. Chicago: Academy Chicago Publishers.

Parmar, Vikram S. 2009. Design framework for developing ICT products and services for rural development: A persuasive health information system for rural India. PhD thesis, Delft University of Technology, Delft, The Netherlands.

Prahalad, C.K. 2005. *The fortune at the bottom of the pyramid: Eradicating poverty through profits*. Delhi: Pearson Education.

Redelinghuys, Christiaan. 2006. Counting the seeds of innovation: The assessment of technological creativity. In *Measuring innovation in OECD and non-OECD countries: Selected seminar papers*, ed. William Blankley, Mario Scerri, Neo Molotja, and Imraan Saloojee, 59. Cape Town: Human Sciences Research Council Press.

Robeyns, Ingrid. 2005. The capability approach: A theoretical survey. *Journal of Human Development* 6(1): 93–117. doi:10.1080/146498805200034266.

Robeyns, Ingrid. 2006. The capability approach in practice. *The Journal of Political Philosophy* 14(3): 351–376. doi:10.1111/j.1467-9760.2006.00263.x.

Robeyns, Ingrid. 2011. The capability approach. In *The Stanford Encyclopedia of Philosophy*, ed. Edward N. Zalta. Stanford: The Metaphysics Research Lab, Stanford University.

Rogers, Everett M. 1995. *Diffusion of innovations*. 4th ed. Aufl. New York: The Free Press.

Roozenburg, N.F.M., and N.G. Cross. 1991. Models of the design process: Integrating across the disciplines. *Design Studies* 12(4): 215–220.

Roozenburg, N.F.M., and J. Eekels. 1998. *Productontwerpen; Structuur en Methoden* [Product design: Fundamentals and methods]. Utrecht: Lemma.

Schumpeter, Joseph A. 1983. *The Theory of Economic Development: An Inquiry into Profits, Capital, Credit, Interest, and the Business Cycle*. Trans. Redvers Opie. New Brunswick: Transaction Publishers.

Sen, Amartya. 1995. *Inequality reexamined*. Oxford: Oxford University Press.

Sen, Amartya. 1999. *Development as freedom*. New York: Oxford University Press.

Simon, Herbert Alexander. 1996. *The sciences of the artificial*. 3rd ed. Aufl. Cambridge, MA: MIT Press.

Sklar, Aaron, and Sally Madsen. 2010. Global ergonomics: Design for social impact. *Ergonomics in Design* 18(2): 4–5, 31. doi:10.1518/106480410x12737888532921.

Skogstad, Philipp, and Larry Leifer. 2011. A unified innovation process model for engineering designers and managers. In *Design thinking: Understand – Improve – Apply*, ed. Hasso Plattner, Christoph Meinel, and Larry Leifer, 19–43. Berlin: Springer.

Suber, Peter. 1999. Paternalism. In *The philosophy of law: An encyclopedia*, ed. Chrisopher Berry Gray, 632–635. New York: Garland Publishing.

Thackara, J. 2005. *In the bubble: Designing in a complex world*. Cambridge, MA: MIT Press.

Thomas, Angharad. 2006. Design, poverty, and sustainable development. *Design Issues* 22(4): 54–65.

Toboso, Mario. 2011. Rethinking disability in Amartya Sen's approach: ICT and equality of opportunity. *Ethics and Information Technology* 13(2): 107–118. doi:10.1007/s10676-010-9254-2.

Ulrich, Karl T., and Steven D. Eppinger. 2004. *Product design and development*, 3rd ed. New York: McGraw Hill.

UNDP (United Nations Development Programme). 2010. *Human development report 2010: The real wealth of nations: Pathways to human development*. New York: United Nations Development Programme.

UNDP (United Nations Development Programme). 2012. *Multidimensional poverty index (MPI)*. http://hdr.undp.org/en/statistics/mpi. Accessed November 2012.

Unger, Darian, and Steven D. Eppinger. 2010. Improving product development process design: A method for managing information flows, risks, and iterations. *Journal of Engineering Design* 22(10): 689–699.

van den Waeyenberg, Sofie, and Luc Hens. 2008. Crossing the bridge to poverty, with low-cost cars. *Journal of Consumer Marketing* 25(7): 439–445. doi:10.1108/07363760810915653.

Verein Deutscher Ingenieure, Richtlinie 2221. 1993. Methodik zum Entwickeln und Konstruieren technischer Systeme und Produkte. [Systematic approach to the development and design of technical systems and products]. In: VDI-Handbuch Konstruktion, ed.Verein Deutscher Ingenieure. Berlin.

Veryzer, Robert W. 2004. The roles of marketing and industrial design in discontinuous new product development*. *The Journal of Product Innovation Management* 22(1): 22–41.

Viswanathan, Madhubalan, and Srinivas Sridharan. 2012. Product development for the BoP: Insights on concept and prototype development from university-based student projects in India. *Journal of Product Innovation Management* 29(1): 52–69. doi:10.1111/j.1540-5885.2011.00878.x.

Viswanathan, Madhubalan, Ali Yassine, and John Clarke. 2011. Sustainable product and market development for subsistence marketplaces: Creating educational initiatives in radically different contexts. *Journal of Product Innovation Management* 28(4): 558–569. doi:10.1111/j.1540-5885.2011.00825.x.

Wheelwright, Steven C., and Kim B. Clark. 1992. *Revolutionizing product development. Quantum leaps in speed, efficiency, and quality*. New York: Free Press.

Williamson, John, and Malia Robinson. 2006. Psychosocial interventions, or integrated programming for well-being? *Intervention* 4(1): 4–25.

Chapter 9
Conceptualizing Responsible Innovation in Craft Villages in Vietnam

Jaap Voeten, Nigel Roome, Nguyen Thi Huong, Gerard de Groot, and Job de Haan

Abstract Previous research by the authors has explored small-scale innovations in poor craft producers' clusters in villages in the Red River Delta in northern Vietnam. Although these innovations resulted in value creation and increased incomes, they also often gave rise to negative environmental or social consequences that were in conflict with broader development goals, such as poverty alleviation and sustainable development. Innovation that meets these goals is broadly termed *responsible innovation* and increasingly made explicit in western innovation debates. This chapter seeks to conceptualize responsible innovation in a very different context; that of informally organized small producers' clusters in a developing country

---

The study was made possible by a grant for a research programme on responsible innovation from the Dutch Organization for Scientific Research (NWO) that was jointly carried out by Tilburg University and Hanoi University of Technology (Vietnam). Submitted version 22 Sept. 2012.

J. Voeten (✉) • G. de Groot
Development Research Institute (IVO), Tilburg University, Warandelaan 2, PO Box 90153, 5000 LE Tilburg, The Netherlands
e-mail: j.voeten@uvt.nl; G.A.degroot@uvt.nl

N. Roome
Governance & Ethics Management Domain, Vlerick Leuven Gent School of Management, Vlamingenstraat 83, 3000 Leuven, Belgium
e-mail: nigel.roome@vlerick.com

N.T. Huong
School of International Education (SIE), Hanoi University of Science and Technology, N1, Dai Co Viet Road, Hanoi, Vietnam
e-mail: nguyenthihuong.hut@gmail.com

J. de Haan
Tilburg School of Economics and Management, Tilburg University, Warandelaan 2, PO Box 90153, 5000 LE Tilburg, The Netherlands
e-mail: j.a.c.dehaan@uvt.nl

J. van den Hoven et al. (eds.), *Responsible Innovation 1: Innovative Solutions for Global Issues*, DOI 10.1007/978-94-017-8956-1_9,

(Vietnam). We employed grounded theory to investigate the outcomes of innovations with a view to developing a set of objective operational criteria for evaluating responsible innovation. However, we found that such an 'outcomes' approach posed epistemological problems when it came to defining criteria and objectively measurable threshold values. We also found that objective measurements imposed a normative framework on the communities we were studying: one that villagers did not necessarily recognize or concur with. As an alternative, we came to conceptualize responsible innovation from a behavioral perspective and modeled it as a dynamic societal process that involves innovators acknowledging responsibility in the resolution of societal conflicts resulting from the harmful outcomes of innovation. This model enabled us to differentiate between responsible and what it is not at the village level.

9.1 Introduction

9.1.1 Background and Research Question

Many economists, politicians and economic actors consider innovation to be a key for competitiveness and economic development. Although this viewpoint is generally accepted in economic circles in relation to developed economies, there is a debate whether innovation are applicable is accessible and relevant for all businesses in every economic context (Schmitz 1999; Kaplinsky 2000). With this in mind, we revisited our past research into new economic dynamics among small business clusters in several craft villages in the Red River Delta (northern Vietnam), where poor small producers introduced new technology, new products and new ways of doing business. These new ways of creating value and improving competitiveness resulted in economic development – a view shared by the innovators, the villagers and local officials. Below we provide several examples of the innovations that we discovered.

Duong Lieu cassava noodle village. Groups of households traditionally processed cassava tubers into starch (as an intermediate product), and sold it to other groups of households producing noodles. Recently, several households switched to making new end products from the starch: children's sweets, medicine pills and soft drinks. These small producers have invested in small machinery, add more value and now sell their products to more profitable outlet channels in Hanoi and beyond. As a result they enjoy higher family incomes.

Bat Trang ceramics village. In the old days, small producers in the cluster baked ceramic products in traditional pottery kilns, fired with wood and charcoal. Over the last 10 years, small producers have begun to fire their kilns with Liquefied Petroleum Gas (LPG). This has resulted in better quality ceramics higher production volumes and new designs for the ceramics. The village has become a ceramics hot spot in northern Vietnam. Small producers have established ceramics shops for the many tourists that now visit the village and have concluded export contracts with buyers from Japan, Europe and the USA.

Van Phuc *silk village. The silk industry village was collectivized for a long period in the socialist command economy. After the introduction of the free market economy in Vietnam, some members of the diluted collective established retail shops in the village and introduced new marketing practices. This resulted in an increase of home-based silk production, many new clients and tourists visiting the village and economic prosperity in the village.*

Phu Vinh *rattan and bamboo village. For decades the village has produced traditional bamboo and rattan articles for the domestic market. Small producers in the village have special skills in weaving rattan and bamboo. Some 10 years ago, a number of export companies were established around the village and successfully initiated export operations to the USA and Europe. The export companies outsource the orders to middlemen in the village who subsequently engage small producers for the actual production. There has been a significant shift to producing higher quality and more expensive rattan products with a large increase in value created.*

In line with our interest in poverty alleviation in developing countries, we wondered whether the new practices could be understood as 'innovation' in the economic sense; entrepreneurs themselves initiating new business practices, acquiring or developing new technologies and making new products thereby improving their business and competitiveness and ultimately increasing their incomes. This was not an easy question to answer, as the understanding of innovation is strongly rooted in advanced hi-tech western economic systems (Tether 2003), very different from the largely informal economic context that prevails in Vietnam and other developing countries. We sought to conceptually clarify innovation in the Vietnamese context by taking theoretical insights from various fields of social sciences including economics, sociology and business administration. This led us to develop a generic definition of innovation 'as the process of introducing something new that creates value' (Voeten et al. 2011) from which we derived and develop an innovation assessment instrument; a criteria checklist with threshold values. With this instrument we identified a number of cases of cluster-level innovation in northern Vietnam (ibid.).

While these new practices generated the economic advantages (e.g. value creation and improved competitiveness) often associated with innovation (Porter 1990), we also noticed that the innovations led the villages to experience other environmental and social consequences, some negative, others positive. Some examples are listed below.

In Duong Lieu, increased cassava starch production, an intermediary product for the newly introduced products, has created significantly higher amounts of organic waste which is dumped into the open sewage system. Universities and NGOs have reported on alarming levels of soil and water pollution and associated health problems.

In the past the smoke emissions from traditional charcoal kilns produced a lot of air pollution in Bat Trang village. The introduction of the new LPG technology resulted in a much better air quality which was recognized and appreciated by the majority of villagers.

The increase in silk production and introduction of new fashionable products in Van Phuc village resulted in the use of toxic chemicals in the dying process. Severely polluted waste water is discharged into the sewage system and the surface water around the village became very dirty.

In Phu Vinh, business has been good for the export companies and the middlemen but not so good for the small and poor household producers in the village. To their dissatisfaction, the export companies have repeatedly negotiated lower unit prices, paying the workers less per unit produced. These days more family members (including children and elderly people) do the actual rattan weaving these days, yet despite this poverty levels have increased.

These consequences caught our attention and led us to question whether the societal impacts of these small-scale innovations were in line with current notions of poverty alleviation, which extends beyond the narrow economic focus of measuring poverty in terms of incomes and consumption (Wagle 2002; London 2007). The true meaning of poverty in developing countries has become a subject of intense debate over the last few years (Jitsuchon 2001). Development practitioners, scientists and policy makers have developed various 'alternative' approaches for defining and measuring poverty. For instance, the basic needs approach includes defining households as being poor if their food, clothing, medical, educational and other needs are not being met (Glewwe 1990). Others have viewed poverty as a function of the lack of individual capabilities, such as education or health, to attain a basic level of human well-being. Sen (1999) proposed that measures of poverty should include the physical conditions of individuals and their capabilities to make the most of the opportunities they have. Alkire (2007) has suggested adding participation, highlighting the importance of the notion of inclusive development (World Bank 2008). Among these many approaches and notions, there is general agreement that poverty in developing countries is a complex, multidimensional, issue and that assessments of poverty need, above all, to take contextual environmental and social aspects into consideration. As such, it is no coincidence that the poverty debate has been increasingly interconnected and merged with the debate on *sustainable development* (Hopwood et al. 2005). Although the term sustainable development has evolved into a widely used (some might say over-used) phrase and idea that conveys many different meanings (Hopwood et al. 2005), it has become inextricably bound up with poverty alleviation.

Brundtland (1987) first articulated the term sustainable development in the report 'Our Common Future' which governments, multilateral organizations and civil society further consolidated into Agenda 21 (United Nations 1992). These documents advocated forms of development that would meet the needs of the present generation without compromising the ability of others around the planet and future generations to meet their needs. The concept of sustainable development is the result of the growing awareness of the global links between mounting environmental problems, socio-economic issues and inequality and an expression of concerns about a healthy future for humanity (Hopwood et al. 2005). The challenge of sustainable development is to align the local interests of today with interests elsewhere on the planet and in the future. It has both temporal and spatial dimensions; Sayer and Campbell (2004) state that local livelihoods and actions should be connected with the global environment and with the future outcomes.

Early poverty alleviation debates questioned the post-war claim – that still dominates much mainstream economic policy – that international prosperity and human well-being can be best promoted through increased global trade and industry (Reid 1995; Hopwood et al. 2005). Critics often point out that such growth-led models have failed to eradicate poverty, either globally or within developing countries. This pattern of growth has also damaged the environment, creating a 'downward spiral of poverty and environmental degradation' (Brundtland 1987). Against this background, sustainable development is explicitly concerned with poverty alleviation, advocating policies such as safeguarding equity, equitable distribution and access to resources, clean water, sanitation, a healthy environment, gender equality, political freedoms and preserving cultural heritage, to name but a few (Sen 1999: World Bank 2000).

While sustainable development was initially exclusively a concern for governments and civil society, business has increasingly come to play a prominent role in the sustainability discourse. The formation in 1995 of the Business Council for Sustainable Development marked the formal emergence of business involvement in sustainable development (Najam 1999). This idea flourished further, driven partly by a series of incidents (such as Brent Sparr, Exxon Valdez, Enron and Bhopal) in the 1990s that created led social actors to question the actions of business. Public awareness and outrage about these events increased as a result of the widespread diffusion of information technologies that facilitated the spread of information about the societal impacts of businesses. Leading business management authors of that time that businesses were better equipped to lead the drive to sustainable development than governments or civil society (Elkington 1999; Hart 2007; Roome and Boons 2005). *Sustainable business* developed the catchphrase People-Planet- Profit, which expresses a new and expanded spectrum of values and criteria for measuring the organizational and societal achievements of business. Key aspects of the concept of sustainable business concept include anticipation, precaution and the recognition that, when there is a plausible risk, business has a responsibility to protect the public from exposure to harm and to avoid conflict (O'Riordan and Cameron 1994). The concept of sustainable business includes an economic aspect (an economically sustainable system that produces goods and services on a continuing basis); a social aspect (a socially sustainable system that provides distributional equity, adequate provision of health, education and other social services, gender equity, political accountability and participation) and an environmental aspect (an environmentally sustainable system that maintains a stable resource base and avoids the over-exploitation of renewable resource systems or environmental functions) (Harris et al. 2001).

We used these ideas about sustainable business to reflect back on our initial observations regarding the environmental and social consequences of innovation in Vietnam. We became interested in exploring whether the innovations made by these small producers was in line with the concept of sustainable business; innovation initiated and owned by poor people – who enjoy value creation and raise their incomes through applying the principles of people planet profit. This interest coincided with western debates on the broader societal impacts of innovation, on ethical issues and sustainability, often captured in the phrase *responsible innovation*, which describes innovations that are accompanied by concerns about societal and environmental consequences (NWO 2008; Douglas and Papadopoulos 2012; Ubois 2010[1]). We discovered that there was very little literature into what responsible innovation means in developing economies and in informal small-scale industrial settings. This led to our research question: how can we understand and conceptualize responsible innovation among small clusters of producers in Vietnam? Aside from posing a theoretical challenge, this question also has practical implications. The ability to distinguish responsible innovation (from 'irresponsible innovation') might also offer a means for operationalizing the concept within policies and programmes aimed at poverty alleviation and sustainable development.

9.1.2 Research Approach

Our initial and open-view literature explorations revealed numerous theoretical associations, leads and clues including α-sciences (humanities, ethics, responsibility, corporate social responsibility – CSR – and morality), β-sciences (technology, engineering and the environment) and γ-sciences (economics, business administration, management, sociology). We also observed that most leads and clues emerged from western empirical contexts; we found few references to informally organized small producers in developing countries. In terms of our intended research approach, these explorations did not provide us with a solid basis to select and defend one theoretical arena or discipline -or set of theories –to the exclusion of others. We initially had a deductive research approach in mind: that would involve constructing an initial conceptualization of responsible innovation within a defined theoretical framework identified through a desk study and subsequently validating and refining it in the field.

As an alternative, we choose to use and analyze empirical data without too much reliance on preconceived theories as a way of arriving at a conceptualization of responsible innovation. To do this we applied *grounded theory,* an inductive systematic research approach that involves the development of theory through empirical explorations using an iterative process that also draws on wider theory analysis (Glaser and Strauss 1967). The initial exploration of the literature helped

[1] http://issuu.com/fondazionebassetti/docs/jeff-innovation-aaa-2010-2

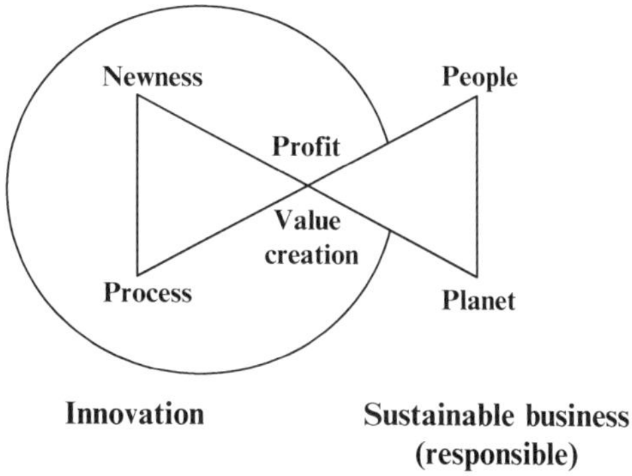

Fig. 9.1 Innovation and sustainable business: responsible innovation

us to frame the empirical and theoretical analysis resulting in us combining two (not exclusively distinct) concepts: innovation and sustainable business. Using a matrix approach, based on Fig. 9.1, we systemically and iteratively examined the literature and empirical material on the social (people) and environmental aspects (planet) of sustainable business against the three key elements of innovation: newness, process and value creation (Voeten et al. 2011).

In the following section, we present these theoretical explorations. In Sect. 9.3 we introduce the empirical material in the form of case studies of four small producers' clusters. In Sect. 9.4 we discuss the conceptualization based on responsible outcomes and responsible behaviour by innovators. In Sect. 9.5 we conclude that responsible innovation is best conceptualized in terms of societal processes (rather than outcomes) through which innovators acknowledge responsibility any societal conflicts that might arise. In this section we also review the operational applicability of this concept, the contribution it makes to theory, the limitations of the model and identify issues for further research. Although the sequence of the sections in this chapter suggests a straightforward research routine, running from theory to data to interpretation, in reality the research involved an iterative process of and going back and forth between theory and empirical analysis (in line with the grounded theory approach).

9.2 Theoretical Explorations

9.2.1 Newness

In an earlier paper we previously defined innovation as the process of introducing newness that creates value. The innovation literature classifies newness in terms of several types of outcomes (Schumpeter 1934; Dosi 1988; OECD 2005) such as new product, new production process/technology, new markets, new inputs etc. The environmental aspects of newness links up with various discussions in the literature on sustainable business, particularly in terms of the environmental impacts of using and disposing of new products (Roome and Hinnells 1993), as well the length of the product use cycle (Davis and Blomstrom 1975). Similar theoretical associations can be found in literature about 'greening' new production processes. The introduction of clean technologies has also received much attention (Blackman and Bannister 1998; Chang 2011; Hart 2007; Fiksel 1996). 'Cradle to cradle' is another concept that has evolved in the past decade (McDonough and Braungart 2002). This calls for reusing all the materials involved in industrial or commercial processes. Bansal and Roth (2000) explored 'ecological responsiveness' and used this concept to explain why businesses 'go green'. There have also been environmental debates about accessing new markets and the environmental impacts of distributing materials along global production chains. The effects of new sources of supply and inputs raise on natural resource depletion have been examined by Vachona and Maoc (2008). Barbier and Homer-Dixon (1996) have discussed resource scarcity and technical innovation in developing countries.

The literature on sustainable business concerning the social aspects of newness in innovation mostly discusses ethical issues. One of discussed social aspects of new products is their ethical acceptability and whether or not they are good for consumers (Earle and Earle 2001). Examples include the introduction of unhealthy foods, extremely violent computer games or children's toys that contain toxic materials (Bhattacharyya and Kohli 2007). Bezençon and Blili (2010) have explored consumer involvement in the design and development of new ethical products. Investigations into the social aspects of new production processes and technology have focused on social issues such as labour conditions and workers' rights (Ewing 2006) and the creation or loss of employment (Pianta 2005). Other discussions about a new sustainable business model, have focused around ethical standards (Drucker 1981) and the economic involvement (or exclusion) of marginal groups. Others have explored social aspects in responsible organizational and management innovation, looking at ethical questions such as human resource management policies and practices (Birkinshaw et al. 2008; Hamel 2006; Trott 2005).

9.2.2 Value Creation

With regard to the environmental aspects of value creation in innovation, there have been discussions about internalizing the real environment costs of production (Rabl 1996) and paying to repair (or avoid) environmental damage (e.g. planting trees or installing pollution control technology). An environmental impact assessment can be used to explore the environmental (and social) impact that a (proposed) project is likely to have on a locality and ways that adverse impacts could be minimized or compensated for (Bartlett and Kurian 1999). Value creation and environmental impacts is reflected the discussion about product life cycles, where Keoleian and Menerey (1994) observe that most design and product life cycles do not follow a sustainable path. *Design for Environment* is an emerging systematic approach for integrating environmental issues into design. The discussions about sustainable consumption and product life cycles follow similar lines (Hertwich 2005). One important issue is the impact that transportation costs in long-distance trade have on globalization and innovation processes. The combination of globalisation and innovation have contributed to a massive increase freight tonnage-miles and the amount of fossil fuel consumed, which has been exacerbated by an increasing tendency to ship freight via more polluting transport modes, i.e. air versus sea and road versus rail (Curtis 2003).

The literature concerning the social aspects of value creation includes discussions about the ethics and principles surrounding the creation of monetary value (Lindfelt and Törnroos 2006). There has been a debate about the just and fair distribution of the benefits (and costs) of the consequences of innovation within communities (Bigsten and Levin 2000). A key question in these debates is whether those who took the risks and invested their time and money shared the benefits in an equitable or even-handed way; excessive appropriation of benefits by a minority is often seen as a cause of poverty in developing countries. Value creation and value appropriation are both necessary to maintain a competitive advantage (Mizik and Jacobson 2002). Value creation also occurs when businesses plough back some of their profits into the community by supporting social activities or study scholarships for instance. Hart (2007) identify different methods for 'sustainable social value creation.' They stress the importance of innovation to a business of (i) managing today's business while simultaneously creating tomorrow's technology, markets and opportunities and; (ii) nurturing and protecting their internal organizational skills, technologies and capabilities while also being receptive and infusing the firm with new external perspectives (e.g. knowledge from outside stakeholders). Economists, the business community and ethicists have also extensively debated the relative merits of creating consumer surplus for a small rich group of consumers (via premium quality and high-priced services and products) or providing lower priced products for those with lower incomes at the Bottom of the Pyramid (Prahalad 2005). Prahalad (2005) stressed the poverty alleviation aspect and suggested that sustainable business offers opportunities for large companies as well as for the four billion poor people at the bottom of the pyramid. Fair Trade provides a working example. It is now a large and growing social movement based on a market-based approach that aims to help improve the conditions and remuneration of producers in developing countries and promote sustainability. The movement pays higher ('fair') prices to producers and promotes better working social and environmental standards (Nicholls and Opal 2004; Stigliz and Charlton 2005).

9.2.3 Process

The innovation process is generally described as consisting of three stages: idea generation, testing and implementation (Tether 2003; Dosi 1988). It is not, however, a linear process, but a rather chaotic one of learning through feedback loops and interactions with other innovators (Kline and Rosenberg 1986). Various strands of the literature examine how, in the innovation process, entrepreneurs and innovators manage and respond to emerging social and environmental consequences. The background paper of the 'Responsible Innovation thematic program of the Dutch Organization of Scientific Research (NWO 2008) describes responsible innovation as an innovation process that includes societal values, interests, needs, rights and welfare, and involves discussions and interactions between the developers of technology, individual actors and others. This view stresses the need to adopt a responsibility for the broader societal consequences of innovation throughout the innovation process. It addresses the morality of decision-making, and recognizes that choices made today that will affect others and future generations. This is in line with anticipation and the precaution principle of sustainable development discussed above.

Acknowledging responsibility in the innovation process involves subjective and interpretative ethical issues that are embedded in the moral responsibility of individuals. Philosophers since the ancient Greeks have explored responsibility, moral behaviour and the extent to which individuals are (and should be) responsible for their actions. It was argued that the good will and adherence to a rule, was the highest good –a view also known as the deontology or the merit-based view (Kant 1786, 1787 [1781]). This view sees the intention behind an action (rather than its consequences) as defining whether or not an action is good and the individual has acted responsibly. From this perspective, responsible behaviour involves a willingness to innovate in a sustainable way. However, what people perceive as responsible behaviour – aimed at making the world more sustainable - may not always produce the intended effects. The consequentialist view (Bentham 1948 [1789]; Mill 1859, 1979 [1863]) considers the rightness of an action in terms of its consequences. From this standpoint, a morally right action is one that produces a good outcome.

These older philosophical debates have influenced a vast amount of literature covering CSR (Frederick 1960), business ethics (Drucker 1981; Bowie 1999) and stakeholder theory (Freeman 1984). Ideally, CSR would function as a built-in, self-regulating mechanism through which businesses monitor and ensure their adherence to the law, ethical standards and international norms. As an example, Siemens[2] has formulated a business ethics code with rules covering integrity, anti-corruption, occupational health and safety and human rights. Following the principles of CSR and stakeholder theory, a number of reporting guidelines or standards have been developed to serve as frameworks for social accounting, auditing and reporting (Henriques and Richardson 2004). Elkington (1999) operationalized the People-Planet-Profit concept through Triple Bottom Line accounting (TBL). All these approaches involve reporting on criteria and threshold values for measuring economic, ecological, and social costs and benefits. CSR has now become a 'projectified' idea within the business community, expanding the traditional reporting

[2]http://www.siemens.com/annual/09/pool/en/downloads/siemens_ar09_integrity.pdf

framework to take social and environmental performance into account in addition to financial performance. Typically this is done through operationalized impact checklists, scoreboards, etc. (Spreckley 1981; Elkington 1999). Although these ideas have gained ground in recent years, there are still ambiguities and discussions about the underlying concepts. The theories assume a certain degree of altruism among entrepreneurs, managers and other business decision-makers, an assumption which some question (Stieb 2009). Critics and proponents of CSR often disagree about the actual and desirable nature and scope of CSR, partly because they have different perceptions and understandings of the role and purpose of business in society (Idemudia and Ite 2006). These tensions mean that CSR has become a disputed concept that combines elements of sustainability, corporate governance and corporate accountability to stakeholders.

9.3 An Empirical Exploration of Responsible Innovation: Outcomes and Behaviour

During our examination of the social and environmental aspects of innovation newness, value creation and process, we came to see two perspectives for viewing responsible innovation. It can be viewed in terms of innovation *outcomes* and whether they are responsible – such as clean technology, ethical products, equal distribution of benefits to name a few – or from the perspective of *responsible behaviour*. The latter occurs when innovators acknowledge their responsibility and act upon it to address any (unexpected) societal consequences of their innovations. We initially chose to investigate the social and environmental outcomes of innovation activities through empirical analysis at the village level. We envisaged that this approach would allow us to identify patterns and develop criteria for responsible innovation in an informal economic context. This, we thought, would provide an objective basis for the conceptualization and allow us to develop a tool akin to the Triple Bottom Line accounting practices. We also explored responsible behaviour, investigating individual perceptions, attitudes and responses to the harmful outcomes of innovation. This led us to explore how social interactions can lead innovators to acknowledge responsibility for emerging societal conflicts that are the consequences of an innovation.

From an epistemological perspective there is an important difference between research into responsible outcomes and research into responsible behaviour. The first requires a positivist approach (which sees one objective reality out there), while the second concerns the subjective construction of mental models, interaction and behaviour, epistemologically related to post-modernism and constructivism.

A team of Dutch and Vietnamese researchers carried out the fieldwork which consisted of two data collection missions of 2 weeks each in the four Vietnamese villages in 2010. The case study data were collected through observations and 15–20 open interviews with relevant actors in each village. The interviews explored the perceptions of harmful innovation outcomes within the community and the responsibility of the innovators in preventing undesirable or promoting desirable consequences. Additional data collection involved reviewing the previously collected data and secondary context-specific data on social and environmental changes in the four locations in Vietnam and materials from research institutes, NGOs and government agencies.

9.3.1 Duong Lieu Cassava Products Village

Duong Lieu is a cassava starch and noodle-producing village in the Red River Delta in northern Vietnam, 30 km southwest of Hanoi. One group of the economically-active village population is involved in home-based cassava starch production, an intermediate product which another group of small producers traditionally used to produce noodles. In more recent years several of these small producers have introduced new end-products which add more value. These include medicine pills, soft drinks, boxes and candy. These small producers have invested in small machinery, add more value and now sell their products to more profitable outlet channels in Hanoi and beyond. Candy production has been quite a success story for households, giving them better incomes. It also involves much lighter and quieter work, in contrast to the harder and dirtier tasks associated with starch and noodle production. But candy production requires some investment and so is not an option for the poorest of the poor. This said the poor can benefit from the new product as the candy industry creates extra employment and there is a shortage of workers in Duong Lieu.

There is also an emerging pollution problem in the village. New end products have increased the demand for starch, resulting in more organic waste being discharged into the open sewage system. The starch producers in the village discharge vast amounts of organic solid waste into the open sewage system, which is becoming an increasing source of debate and conflict. Several government research centres and NGOs have carried out environmental impact studies in Duong Lieu which indicate a worrying environmental situation. There have been some adverse reports in the media and local people are somewhat irritated about this as they think it will have a negative impact on demand for their products.

Many of the villagers – particularly the starch producers – ignore the problem and consider it as a trade-off for their livelihoods; they do not acknowledge responsibility. But more and more villagers are bothered by the pollution and concerned about the health impacts and link the pollution to several diseases that have recently become more common. Some people in the village link high rates of cancer cases to the pollution.

The household enterprises involved in producing the new products consider the waste issue to be the problem of the starch producers and do not see that they have any responsibility or role to play in addressing this issue. They ignore the potential to allocate some of the wealth they create by producing candy to pay for the environmental damage it causes.

9.3.2 Bat Trang Ceramics Village

Bat Trang is a traditional ceramics village in the Red River Delta in northern Vietnam, 15 km east of Hanoi. The village has 1,020 μ and small household enterprises[3] producing ceramics. Over the past 8 years, small producers have introduced LPG kilns for baking ceramics. The new technology enabled higher production volumes, higher quality ceramics (which can be exported) and saves on energy costs. Small producers acknowledge that the innovations have increased household incomes and improved the quality of life. There is less air pollution and the working conditions in the workshops are greatly improved. The innovators have created surplus value in the village and new employment opportunities for poorer people. The improved competitive advantage made it possible to access new (international) markets. Poverty was common in Bat Trang 20 years ago, but today poverty rates are below average for the province and far below the national average. According to the village's administration, unlike other craft villages in the Red River Delta, the gap between rich and poor in Bat Trang has not widened.

A number of poor household enterprises in the village – lacked the means or capability to purchase the new LPG technology and had to close down their business. However, this is not perceived as a major issue since many of them found employment in the innovating enterprises. The production of ceramics – both traditional and in gas ovens – requires special skills and experience and there is a shortage of skilled employees in the village.

The new production process has led to a significant improvement in the village's living environment. The LPG kilns emit less pollution than the charcoal kilns. People believe that the smoke and air pollution from traditional charcoal oven in the past were responsible for many cases of respiratory diseases, particularly among older people. Today the air is much cleaner and there are fewer dirty storage areas for charcoal in the streets. According to the villagers, the village is now a greener and a more pleasant place to live. Early innovators mention that personal profit was not the only reason for developing the technology. They also took the environmental situation into account and wanted to promote the image of Bat Trang as a ceramics village, based on family traditions. The majority of villagers, and particularly those involved in the ceramics industry, see that the introduction of LPG technology has brought a variety of positive outcomes.

Villagers report, with satisfaction and a certain pride, that the village is now much cleaner and greener. Over the years a collective process of becoming more environmentally aware has been underway. Although the profit argument may have been dominant, the small-scale producers also mention that they took environmental considerations into account. They see it as their responsibility. Having seen the benefits of the LPG kilns in past years, they are convinced that they have made

[3]Micro and small entrepreneurs in Bat Trang typically have a home-based workshop, with between 1–5 (micro) or 5–20 (small) employees, often family members employed under informal contracts.

a difference in creating a cleaner environment for themselves. The Ceramics Association, established in 2002, has played a prominent role in the introduction of LPG oven technology. Virtually all the small-scale producers in Bat Trang are members of the association. The association functions as a discussion and exchange platform and actively promotes LPG kiln technology, highlighting the environmental arguments as one of the benefits. These discussions about the societal implications have come naturally as the inhabitants of Bat Trang feel strongly connected through family ties and their shared history in ceramic production. In this sense the innovation process was a collective process and the villagers recognized their responsibility, rather than looking to the government for a solution.

9.3.3 Van Phuc Silk Village

Van Phuc is a traditional silk craft village in Ha Tay province, 10 km west of Hanoi where a cluster of 785 small, home-based, producers is engaged in silk weaving, tailoring and sales. Over the past 10 years, many of these small producers have established retail shops in the village's main street, offering a much broader range of products. Silk weaving families opened retail shops in the village's main street and benefitted from the growing demand for silk products, spurred by the increasing number of domestic and foreign tourists coming to the village. By and large the village has benefitted from advantage of the new marketing practices although some actors in the value chain claim that the distribution of benefits is unfair. The silk weavers and silk dye workshops in the village enjoy higher and more stable incomes than before, but not to the same extent as the shop owners.

Competition is increasing and the shops are having to compete more on price and lower their quality standards. This implies the need for higher production volumes per business in order to survive. At the same time the shop keepers are sourcing products from outside the village (including from China) One-loom households are no longer viable and those that could not expand and increase production volumes had to close down. However most of these people have been able to find new employment in an expanding weaving workshop.

Although the silk shops do not affect the environment directly, increased silk production in Van Phuc has caused serious environmental problems, particularly water pollution. The weaving workshops and shop owners outsource the dying to several specialized dying workshops in the village. The latter use more toxic chemicals for the dyeing process to obtain fashionably bright colours. The waste water from this process is discharged directly into the sewage system and river without any treatment. According to many villagers, this results in severe pollution, black river water and new and more health problems.

Research institutes measured alarming levels of toxic pollution in the ground water and the river bordering the village. Villagers link the pollution to the increased silk production volumes and increased use of toxic chemicals, particularly in the dyeing process. However, not everybody in the village is convinced about the link; pollution is also associated with the newly established textile-related companies further along the road and around the village. There is no agreement about the precise sources of the pollution – from the village and factories around or the impacts on human health. But more villagers, research institutes and the local administration express great concern about the surface water quality in and around the village. A growing minority of people in Van Phuc consider the pollution as a serious threat to the village and associate it with the occurrence of more serious and fatal diseases. Research institutes have examined the pollution and its impacts and produced several scientific articles on the matter. However, the often detailed research outputs are sometimes conflicting; the villagers do not have access to straightforward and practical information about the origin and effects of the pollution or possible solutions to the problem.

There is a growing sense among the villagers that the pollution of the river is a problem and might affect human health. However, the general attitude among small producers and shop owners is that the problem is an acceptable trade-off for increased economic prosperity. Most dye workshops owners are less concerned and see the pollution as a fact of life and an acceptable consequence of making money in the silk industry. As individuals, they consider themselves as small players in a larger complex. The villagers feel some sympathy for the dye workshop producers and do not blame them for the pollution. They recognize that these workshop owners are poor and trying to survive. The richer shop owners do not see themselves as having a responsibility to solve the problem. The main street – where they have their shops – is some distance from the polluted areas. However, the rich shop owners do see that the pollution will eventually have an adverse effect on tourists coming to visit the village and that does worry them because this might impact their own prosperity. Small producers and some villagers are looking to the local government for a solution. The village administration is assuming responsibility and developing plans to move the polluting workshops to a location just outside the village where they will be concentrated and provided with a waste water purification plant.

9.3.4 Phu Vinh Bamboo and Rattan Village

Phu Vinh is a traditional bamboo and rattan weaving village 30 km southwest of Hanoi. After the introduction of the free-market economy in Vietnam, entrepreneurs established export companies just outside the village. When they sign an overseas contract they outsource the actual production to middlemen in the village who in turn sub-contract the order to household enterprises scattered around the village. The small producers do the weaving and deliver the semi-finished rattan and bamboo products to the middlemen and export companies who then do the final colouring and varnishing, as the last step before shipment overseas.

For the export companies and middlemen it is very profitable business. However, the innovation has worked to the disadvantage of the small household enterprises. They get a lower unit price, have to work harder and more family members are now involved in the production work (including children who work after school and old people) and they still earn less than before. These changes are driving the small producers into poverty and making them feel marginalized.

New environmental problems have also emerged. To meet international quality standards and design requirements, small producers, middlemen and export companies now use more chemicals to whiten, soften, colour and dry the bamboo and rattan. The waste water – containing harmful concentrations of chemicals – is usually discharged untreated into the surface water with no consideration about the effects. Nobody knows the exact level of pollution or what health impacts can be expected. Another emerging problem is the depletion of rattan and bamboo as a result of the increased production volumes of recent years.

In the past the rattan/bamboo products used to have a practical use (as household utensils). Now they have become a more luxury decorative product. Products today are sold to a higher segment of the market, particular the western export market. Consumers in Europe and US enjoy the consumer surplus value but these products are no longer affordable for traditional clients in the domestic market. After the introduction of the free-market economy in Vietnam, entrepreneurs established export companies just outside the village. When they sign an overseas contract they outsource the actual production to middlemen in the village who in turn subcontract the order to household enterprises scattered around the village. The export companies have enjoyed handsome profits the system has brought less prosperity for the small-scale producers. To maximize their profit in a free market system, the export companies have increasingly imposed lower unit prices on the small producers. This has created conflict between villagers and the export companies. The producers complain about the lower unit price and increasingly suffer from poverty.

The export companies take a hard-line business attitude and do not see that they have a role to play, or a responsibility to modify unit prices to reduce poverty. They see this as the government's role. The small-scale producers have a different view and blame the export companies for offering such low prices, arguing that they could share more of their profits.

The village administration recognizes and sympathizes with the problems of poverty faced by the small-scale producers, yet is unable to interfere with the economic process and the free market price setting mechanism. In addition, they are closely connected – through family ties – to the export companies. In recent years, the export companies have helped the local authorities to construct a school and a medical clinic, have planted trees and provided tables and computers for the administration's offices. The local government has facilitated the procedures for renting land and completing export license procedures.

9.4 Interpretation of the Case Studies

9.4.1 Innovation Responsible Innovation Outcomes

The innovations in these Vietnamese villages have led to a diverse range of social and environmental consequences, as summarized in Table 9.1. The first column of the table lists the most notable issues we identified within each particular village, through field observations and villagers' reports. The second column lists our assessment and judgement of whether these outcomes were positive or negative to be validated later in subsequent discussions with the villagers. The last column categorizes these empirical observations in terms of the theoretical clues in the preceding section focusing on environmental and social outcomes; newness, value creation and process.

Some outcomes reflect environmental issues such as air and water pollution, others relate to social issues: labour conditions, income disparity and health. Some were the consequence of newness; others the result of innovative ways of value creation. At first glance, the outcomes and consequences of innovation in Bat Trang are all positive. By contrast in Van Phuc and Duong Lieu there were several negative outcomes alongside the positive ones. In Phu Vinh the consequences were mostly negative. This highlights two issues that are relevant to our initial research question regarding the conceptualization of responsible innovation in the Vietnamese small producers' reality. Firstly, it suggests that one village – Bat Trang – might be categorized as pursuing a path of responsible innovation, while the other villages do not. Secondly, it leads us to ask if the identified issues can help us to develop a checklist of generic criteria to which threshold values to distinguish responsible innovation from what it is not might later be added.

In regard to the first issue, from our assessment as researchers, we were inclined to assert that Bat Trang village could be labelled as experiencing responsible innovation. During our discussions in later rounds of validating our tentative field assessments, we were confronted with the views of innovators and villagers in the other villages who had a different judgement than us about the whether the outcomes were negative or positive. In Duong Lieu and Van Phuc the villagers considered the emerging pollution problem as an acceptable trade-off for the benefits of the innovation. In Phu Vinh, our normative framework reflecting universally agreed ILO conventions saw some practices as child labour, a view not shared by the villagers. Any attempt that we – as western researchers, not living in the village – might make to define threshold values for these criteria, would involve imposing our normative framework about what is acceptable and what is not. This was particularly critical in the qualitative outcomes include labour conditions, the quality of products and the living environment, the position of employees and the consequences of innovation for cultural and traditional values. It is difficult to measure these criteria in an objective positivist fashion, as they are largely socially constructed, context specific (Adcock and Collier 2001). We faced an epistemological challenge that is inherent to the positivist approach which assumes 'one truth' and seeks to establish one set of

Table 9.1 Summary of economic, social and environmental outcomes in the cases

Cases	Outcomes and consequences of innovation	Assessment	Theoretical category
Bat Trang (Ceramics)	Cleaner air	+	Environmental – newness (Blackman and Bannister 1998; Chang 2011)
	Better labour conditions in the workshops	+	Social – newness (Ewing 2006)
	Employment creation	+	Social – newness (Drucker 1981; Pianta 2005)
	Equitable distribution of value; poverty alleviation	+	Social – value creation (Bigsten and Levin 2000; Hart 2007)
	Better quality products with longer product use cycle	+	Environmental – value creation (Bartlett and Kurian 1999; Hertwich 2005)
Van Phuc (Silk)	Increased water pollution due to chemical use	–	Environmental – newness (Roome and Hinnells 1993)
	Increased sales, benefiting many in the village	+	Social – value creation (Bigsten and Levin 2000; Hart 2007)
	Employment creation	+	Social – newness (Drucker 1981; Pianta 2005)
	Uneven distribution of value creation (favouring the shop owners)	–	Social – value creation (Bigsten and Levin 2000)
	Lower quality products, shorter product use cycle	–	Environmental – value creation (Bartlett and Kurian 1999; Hertwich 2005)
Dong Lieu (cassava candy)	New income and employment	+	Social – value creation ((Lindfelt and Törnroos 2006)
	Sweets are not healthy food for children	–	Social – newness (Bhattacharyya and Kohli 2007)
	Good business accessible to villagers	+	Social – value creation (Pianta 2005)
	Better labour conditions in the workshops	+	Social – newness (Ewing 2006)
	Severe water pollution from increased starch production	–	Environmental – newness (Roome and Hinnells 1993)
Phu Vinh (Rattan)	Older people and children do a significant amount of work	–	Social – newness (Ewing 2006)
	Increased pollution from chemicals used for whitening the rattan	–	Environmental – newness (Roome and Hinnells 1993)
	New poverty in the village; small producers earn less	–	Social – value creation (Bigsten and Levin 2000; Hart 2007)
	More transport environmental costs due to remote markets	–	Environmental – newness (Curtis 2003)
	New products only for export. No BOP products	–	Social – value creation (Prahalad 2005)

universally applicable threshold values. In reality, perceptions of the relevance and legitimacy of these thresholds may vary considerably, according to the situation of the people concerned. Given the growing view that sustainability should be owned by people; that it should be participatory (Bell and Morse 2003), it is essential to include the judgments of the actors involved (in this case the villagers). This made it problematic to conclude that the innovations in Van Phuc, Duong Lieu and Phu Vinh should not be viewed as responsible innovation.

In regard to the second issue, although the villages appear comparable in terms of their innovative activities, the social and environmental consequences of innovation in the four villages were very different. A diverse variety of issues emerged in the different villages. We tried to translate these and reduce this multi-dimensional reality into a simplified and comparable set of criteria. In so doing we faced the problem of which criteria to include and which to exclude and what weight to give to each criteria, that is which ones were critical and which were less essential. A long list of criteria would not contribute to conceptual clarity, let alone provide practical and feasible measurement tools. There was also the issue of who will decide which criteria to include and exclude– clearly a potentially political issue.

We also faced the problem of addressing the space and time aspects of sustainability. In spatial terms human activities such as innovation can have impact on sustainability that stretches from the local to the global. Sayer and Campbell (2004) point out that fragmented analysis over extremely small spatial scales may be meaningless in terms of tacking bigger problems of both a local and trans-local nature. In temporal terms, peoples' perceptions change: we witnessed this ourselves when recording perceptions about the consequences of innovation during the years of working on this matter in Vietnam. Sustainability has several temporal dimensions: physical and perceptual changes occur at different speeds. Some consequences of innovation may be immediately visible, while others may take much longer to become apparent. As a result some processes and their impacts may be studied over short time frames, while others need to take into account a period of decades or even, in the case of climate change, centuries and the prospects of future generations (Sayer and Campbell 2004).

In sum, our interpretation of the data that we collected and the associated serious methodological and measurement problems discussed above led us to doubt the feasibility of developing a defendable responsible innovation outcome criteria checklist that could be used a basis for conceptualizing (or operationalizing) responsible innovation. Although we do not exclude the possibility that such a positivist generic checklist might be developed in more thorough analysis, we could not see a feasible way of doing so with the empirical material we had available from our research.

9.4.2 Interpretation of Responsible Innovation Behaviour

The case descriptions offered another perspective: on responsible behavior from innovators. This concerned how innovators considered their responsibility in resolving any societal conflicts arising from harmful outcomes of innovation. The cases showed essential differences in that respect. By comparing various aspect of the perceptions and attitudes of the villagers and innovators in the four cases we identified patterns that we modeled into a five-stage societal process towards responsible innovation. The detailed development of the model is based on Voeten et al. (2012). Here we summarize the essential points of the model, the stages it follows add and combine this with our earlier considerations from the literature.

9.4.2.1 Stage 1: Perception of Societal Change

The cases show that there were differences in the initial recognition by the villagers of an environmental or social change and how different actors in the communities perceived these outcomes. In the rattan bamboo case villagers are clearly aware of increasing poverty levels among some parts of the community over recent years and consider this to be unacceptable. In the ceramics case more or less all the villagers remember the bad air quality caused by the charcoal kilns. By contrast with these two cases, the social and environmental consequences in the silk and cassava cases are less clear-cut and not commonly agreed upon. Some community members in these villages see increased pollution and see serious health problems emerging, while others do not.

Several theoretical insights might help to explain this. Simon (1957) argued that the rationality of individuals is limited by available information, the finite amount of time that people have and their cognitive limitations in interpreting the complex environments in which they operate. In the sustainability debate it is acknowledged that it sometimes takes a while before harmful outcomes materialize. The temporal dimension of sustainability is complicated because different processes take place at different speeds (Sayer and Campbell 2004). Individuals, groups, or organizations perceive and react to changes in their environment through a learning process, identified by Argyris and Schön (1978) as single loop learning. Experiential learning – the process of making meaning *through the transformation of direct experience* (Kolb 1984) – was particularly relevant in the daily reality in the villages. Experiential learning can be both an individual, and a joint, process. It is referred to in the literature as a social learning process; since it is often beyond the capacity of any single actor to understand the nature of these emerging societal problems (Pahl-Wostl 2006; Beers et al. 2010). The cases showed differences in the extent to which the communities developed, or failed to develop, a “critical mass” of common perceptions. The literature refers to such an accumulation of perceptions as a “tipping point”. This concept was introduced by Gladwell (2000) who defined it as “the moment of critical mass, the threshold, the boiling point—the levels at which the momentum for change becomes unstoppable”.

9.4.2.2 Stage 2: Linking Innovation with Societal Change

The next stage is whether and how the actors relate the social or environmental problem to the innovation, the innovators or those who benefit from the innovations. In some cases the link is clear to everybody while in other cases is difficult or impossible to establish a causal relationship. In the rattan case, the community has no doubts that the lower prices offered by the export companies to local producers have resulted in more poverty among small-scale producers, a negative societal outcome. Similarly in the ceramics case, there is general agreement in the community that cleaner air is a result of adopting a new technology (charcoal burning created pollution and was harmful). Conversely, the links are less obvious in the silk and cassava candy villages. The scattered workshops over the villages and other local sources of pollution make it difficult to trace who is contributing to the increased pollution and to what extent. Moreover, the innovators in the silk and cassava cases are not actually producing the pollution themselves. The cassava starch producers in Duong Lieu – who themselves did not innovate – pollute more due to the increased demand by the innovative households making new products. This is also the case in Van Phuc where the silk dye workshops pollute more due to increased demand by shop owners, the actual innovators. The innovators only indirectly affect the environment.

In reality, it is often difficult to understand the causality between an innovation and societal change because social and economic phenomena are complicated and intertwined. They often overlap and this makes it difficult to establish a cause-effect relation. Waller and Felix (1989) argue that the causality principle can successfully be applied to cases in which one has complete information on the situation (ceteris paribus). However, causality becomes problematic in a situation where limited information is available (Eve et al. 1997). Complete information needed to establish a causal relation. However, in the societal context of small producers in Vietnam complete information does not exist. The clusters are complex systems in which numerous independent elements continuously interact and spontaneously organize and reorganize themselves (Valle 2000). This is particularly the case in complex value chains, where the innovation may be initiated by one actor in the chain, and the societal harmful consequences are produced by other actors. This complicates the establishment of a cause-effect relation.

There is no clear agreement about the exact causes of the pollution because of the complexity of these environmental pathways. As with perceptions of societal change, learning within the community is instrumental to developing an understanding of links between innovation and societal change. Learning may involve developing new insights into the origin of societal changes. The community has to question the issues that gave rise to the societal changes; if they are able to understand that these changes are related to an innovation (or another recognizable cause), then second-order or double-loop learning has taken place (Argyris and Schön 1978).

9.4.2.3 Stage 3: Dissatisfaction with the Trade-Offs; Emerging Conflict

Once an innovation has been linked to a societal change a community can respond in different ways, as our case studies show. This is particularly evident in the different ways in which communities weighed the harmful changes against benefits, such as prosperity, income, employment and stability, which is in fact an informal form of social cost-benefit analysis. For instance, in Phu Vinh (rattan), the small-scale producers find the new problem of poverty unacceptable and do not see any compensatory benefits. The result is dissatisfaction and an emerging conflict with the export companies. On the other hand, in Van Phuc (silk) there is a common perception that the new problems of pollution are sufficiently compensated for by the economic benefits of innovation and the community sees the pollution as an acceptable trade-off. A similar story emerges in the cassava candy case, where no overall conflict of interests has emerged about the harmful environmental consequences, which are sufficiently compensated by the economic outcomes. In the third stage we can also see 'dissatisfaction with the trade-offs'.

When harmful societal consequences are not compensated for by benefits, conflicts can emerge among people with differing interests and resources (Mills 1959). They may create social groupings that reflect the unequal distribution of power and resources in the community. In practical terms, these conflicts stem from the perception that one's own needs, interests, wants, or values are incompatible with someone else's (Mayers 2000). They create a situation in which (at least) two groups of actors are striving to acquire a set of scarce resources at the same time. Dissatisfaction provides the potential for conflict, also known as the "latent phase" in the process towards conflict (Brahm 2003). Glasl (1999) shows how parties in a conflict lose the ability to cooperate in a constructive manner, as they share fewer common and mutually beneficial experiences and lose the links that used to bind them in the past. He identifies several "points of no return" which contribute decisively to this escalation.

9.4.2.4 Stage 4: Escalating Conflict; Opportunism or Altruism

When actors in the community feel disadvantaged and conflict arises, innovators can react to these concerns in different ways. The innovators might be sensitive and exhibit altruism or feel a sense of responsibility for the outcomes and arrange for some form of compensation within the community (Schacter and Marques 2000). Internal mechanisms within the community could push the innovators to acknowledge their responsibility in resolving the emerging conflict. In Bat Trang, where pollution was recognized to be a problem, the small-scale producers included environmental considerations when deciding whether to invest in LPG-fired kilns. On the other hand, the innovators could intentionally not take responsibility, act opportunistically and selfishly take advantage of circumstances with little regard for principles or the welfare of others. Such behavior involves taking advantage of exploiting information asymmetries by seeking *self-interest with guile* (Williamson

1986). This situation can often escalate into a conflict. In the rattan case, the export companies lowered the price they offered and behaved opportunistically, following the principles of the free market game. Altruism and opportunism need to be discussed within the context of morality. Frederiksen (2010) distinguishes several moral frameworks that inform CSR. These include moral egoists, libertarians (who believe in not violating anyone else's rights), utilitarians (who promote the best possible outcome) and supporters of 'common-sense' morality. Most literature about CSR and stakeholders contains the assumption that an entrepreneur (as an individual or organization) has a social interest and is willing to accept responsibility for his or her behaviour and to redistribute benefits and important decision-making powers to stakeholders (Stieb 2009). These strands of literature assume a variety of motives among dominant actors, once they have acknowledged responsibility for the consequences of their actions. While there are well-intentioned innovators who are willing to compensate others for harm caused, the scale and complexity of the problems, uncertain causality and bounded rationality may all make it difficult to know how to do so. Even if the causes are known it may still be difficult to establish the appropriate level or method of compensation. When there is proximity between the actors, as in the clusters in Vietnam, it should be easier for the innovators to recognize and arrive at acceptable compensation arrangements for the community.

9.4.2.5 Stage 5: Enforcement of Responsibility by Third Party

In cases where there is no internal settlement of the conflict or voluntary compensation and the conflict remains unsolved or escalates, innovators could be pushed by an external force or a new institutional arrangement to acknowledge their responsibility and offer some form of resolution. In the silk village, the local administration sees that it has a responsibility to address the pollution problem. If this does not happen the community is likely to end up with an unresolved and escalating conflict (as witnessed in the rattan case). A third party, e.g. a court of law, can also be called in to intervene and to act as an arbiter. However in many developing countries, there is limited awareness of, or access to, *de jure* rights and poor people are often excluded from the formal legal system (Barendrecht 2009). There are no or few formalized processes for local actors to claim their rights as a result of failing laws, judiciaries and other legal mechanisms (Buscaglia and Ratliff 2000). An alternative is that resolutions might be arrived at through informal and multi-actor conflict resolution arrangements (Crowfoot and Wondolleck 1990). One difficulty here is that disadvantaged parties may lack the courage to fight a claim and voluntarily assume a position of powerlessness. This will partly depend on the cultural patterns within the community. For example intimidation might play a key role in some countries. In the case of Vietnam, social mores about harmony, not complaining and accepting one's destiny are likely to be more decisive (Warner 2003). Some societies stress tolerance and harmony and are less inclined to engage in behaviour that is seen as creating conflict.

In sum, based on the evidence developed from our cases we advance a conceptualization of responsible innovation. From the empirical material and theoretical insights drawn from the literature, we propose the five-stage model shown in Fig. 9.2. This shows a process that either ends in an unresolved conflict or moves into the zone of what can be termed responsible innovation – innovation that takes account of social, environmental or distributive issues and is acceptable for the community concerned.

9.5 Concluding Remarks

This chapter has attempted to conceptualize responsible innovation in clusters of poor small producers in Vietnam. Earlier research discussed small producers in clusters introducing new technology, products and business practices, such as linking up with global value chains and international markets, innovations which sometimes bought about harmful environmental and social consequences within their villages. Through an inductive grounded theory approach we attempted to develop a positivist understanding of responsible innovation based on innovation outcome criteria. However, due to the variety of facets and different normative frameworks we could not produce a convincing set of objectively measurable criteria as a basis for the conceptualization. Instead, we examined innovators' behaviour in acknowledging responsibility for societal conflicts resulting from innovation outcomes and arrived at a model of responsible innovation that represents it as a five-stage societal process. This process is situated in a community where innovators and community members closely live together and have some link with the innovation process. We view responsible innovation as a situation – which we term the responsible innovation zone – where societal conflicts in the community resulting from perceived harmful innovation outcomes are resolved either by community members considering the consequences of a trade-off, by innovators acknowledging responsibility by themselves or being forced by third parties to act appropriately. A central aspect of the model is the acknowledgement of responsibility over conflicts resulting from environmental and social outcomes. Communities below the line demarcating the responsible innovation zone are in a situation of an ongoing latent or escalating conflict.

An essential element of the model is that the normative framework employed for judging the innovation outcomes is based on the values and perceptions of the community itself. As we argued earlier in the chapter, imposing an external – western – normative framework is not viable, since the perceptions and evaluations of environmental and social outcomes are subjective, context specific and subject to constructivism and multiple realities. Another element is that human perceptions and interactions are dynamic, a reality that is accommodated within the model. For example, a village may be in the responsible innovation zone at one point in time, but new insights of individuals, social learning or institutional changes may later move that village outside the innovation zone. The model also adopts a consequentialist

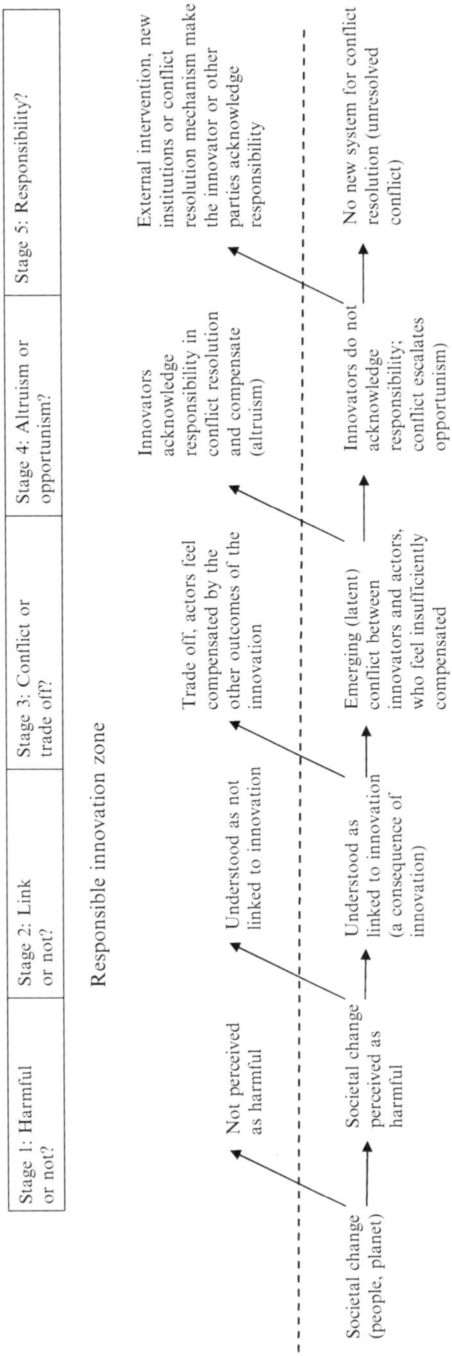

Fig. 9.2 The societal process towards acknowledging responsibility (Adapted from Voeten et al. 2012)

perspective; the consequences of the innovators' behaviour – as perceived in the community – form the basis for the judgment about the rightness of that conduct. This ensures that the prevailing societal values, interests, needs and rights become the yardstick through which the outcomes of the innovation process are measured.

The societal process surrounding innovation is a distinctive characteristic in our conceptualization of responsible innovation and this is in contrast to the projectified approach of CSR (Voeten et al. 2012). The projectified approach assumes a predictable process capable of anticipating the outcomes. In this respect, CSR is in line with the precautionary principle of sustainable development that relies on anticipation, the precautionary principle and the recognition that, when there is a plausible risk, businesses have is a social responsibility to protect the public from exposure to harm and to avoid conflict (O'Riordan and Cameron 1994). However, the societal processes analyzed in this research were far from predictable. They were open-ended, responsive and unplanned processes steered by incidental human interactions, actions and reactions involving a multitude of actors, some of whom come and go. In effect what we have done is to shift attention from the *quality of the outcomes* to the *quality of the process*. This is line with the growing attention being paid in western business management practices to total quality management, process control and quality assurance, which are increasingly focused (together with process improvement and benchmarking) on businesses' search for sustainable competitive advantage (Powell 1995).

With regard to the connection to global and future impacts signaled in introductory discussion on sustainable development (Sayer and Campbell 2004), it should be noted that the model is based upon the innovation outcomes directly perceived by the community members at the village level it does not give much weight to temporal and (broader) spatial issues. This does limit the scope of the model, because even the relatively simple local innovations taking place in the Vietnamese villages can lead to indirect global impacts effects, such as CO_2 emissions and water pollution. The model is, however based on a cluster-level situation and the innovators and villagers experiencing the consequences live close together. In the cases we have described, innovators did not consider the outcomes and impacts of 'downstream stakeholders' or future generations. In one sense this is due to the unplanned nature of the innovations. It is also, to a large part, due to their bounded rationality and the lack of information available to them about global and future impacts (step 1 in the model) and clarity about any causal links with their innovations (step 2). Then again, even if it was possible to provide complete information and 'hard evidence', the small producers may ignore this, such harmful outcomes might appear too distant in location or time (discounting). Geographical proximity and firsthand experience were key factors in shaping entrepreneurs' perceptions and encouraging them to behave responsibly. Lähdesmäk and Suutari (2012) have examined relationships between business and local communities and identified the importance of reciprocity, suggesting that CSR at the local level is an expression of reciprocal community relations between the owners and managers of small businesses and the local community in which they live. It would be informative to test the model further, for example through experiments in game theory that could incorporate behavioral economics and various levels of geographical proximity. This might illustrate the extent to which extra available information about global and future impacts would influence behavior and whether model will work in the same way.

But even so, the serious issue of bounded rationality would remain: the difficulty of having available accurate, objective and complete information about the impacts of an innovation and an indication of causality. In the empirical part of this chapter, we found how difficult it is – even at village/micro level – to develop and agree on a list of harmful innovation outcomes and attribute causality. For investigating global and future impacts at macro level this is far more difficult. Many academics, CSR managers and sustainable development practitioners have struggled to develop positivist impact evaluations on a global level. Despite the many checklists, scoreboards, TBL and certification systems, nobody working at a reasonable scale in the domain of sustainable development has been able to develop a comprehensive approach to collecting and providing the complete information that would be required to overcome bounded rationality. Attempts at doing so have led to disagreements about the measurements and indicators that should be included (and excluded) and the normative framework of benchmarking (Bell and Morse 2003). The latter has been a particular problem in the ongoing global debate about climate change, which severely hampered progress at Kyoto and Copenhagen (Hasan and Dwyer 2010).

Against this background, the model and its focus on the quality of process might provide a useful additional tool for understanding and promoting sustainable development. It contains the underlying concept that a good process leads to good outcomes. Another positive aspect of our focus on societal processes is it allows sustainable development to be framed in a participatory manner, which accepts and recognizes different normative frameworks. This is in line with the growing view that sustainability should be owned by people; *the very soul* of sustainability is that it is participatory (Bell and Morse 2003).

In conclusion, it is possibly true that it may be more difficult to operationalize this dynamic model than to employ a static, objectively verifiable, outcome criteria check list. However, this method of conceptualizing responsible innovation contributes to our understanding of the underlying factors and societal process that shape responsible innovation. This understanding may in turn be used to develop policies and programs that can promote responsible innovation in informal economic situations in developing countries.

References

Adcock, Robert, and David Collier. 2001. Measurement validity: A shared standard for qualitative and quantitative research. *American Political Science Review* 95(3): 529–546.

Alkire, Sabina. 2007. *Choosing dimensions: The capability approach and multidimensional poverty*. Working Paper No. 88. Chronic Poverty Research Centre. doi:http://dx.doi.org/10.2139/ssrn.1646411.

Argyris, Chris, and Donald Schön. 1978. *Organizational learning: A theory of action perspective*. Reading: Addison Wesley.
Bansal, Pratima, and Kendall Roth. 2000. Why companies go green: A model of ecological responsiveness. *The Academy of Management Journal* 43(4): 717–736.
Barbier, Edward, and Thomas Homer-Dixon. 1996. *Resource scarcity, institutional adaptation, and technical innovation: Can poor countries attain endogenous growth?* Occasional Paper Project on Environment, Population and Security. Washington, DC: American Association for the Advancement of Science and the University of Toronto.
Barendrecht, Marinus. 2009. *Best practices for an affordable and sustainable dispute system: A toolbox for microjustice*. Working Paper No. 003/2009. Tilburg: Tilburg University Legal Studies.
Bartlett, Robert V., and Priya A. Kurian. 1999. The theory of environmental impact assessment: Implicit models of policy making. *Policy and Politics* 27(4): 415–433.
Beers, Pieter J., Jifke Sol and Arjen Wals. 2010. *Social learning in a multi-actor innovation context*. Ninth European IFSA symposium, building sustainable rural futures—The added value of systems approaches in times of change and uncertainty, Vienna, Austria, 4–7 July 2010.
Bell, Simon, and Stephen Morse. 2003. *Measuring sustainability: Learning by doing*. London: Earthscan.
Bentham, J. 1948 [1789]. *An introduction to the principles of morals and legislation*. Oxford: Hafner Press.
Bezençon, Valéry, and Sam Blili. 2010. Ethical products and consumer involvement: What's new? *European Journal of Marketing* 44(9/10): 1305–1321.
Bhattacharyya, Rita, and Sangita Kohli. 2007. *Target marketing to children – The ethical aspect*. International marketing conference on marketing & society, 8–10 Apr 2007, IIMK.
Bigsten, Arne, and Jörgen Levin. 2000. *Growth, income distribution, and poverty: A review*. Working Papers in Economics 32. Göteborg: Department of Economics, Göteborg University.
Birkinshaw, J., G. Hamel, and M. Mol. 2008. Management innovation. *Academy of Management Review* 33(4): 825–845.
Blackman, A., and G. Bannister. 1998. Community pressure and clean technology in the informal sector: An econometric analysis of the adoption of propane by Traditional Mexican Brickmakers. *Journal of Environmental Economics and Management* 35(1): 1–21. doi:http://dx.doi.org/10.1006/jeem.1998.101.
Bowie, Norman. 1999. *Business ethics, A Kantian perspective*. Oxford: Blackwell Publisher.
Brahm, E. 2003. Conflict stages. In *Beyond intractability*, ed. G. Burgess and H. Burgess. Boulder: Conflict Research Consortium, University of Colorado.
Brundtland, G. 1987. *Our common future: The world commission on environment and development*. Oxford: Oxford University Press.
Buscaglia, Edgardo, and William Ratliff. 2000. *Law and economics in developing countries*. Palo Alto: Hoover Institution Press.
Chang, Ching-Hsun. 2011. *Green innovation performance: Antecedent and consequence*. Technology Management in the Energy Smart World (PICMET), proceedings of PICMET'11, 1–8. Portland (USA): Portland State University.
Crowfoot, James, and Julia Wondolleck. 1990. *Environmental disputes: Community involvement in conflict resolution*. Washington, DC: Island Press.
Curtis, Fred. 2003. Eco-localism and sustainability. *Ecological Economics* 46(1): 83–102.
Davis, Keith, and Robert Blomstrom. 1975. *Business and society: Environment and responsibility*. New York: McGraw-Hill.
Dosi, Giovanni. 1988. Chapter 10: The nature of the innovation process. In *Technical change and economic theory*, ed. G. Dosi, C. Freeman, R. Nelson, G. Silverberg, and L. Soete. London/New York: Pinter Publishers.
Douglas, D., and G. Papadopoulos. 2012. *Citizen engineer*. New York: Prentice Hall.
Drucker, Peter. 1981. What is business ethics? *The Public Interest Spring* 63: 18–36.

Earle, Mary, and Richard Earle. 2001. *Creating new foods the product developer's guide – A systematic approach to managing the development of commercial food products*. London: Chadwick House Group Ltd.
Elkington, John. 1999. *Cannibals with Forks, The triple bottom line of the 21st century business*. Oxford: Capstone Publishing Limited.
Eve, R.A., S. Horsfall, and M.E. Lee (eds.). 1997. *Chaos, complexity and sociology. Myths models and theories*. Thousand Oaks: Sage.
Ewing, Keith. 2006. International labour standards. In *Global industrial relations*, ed. M. Morely, P. Gunnigle, and David G. Collings. London: Routledge.
Frederick, William. 1960. The growing concern over business responsibility. *California Management Review* 2(4): 54–61.
Frederiksen, Claus. 2010. The relation between policies concerning corporate social responsibility (CSR) and philosophical moral theories – An empirical investigation. *Journal of Business Ethics* 93(3): 357–371. doi:10.1007/s10551-009-0226-6.
Freeman, R. Edward. 1984. *Strategic management: A stakeholder approach*. Englewood Cliffs: Prentice Hall.
Fiksel, Joseph. 1996. *Design for environment: Creating eco-efficient products and processes*. New York: McGraw-Hill.
Gladwell, Malcolm. 2000. *The tipping point: How little things can make a big difference*. Boston: Back Bay Books.
Glaser, Barney G., and Anselm L. Strauss. 1967. *The discovery of grounded theory: Strategies for qualitative research*. Chicago: Aldine Publishing Company.
Glasl, F. 1999. *Confronting conflict: A first-aid kit for handling conflict*. Gloucestershire: Howthorn Press.
Glewwe, Paul W. 1990. Identifying the poor in developing countries: Do different definitions matter? *World Development* 18(6): 803–814.
Hamel, Gary. 2006. Why, what, and how of management innovation. *Harvard Business Review* 84(2): 72–84.
Harris, Jonathan M., Timothy A. Wise, Kevin P. Gallagher, and Neva R. Goodwin (eds.). 2001. *A survey of sustainable development: Social and economic dimensions*. Washington, DC: Island Press.
Hart, Stewart. 2007. *Capitalism at the crossroads – Aligning business, earth and humanity*, 2nd ed. Upper Saddle River: Wharton School Publishing.
Hasan, H., and C. Dwyer. 2010. *Was the Copenhagen summit doomed from the start? Some insights from Green IS research*. Available at http://csis.pace.edu/~dwyer/research/HasanDwyerAMCIS2010.pdf. Accessed 15 Aug 2010.
Henriques, Adrian, and Julie Richardson. 2004. *The triple bottom line – Does it add up? – Assessing the sustainability of business and CSR*. London: Earthscan Publications Ltd.
Hertwich, Edgar. 2005. Life cycle approaches to sustainable consumption: A critical review. *Environmental Science and Technology* 39(13): 4673–4684. doi:10.1021/es0497375.
Hopwood, Bill, Mary Mellor, and Geoff O'Brien. 2005. Sustainable development: Mapping different approaches. *Sustainable Development* 13: 38–52. doi:10.1002/sd.244.
Idemudia, U., and U. Ite. 2006. Corporate–community relations in Nigeria's oil industry: Challenges and imperatives. *Corporate Social Responsibility and Environmental Management* 13(4): 194–206.
Jitsuchon, Somchai. 2001. What is poverty and how to measure it? *TDRI Quarterly Review* 16(4): 7–11.
Kant, I. 1786. *Groundwork for the metaphysics of morals* (original German title: Grundlegung zur Metaphysik der Sitten). Oxford: Oxford University Press.
Kant, I. 1787 [1781]. *Critique of pure reason* (original German title: Kritik der reinen Vernunft). New York/Toronto: St. Martin's Press/Macmillan.
Kaplinsky, Ralph. 2000. Spreading the gains from globalisation: What can be learned from value chain analysis? *Journal of Development Studies* 37(2): 117–146.

Keoleian, Gregory A., and Dan Menerey. 1994. Sustainable development by design: Review of life cycle design and related approaches. *Air and Waste* 44: 645–668.

Kline, S.J., and N. Rosenberg. 1986. An overview of innovation. In *The positive sum strategy: Harnessing technology for economic growth*, ed. R. Landau and N. Rosenberg. Washington, DC: National Academies Press.

Kolb, D. 1984. *Experiential learning*. Englewood Cliffs: Prentice Hall.

Lähdesmäk, Merja, and Timo Suutari. 2012. Keeping at arm's length or searching for social proximity? Corporate social responsibility as a reciprocal process between small businesses and the local community. *Journal of Business Ethics* 108: 481–493. doi:10.1007/s10551-011-1104-6.

Lindfelt, L.-L., and J.-Å. Törnroos. 2006. Ethics and value creation in business research: Comparing two approaches. *European Journal of Marketing* 40(3/4): 328–351.

London, Ted. 2007. *A base-of-the-pyramid perspective on poverty alleviation*. Growing Inclusive Markets Working Paper Series. Washington, DC: United Nations Development Program.

Mayers, J. 2000. Company–community forestry partnerships: A growing phenomenon. *Unasylva* 51(200): 33–41.

McDonough, William, and Michael Braungart. 2002. *Cradle to cradle: Remaking the way we make things*. New York: North Point Press.

Mill, J.S. 1859. *On liberty*. London: John W. Parker and Son.

Mill, J.S. 1979 [1863]. *Utilitarianism*. London: Collins.

Mills, C.W. 1959. *The sociological imagination*. London: Oxford University Press.

Mizik, N., and R. Jacobson. 2002. *Trading off value creation and value appropriation: The financial implications of shifts in strategic emphasis*. Working Paper [02–114]. Cambridge, MA: Marketing Science Institute (MSI).

Najam, Adil. 1999. World Business Council for Sustainable Development: The greening of business or a greenwash? In *Yearbook of international co-operation on environment and development 1999/2000*, ed. Helge Ole Bergesen, Georg Parmann, and Øystein B. Thommessen, 65–75. London: Earthscan Publications.

NWO. 2008. *Responsible innovation, description of thematic programme*. Den Haag: Netherlands Organisation for Scientific Research (NOW).

Nicholls, Alex, and Charlotte Opal. 2004. *Fair trade: Market-driven ethical consumption*. London: Sage.

OECD. 2005. *The measurement of scientific and technological activities – Proposed guidelines for collecting and interpreting technological innovation data*. Paris: Organization for Economic Co-operation and Development (OECD).

O'Riordan, T., and J. Cameron. 1994. *Interpreting the precautionary principle*. London: Earthscan Publications Ltd.

Pahl-Wostl, Claudia. 2006. The importance of social learning in restoring the multifunctionality of rivers and floodplains. *Ecology & Society* 11(1): Art.10.

Pianta, Mario. 2005. Innovation and employment. In *The Oxford handbook of innovation*, ed. J. Fagerberg, D. Mowery, and R. Nelson, 568–598. Oxford: Oxford University Press.

Porter, Michael. 1990. *The competitive advantage of nations*. London: Macmillan.

Powell, Thomas C. 1995. Total quality management as competitive advantage: A review and empirical study. *Strategic Management Journal* 16(1): 15–37.

Prahalad, C.K. 2005. *Fortune at the bottom of the pyramid – Eradicating poverty through profits*. Upper Saddle River: Wharton School Publishing.

Rabl, A. 1996. Discounting of long-term costs: What would future generations prefer us to do? *Ecological Economics* 17(3): 137–145.

Reid, David. 1995. *Sustainable development: An introductory guide*. London: Earthscan Publications Ltd.

Roome, N.J., and F. Boons. 2005. Sustainable enterprise in clusters of innovation. In *Corporate environmental strategy and competitive advantage*, ed. S. Sharma and A. Aragón-Correa. Cheltenham: Edward Elgar Publishing.

Roome, N.J., and M. Hinnells. 1993. Environmental actors in the management of new product development: Theoretical framework and some empirical evidence from the white goods industry. *Business Strategy and the Environment* 2(2): 12–27.

Sayer, J., and B. Campbell. 2004. *The science of sustainable development – Local livelihoods and the global environment*. Cambridge: Cambridge University Press.

Schacter, M., and E. Marques. 2000. *Altruism, opportunism and points in between trends and practices in corporate social responsibility*. Ottawa: Institute on Governance.

Schmitz, Herbert. 1999. Collective efficiency and increasing returns. *Cambridge Journal of Economics* 23: 465–483.

Schumpeter, Joseph. 1934. *The theory of economic development*. Cambridge: Harvard University Press.

Sen, Amartya. 1999. *Development as freedom*. Oxford: Oxford University Press.

Simon, Herbert (ed.). 1957. A behavioral model of rational choice. In *Models of man, social and rational: Mathematical essays on rational human behavior in a social setting*, 241–261. New York: Wiley.

Stieb, J. 2009. Assessing Freeman's stakeholder theory. *Journal of Business Ethics* 87(3): 401–414.

Stigliz, Joseph, and Andrew Charlton. 2005. *Fair trade for all: How trade can promote development*. New York: Oxford University Press.

Spreckley, Freer. 1981. *Social audit – A management tool for co-operative working*. Leeds: Beechwood College.

Tether, Bruce S. 2003. *What is innovation? – Approaches to distinguishing new products and processes from existing products and processes*. ESRC Centre for Research on Innovation and Competition (CRIC) Working Paper No. 12. Manchester (UK): University of Manchester.

Trott, Paul. 2005. *Innovation management and new product development*. Essex: Pearson Education Limited.

Ubois, Jeff. 2010. *Responsible innovation/sustainable innovation*. Paper presented at the annual meeting of the American Anthropological Association, November 2010. Available at http://issuu.com/fondazionebassetti/docs/jeff-innovation-aaa-2010-2. Accesses 15 Aug 2012.

United Nations. 1992. *Rio declaration on environment and development*. Report of the United Nations Conference on Environment and Development, Rio de Janeiro, 3–14 June 1992, UN Report A/CONF.151/26.

Vachona, Stephan, and Zhimin Maoc. 2008. Linking supply chain strength to sustainable development: A country-level analysis. *Journal of Cleaner Production* 16(15): 1552–1560.

Valle, Vicente. 2000. *Chaos, complexity and deterrence*. National War College. Available from http://www.au.af.mil/au/awc/awcgate/ndu/valle.pdf. Accessed 15 Aug 2012.

Voeten, Jaap, Job de Haan, and Gerard de Groot. 2011. Is that innovation? Assessing examples of revitalized economic dynamics among clusters of small producers in Northern Vietnam. In *Entrepreneurship, innovation, and economic development*, ed. Adam Szirmai, Wim Naudé, and Micheline Goedhuys. Oxford: Oxford University Press.

Voeten, Jaap, Nigel Roome, Gerard de Groot, and Job de Haan. 2012. Resolving environmental and social conflicts – Responsible innovation in small producers' clusters in northern Vietnam. In *A stakeholder approach to corporate social responsibility: Pressures, conflicts, reconciliation*, ed. Adam Lindgreen, Philip Kotler, Joëlle Vanhamme, and François Maon. Aldershot: Gower Publishing.

Wagle, U. 2002. Rethinking poverty: Definition and measurement. *International Social Science Journal* 54: 155–165. doi:10.1111/1468-2451.00366.

Waller, W.S., and W.L. Felix. 1989. Judgments under uncertainty: Heuristics and biases. *Science* 188: 1124–1131.

Warner, M. 2003. *Culture and management in Asia*. London: Routledge.

Williamson, Oliver. 1986. *The economic institutions of capitalism*. New York: Free Press.

World Bank. 2000. *World development report 2000/2001 – Attacking poverty: Opportunity, empowerment, and security*. Washington, DC: The World Bank.

World Bank. 2008. *Growth report: Strategies for sustained growth and inclusive development*. Washington, DC: Commission on Growth and Development, The World Bank.

Chapter 10
Values in Development: The Significance that Cultural Transitions have for Development

Jan Otto Kroesen and Wim Ravesteijn

Abstract In an endeavour to deal effectively with tensions between competing values in developing countries this contribution will study the desirability of cultural transitions. The approach draws on various moral theories on distributive justice (equal access), unfolded in sociological literature on cultural differences, the impact of such differences on the role of civil society and the relationship between cultural transitions and technological system change. By adopting this approach, the authors are able to provide a framework for analysis and for policy design. To put this framework to the test two cases analysed. The initial aim is not only to show that – but also how– development involves conflicts and trade-offs between diverging value priorities. In the second place, the values and value priorities at stake will be highlighted and finally the point that such trade-offs require explicit dialogue and negotiation processes if equilibrium between the different value priorities is to be achieved will be discussed. In short, it is the authors' contention that system and regime change in developing countries involves cultural transitions and that such value reorientations should be an integral and explicit part of the development agenda if sustainable results are to be attained.

10.1 Introduction

Sometimes it seems as if all talk and scientific discourse on development is redundant. China and India are developing rapidly and that is not the result of any form of intentional development aid. Africa, despite recent figures pointing to increasing economic growth, is still lagging far behind and so heavy investment in

J.O. Kroesen (✉) • W. Ravesteijn
Faculty of Technology, Policy and Management, Department of Philosophy
Delft University of Technology, Jaffalaan 5, 2600 GA Delft, The Netherlands
e-mail: j.o.kroesen@tudelft.nl; w.ravesteijn@tudelft.nl

J. van den Hoven et al. (eds.), *Responsible Innovation 1: Innovative Solutions for Global Issues*, DOI 10.1007/978-94-017-8956-1_10,

development did not seem to help there. The rapid development of China and India has surprised everybody while intentional intervention in Africa has apparently failed. Observations like this urge development theory to become more reflective. It would seem that certain decisive development factors have not been taken into account in existing theories on development and value creation. Apparently, we still do not know how it really works. In this contribution the authors will not address the matter of why China and India are moving forward, but they will examine why Africa is lagging behind. Why have successive development strategies failed to turn Africa into a productive force?

Since all sorts of approaches have already been tried out the question only becomes even more pressing. In conjunction with the term "governance issues" the poor performance of institutions has been criticized (Calderisi 2006), the rights-based approach or the capabilities-based approach has underscored the importance of specific standards and freedom (Nussbaum 2006; Pogge 2008), and economic approaches may emphasize big plans or large investments (Sachs 2005) or even the withdrawal of such plans or funding (Moyo 2009) yet there remains one important but sensitive issue which all these approaches seem to evade or only mention in passing and that is the issue of the basic value orientation of a given culture. According to many cultural theorists (Hofstede 1997; Trompenaars and Hampden-Turner 1999) the value orientation of a culture is the deepest layer of cultural identity and the most difficult to change. However, it is a sensitive issue which therefore makes it difficult to tackle in the development debate, because it touches on the rights of one's own indigenous cultural identity. It is easy to naïvely invade another culture with one's own value priorities. The damage, however, will be even greater if it is done inadvertently. To give but one example, Samli propagates entrepreneurship as a solution for Africa, but only briefly and almost implicitly explains that the entrepreneurial mentality of initiative and individualism is virtually lacking in Africa and, quite naïvely, he then proposes introducing more of such a mentality (Samli 2009).

This paper will explore the precise role and impact that these basic cultural value orientations have on institutions, social interaction and economic development. It will not do this in a naïve way but rather by putting it explicitly on the development agenda. In this regard, the concept of capacity or the lack of that as something central to development will be a key issue (Chang 2007; Eade 1997). Capacity points to what people can do, thus affirming that it is not what people should be able to do, but rather what they can actually do that becomes decisive for development and economic growth. Capacity, however, does not only entail knowledge and competence, it also embraces attitudes and values. People's attitudes are largely formed by their basic value orientation. What do people expect from life? How do they relate to each other? How is their inner self-experience formed or programmed by their cultural traditions? How are they conditioned by such traditions even if they only observe objective "facts"? All of this goes towards determining people's attitudes and capacities. Such value orientation also determines whether or not particular institutions or policies, introduced from elsewhere, can be adopted and can perform a positive function in the receiving society (De Jong et al. 2003).

In other words, we cannot avoid the question of which value orientations are crucial to the capacity for development; the capacity to participate in the production and reproduction of society.

In this way the present paper aims to overcome the sensitive dilemma between "cultural (western) imperialism" (the West is best) and a detached attitude with regard to culture that places emphasis on respect for indigenous traditions ("rule number one" of Star Trek was: no intervention in different cultures). Never have cultures been hermetically closed off from each other. Whenever the repertoire of any one culture appeared to add up to the capacity to survive and grow, other cultures invariably proved keen to learn from that. In addition, as will be indicated, in rudimentary or embryonic forms all core values from different cultures are present in all cultures anyway. In the emerging global society the cultural values that every culture brings with it may be invested in and become part of a common fund of values from which, from now on, we shall all draw. And these values or, more precisely, these combinations and compositions of values, need to be appraised on the basis of careful empirical research and deliberation.

To make it possible to turn global cultural heritage into a common reserve, a theoretical model, a sort of matrix of cultural repertoires of civilizations, will be put forward. This general framework will help to provide an understanding of diverging cultures as entities specializing in a particular value set which is, in principle, accessible to and present in all cultures. It will furthermore be demonstrated that such cultural repertoires are related to the concept of civil society, which originated in the West, as well as to the affiliated dynamic evolvement of culture and technology. Whether by necessity or accident, these cultural traits emerged in Western history and are now in the process of being adopted by older cultural repertoires all over the world. That does not mean to say that these cultures will simply be replaced by Western culture. It will be more probable; indeed, it is perhaps already the case, that cultures will innovate and reaffirm their original cultural values in the process of taking over the values of Western culture. In the end, such innovation may turn out to be a challenging event for the West. So often next steps in cultural innovation have been taken by what are, on the face of it, "backward" regions (Landes 1998; Rosenstock-Huessy 1993 (original 1938)). After this theoretical model has been described, two cases will be used to corroborate and test this approach to culture and development. Finally several conclusions will be drawn which will, at the same time, form hypotheses for further research.

10.2 A Repertoire of Cultural Value Priorities

The debate about culture and its consequences is to a large extent dominated by the work of Hofstede (1997) and Trompenaars and Hampden-Turner (1999), in which a number of different values are listed as aspects in which cultures may differ. These values – including for instance power distance, collectivism and individualism – represent, as is inherent to values, the different ways in which humans relate to

Fig. 10.1 The four dimensions on the space-time axis of: dialogue, labour division, group belongingness and innovation

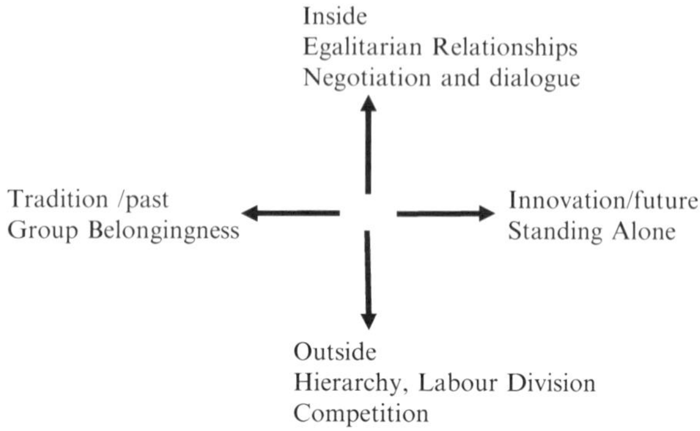

social reality. One can make the lists of such values and cultural repertoires long or short but here we propose that they be deduced from four basic social ways in which human beings relate in all cultures. These four basic types of relationships explain the complicated ramifications of these cultural value sets and make them analytically understandable as variations of one underlying matrix common to all. They are the basic relationships of human beings with each other as first systematically explicated in the vast sociological works of Rosenstock-Huessy (1956, 1958; cf. pp. 131–226). In these works he presents a history of societies of the human race from early times until modernity.

There are four types of human relationships: human relationships can be categorized as consisting of *hierarchical* versus *egalitarian* relationships, and in terms of *belongingness* to a group versus standing *alone* against the group (Rosenstock-Huessy 1981, see Fig. 10.1).

1. *Hierarchical relationships* are present in labour division systems, companies and organizations; in short, everywhere where command and control dominate. They are especially noticeable in production processes and where nature needs to be confronted by labour or struggle. Hierarchical relationships become important in connection with survival. They are indispensable wherever the material or objective side of our existence needs to be coped with. Imperial cultures of the past like those of Egypt, Babel, the Aztecs and China can be categorized as specializing in these sorts of relationships.
2. *Egalitarian relationships* can be found in friendship, marriage and other social and political formations. They are evident everywhere where people are more or less independent of each other and cooperate on a voluntary basis. There people have to create a common horizon of understanding by means of a dialogue between independent subjects opening themselves up to each other. Many old tribes maintained such egalitarian relationships but Western civilization, especially since the time of the French revolution, has specialized in this more than any other culture.
3. *Group belongingness* and identification form a constituent part of any culture or subculture sharing a common history. Tradition and collective symbols turn

the members into a "we", sharing the same past. Tribal societies are the prime cultures to specialize in this type of behaviour, although it can be present in any culture and is always seen where and when group belongingness prevails over independent individual initiative.

4. *Standing alone* against the group is the experience of any person taking up a vacant position or advancing into novel and uncertain areas, finding creative solutions, opening new paths to the future. Hardly any real "civilization" can incorporate this type of human behaviour, because the future remains uncertain and unknown and novelty cannot be predicted, but it shows up in any crisis situation when people are under pressure. A creative foray into the novel tends to be born of deadlock generated by too much tradition, too much repetition and sustained uncertainty avoidance. The Jewish people exhibited this type of behaviour when they gave priority to an uncertain future in preference to an oppressive past by fleeing from Egypt under the leadership of Moses.

This division of relationships can be illustrated in a simple diagram explaining and underpinning the cultural differences researched in much sociological literature whenever endeavours have been made to make sense of and cope with intercultural conflicts and misunderstandings (Fig. 10.1). This diagram makes use of a space and time axis resulting in two spaces (inside and outside) and two periods (past and future). Dialogue between equals creates shared space *inside* the group. Control from above coordinates people's actions and bodies in the world *outside*. A shared *history* turns fragmented individuals into a group with a common identity. *Future* challenges require a daring personal/individual answer.

In principle, each society has to deal with these four types of relationships and the corresponding values and organizational patterns. The usual repertoire of cultural dimensions can be understood either as specializations or combinations of these ways of relating to each other within social reality. Where hierarchy is strong (Hofstede: power distance), there is little room for individualism or egalitarianism so that negotiation and dialogue remain difficult. In a culture in which both hierarchy and group belongingness are highly valued (Hofstede: collectivism), standing alone and showing a critical attitude is experienced as dangerous (Hofstede: uncertainty avoidance), and will be modified. Sometimes, like in the tribes of old, group belongingness can be organized without hierarchy (or simply with little hierarchy), thus making these societies more egalitarian, but within such societies it is difficult to be innovative, since the individuals have to adapt to the group and to tradition. Strong initiation rituals incorporate individuals into the group – that is their main function (Rosenstock-Huessy, 1958).

10.3 Cultural Characteristics and the Role of Civil Society

In societies in which egalitarianism is an important value there will be more space for individual opinion but also for cultural traits (mentioned by Trompenaars) like *universalistic rules* and *neutral roles* as the coordinating mechanisms. Universalistic rules and neutral roles help to create cooperation without the need for the command and control lines imposed by hierarchy because everybody is expected to stick to the rules. In such societies labour usually also becomes a value from which status can be derived instead of one's position in the group or in the hierarchy (Trompenaars). An open civil society in which voluntary cooperation is more common cannot do without such shared rules of behaviour and open communication as coordinating mechanisms. We now call them coordinating mechanisms but these mechanisms are really values in themselves. These values need to be in place because otherwise the egalitarian relationships based on individual preferences will lead to chaos. This is in fact the fear of many hierarchically organized countries (Wiarda 2003). Equality, open communication and universalistic rules that give status, based on achievement rather than on position are indispensable components of a society based on more egalitarian relationships for keeping individualism in check. If that were not the case individualism would lead to anarchy. Here it appears that the Western notion of individualism has an inbuilt anticipation of correction by and dialogue with other individuals and that this game of exchange and mutual adjustment must be played by the rules if a common support base for action is to evolve. If innovation and creative genius are valued more than tradition, uncertainty avoidance is pushed back and independent subjects start viewing themselves as turning points of change, which leads to character traits like *voluntarism* and a *sequential* approach to dealing with time, or planning (also mentioned by Trompenaars). Here an *entrepreneurial attitude* – which would now appear to be value laden in itself – appears on the scene, directed towards the future.

Traditional and agricultural societies generally show a combination of hierarchical and collectivist/traditional relationships. Segregated communities at the bottom (villages, tribes, casts) that were collectivist and traditional in character were only united at the top by means of hierarchical, military, elitist rule (Gellner 1997; Gupta 2007; Van der Pijl 2007). Although even these societies allowed for egalitarian and democratic relationships at the top (aristocratic rule like in the Greek city-states), it is in general the West which made the leap forward towards democratic rule involving all citizens (Ferguson 2011; Rosenstock-Huessy 1993; Gauchet 1985; Winkler 2010; Fukuyama 2011). In these societies, where egalitarianism and innovation were highly valued, as has already been pointed out, new non-traditional and non-hierarchical cooperation mechanisms needed to be installed. In plain language: in these societies people were supposed to cope with a plurality of opinions and views without state intervention. Accordingly they needed the capacity to find a common support base for action at all levels of society, both at the bottom and at the top, not by force or by tradition but by means of free and voluntary effort. This presupposes the capacity to establish a common support base despite a continuous pluralism of opinions and preferences. In essence, this is what the emergence of civil society is all about; finding a support base for action away from the state hierarchy but also away from traditional collectives (Stackhouse 1984). Such creation of a support base for common action involves temporary groupings and re-groupings of individuals and organizations. Accordingly group loyalty and state authority recede into the background (Stackhouse 1984; Rosenstock-Huessy 1993). If such civil society does not develop the society in question may end up deadlocked because the cultural repertoire to create equilibrium between innovation and tradition is lacking. It will appear to be too controlled either by the state or by traditional groups, ethnic loyalties, patrimonial systems, and so on. In Africa both cases apply as collectives, either based on or not based on ethnic loyalties, have encroached upon the state many times thus making it subject to group interests, and to the controlling of the rest of society on that basis.

That said, it should also be noted that a civil society does exist in different ways and to different degrees (Wiarda 2003). Here, too, a path dependent trade-off is possible and desirable between either more control from above or more cooperation from below. For instance, if indeed the United States of America is the most democratic society on earth, such democratic openness will, at the same time, be counterbalanced by the role of the central authority in the form of the presidential powers, which are much greater than presidential powers in, for instance, European countries. What is decisive is the question relating to the direction in which societies develop. Checks and balances in governmental affairs and transparency (Collier 2007), universalistic rule without privileges (Ferguson 2011; Fukuyama 2011) and free competition without state control (Mahbubani 2008) all introduce some civil society traits, even if no representative government is fully in place. Sometimes a civil society only functions at the level of the marketplace (Sen 1999) and only to a very limited extent in politics or not at all, as is the case in China. Western states and donor agencies are often blamed – and rightly so – for merely being focused on free elections as a proof of democratic governance (Collier 2007; Kasfir 1998). There are, however, more and other requirements that need to be met in order to make a society free and open.

10.4 Systems Approach

The negotiation and establishment of different cultural value priorities does, in itself, need to be placed within the framework of technological system innovations and regime shifts in relation to the broader societal context. What is important here, is that large-scale technological innovations have a dynamics of their own so that, ultimately, the entire society is involved and this needs to be taken into account when introducing technology for development (see for examples in the water domain: Bressers and Kuks 2004; Kissling-Näf and Kuks 2004; Kates and Burton 1986). It is specifically, the "socio-technical system" concept that helps one to analyse and understand multiple types of large technological systems and the inherent tensions while also revealing the possibilities and constraints regarding the management of such large systems. Three system mechanisms are of particular importance (Hughes 1983; Ravesteijn et al. 2002):

- The "momentum" or path dependency which constrains the operating space.
- The "load factor" which points to the maximum use the system can deliver and which could lead both to "lock in" and to system transition.

- The "reverse salient" component; the obstacle which prohibits the continuation of the system, translated into one or more "critical problems" that need to be solved in order to open perspectives for the future.

Such systems always operate in some social environment or another and Hughes' (1983) characterization of the interrelationship between the system and the environment is stimulating. On the one hand, he maintains that the system boundaries are determined by the influence of one central actor whilst he shows, on the other hand, that the system environment initially largely shapes the emerging system. In later stages of development it is the system which, in its turn, has a determining influence on the environment, though the environment still enables its existence (Van Vleuten 2004).

What is essential here is that from a systems approach, technological change involves a range of actors and so for that reason a broad support base is indispensable if lasting change is to occur (Ravesteijn and Kroesen 2007). This raises a crucial question: how does a particular society create a support base for large-scale change? If change does not ensue from an all-powerful state, if the state has become just one actor among many, it would appear that somehow the different actors themselves should make the difference through their reciprocal relationships. In other words, a systems change also demands cultural transition; a change in opinions and attitudes at all levels. If different social actors do not succeed in creating bonds of cohesion in support of profound change no progress will be made in the adopting of large-scale technologies. If state initiative cannot bring this about then somehow and to some degree voluntary action on the part of many civil society actors will have to come into play. If this does not succeed, one or more of the actors may turn to violence as a last resort and impose its preferences on the other stakeholders. Violence, however, brings change which is not based on any sort of broad support and is therefore not lasting. In such cases society falls apart.

Two cases will be put forward to corroborate the above analysis and put it to the test, the first is a small-scale project on solid waste management in Nairobi whilst the second is a large-scale project focused on water management in the Senegal River. The present authors maintain that both require cultural transition and system change if they are to be successful.

10.5 Case I: Solid Waste Management in Nairobi

In Nairobi internship students engaged in an internship program at TU Delft studied the operations and business models of a number of solid waste recycling companies, the aim being to find an optimal business plan for a start-up company (Alberts et al. 2010). By conducting in-depth interviews they studied 13 companies involved in solid waste collection in Nairobi. This case description is based on their research findings. It appears that a business involved in the recycling of solid waste has to operate in a challenging and complex environment. It should, however, be noted that this case does not stand on its own. Many other cases drawn from this internship program convey more or less the same message.

The City Council of Nairobi (CCN) charges every area in Nairobi for waste collection. But the CCN does not have the resources to collect all the waste. Since the 1990s the quantity of waste has increased but the number of collection trucks has decreased. There were 1,000 tons of waste per day and 100 trucks in the 1970s and 1980s, but by the 1990s the waste had increased to 1,600 ton per day and the number of trucks had dropped to 40. By the time of the study 2,400 tons of waste was being produced each day. Hence the reason that 80 % of all the waste is currently being picked up either by community-based organizations or by small and medium-sized enterprises.

Although there are many laws and policies related to solid waste management, they are not working. The CCN has divided Nairobi into different waste collection zones. Officially the collectors can only pick up waste in the zones assigned to them, but due to any lack of rule enforcement in practice the collectors can collect wherever they want. Most collectors, especially the smaller ones, simply collect waste in their neighbourhood. Plastic is the most valuable product for recycling. Occasionally rival waste collection firms pick up all the plastic so that the area's normal collectors have to go elsewhere. A lot of areas in Nairobi are only partially served or not served at all by the CCN or any other collection company. Nevertheless competition in the solid waste sector is really stiff, precisely because it is not regulated. Illegal dumping is not prevented or controlled by the CCN and so that is one of the causes of unfair competition. Another reason why companies are afraid to invest is because of the uncertain future in terms of the legislation on waste collection companies and truck requirements. Other complaints about the CCN are that sometimes plastic recyclers have to pay a corruption fee and that the CCN imposes top-down decisions without consulting the collecting companies, which leads to legislation and statistics based on theory instead of on practice and experience. There is no cooperation between the collection companies and the CCN and there is no cooperation between the waste collection companies themselves either.

After collecting, the companies start separating the plastic. Sometimes this is even already done during transport. In order to increase the value, plastics are sorted according to colour and type. After that the plastic is washed in a big tank and cut into pieces by a machine before finally being put into bags. It can be sold in minimum volumes of one and a half tons. There are many companies working with the help of community-based organizations that collect the waste, after which it is taken to the collection depot. Most of the companies also seek to help the community and create jobs. This is the reason why many companies employ boys from the streets.

One of the biggest solid waste management challenges is the lack of organization which means that waste piles up in the streets. In addition there are other challenges for small companies, like delayed payment on the part of customers, the high operation costs, poor infrastructure, CCN harassment, the high price of operation licenses, all the recycling obstacles and insecurity at the dump sites. There is a lack of protective gear as well as a lack of equipment, and space is limited.

At one company, RIREC, the workers received a salary of 200 Ksh per day (around €2), including lunch. The salary was paid per week and the employees only got paid for the days they actually worked. The manager kept careful note of the numbers of days worked. The employers at that company hired 50 street guys to collect plastic. The group was divided into five teams of ten street guys and each team would collect for 1 week and then the next week the other team would collect. Unfortunately most of the collectors were taking drugs which made them unreliable workers. Sometimes they also fell ill and needed to be taken to hospital, all of which made the project expensive. When finally the company got into financial difficulty the street workers were sacked which triggered aggression in front of the compound and the police had to intervene. That company's biggest present problem is the financing. The project is not financially self-sustaining so new sponsors are being sought and in the meantime the project has been suspended.

Another company, Vijana Kwa Mazingira, is an umbrella company which runs several programs. These include tree planting but also waste collection. Vijana has about 320 clients from which waste is collected. Some 300 of them are private residents and they pay about 300–500 Ksh every 2 months. Twenty of the clients are commercial and they pay about 1,000 Ksh per month. They are sent invoices to remind them to pay for the service. The company drives with a rented truck to the Ngong dump. The round trip costs about 500 Ksh. Vijana also buys plastic from the collection points for around 15–18 Ksh per kilogram. Competitors leave deposits at the collection places to make sure that the collectors will save plastic garbage for them, but Vijana is too small for that which puts it at a disadvantage.

There have been problems with the weight of the garbage. The people selling the plastic often use corrupt measuring methods. At the collection place the weight measured might be 500 kg, but once they arrive home they discover that the weight is perhaps only 300 kg. Eventually they were allowed to use their own weighing equipment. Vijana's tariffs were relatively high. Most of the personnel got a salary from the company, but the manager himself complained that he had nothing left at the end of the month. Licences had to be paid, computers had to be purchased, a loan had to be paid back and then there were the monthly costs for electricity and water. Vijana also buys garbage for 15,000 Ksh per week and sells it for 30,000 Ksh per week. So that is 100 % profit.

The students came up with a plan for a start-up company. Their advisory business plan proposed starting on a small scale, not buying land and sorting the waste in the collection truck. To that end, a truck needed to be bought (cheaper than hiring) and apart from the manager and the driver two sorters would be hired. With 85 residential and commercial clients such a company would be able to realize a modest positive revenue. Included in their final recommendations was a small amount of payment for the street boys in the neighbourhood in return for protecting the truck from being damaged and to allow for investments in such things as second-hand but not completely worn out tyres for the truck which would be less expensive and more reliable than buying new tyres, the quality of which could not be guaranteed.

10.6 Case Analysis I

Several features in this case description are striking. We shall mention the following:

1. There appears to be a large *power and status distance* between the CCN and the waste collection companies. The CCN sets standards and introduces regulations without any form of consultation.
2. The CCN is treating companies in an *arbitrary* way. The companies do not know what to expect. As a consequence, they are hesitant to invest. The CCN is even harassing them, it is imposing regulations, which at the same time it is not maintaining and enforcing, and rules based on *particularistic relationships* – call it corruption – so as a result it is highly ineffective.
3. The waste collecting companies themselves do not, however, *cooperate* in any way, they only protect their own interests. They do not keep to the designated areas of waste collection and they cheat by using false measuring equipment etc.
4. They do not use sophisticated technology for the separation of waste and they do not have the *capacity* for any professional form of waste management nor do they seek support to attain that goal, instead they take the existing practice for granted (*traditionalism, uncertainty avoidance*).
5. However, they are often socially motivated to help unfortunate young people to get a job. They work – often supported by community-based institutions or NGOs – with street guys who also lack *capacity*: they are not very motivated or educated, do not work in a reliable way and may even steal equipment if not carefully watched. Sometimes they are even in need of medical care.
6. There are cost recovery problems. In general it appears that the *lack of trust and accountability* at the bottom end of society is reinforced by the *lack of law enforcement and regulation* from above.

In such an uncertain environment the students could only come up with a proposal to start a very small company. The social environment is too uncertain and regulation too weak for a sustainable investment climate to emerge. Even the one truck to be bought has to be one with second-hand tyres. This, once more, is an indication of the *lack of anonymous trust*, thus leading to deception and bad products. In all transactions in Kenyan society this is a recurring problem: it is recommendable upon entering Kenya to get the phone number of a reliable taxi driver from a reliable friend. In general, people do not buy products because they trust a particular company, but rather because they trust an individual who works at the company. This lack of anonymous trust enormously reduces the number of possible transactions. Obviously it is related to the lack of a *civil society*. Collectivist groups, ethnic or not, tend to close themselves off from other collectivist groups with only personal relationships creating an opening. Otherwise the door remains closed. Since government bureaucracy functions in the same way, it becomes a vicious circle: there is no universalistic rule from the side of government, which could alternatively opt to create a society of equal rights and *equitable justice*. In turn, collectivist groups tend to use the state bureaucracy to promote their own group interests (OSSREA 2009). They do not force the state as a countervailing power to become more democratic and open, and less particularistic, as Western donors would so often wish (Eberly 2008). Although the situation is slowly changing, and although many Africans do try to make their leaders more accountable, the old mechanisms are stubborn and recalcitrant. In terms of our diagram above one might argue that *hierarchy* and *collectivism* do to a large extent overrule openness towards *change* and *egalitarian cooperation*. In terms of system dynamics, it is difficult in these circumstances to create a large support base for system change. The *reverse salient* consists in a culture with too much collectivism (in reality a plurality of collectivisms closing themselves to each other) and too much hierarchy: the combination of both hampers further development. The *critical problem* that needs to be solved is that of how to create a civil society in which people can regroup and alter membership at the bottom whilst having this sanctioned by universalistic rule and equal access/treatment from the top.

10.7 Case II: The Senegal River, Modernization without Cultural Change

After an intense drought in the 1970s the *Organization to Enhance the Senegal River* (OMVS), including Mauritania, Mali, and Senegal, launched an integrated development programme to improve the management and exploitation of the Senegal River basin. This led to the construction of two big dams in the Senegal River, one more than a thousand kilometres inland, the Manantali Dam, that was completed in 1988 and the other 25 km from the sea, that was finished in 1986, the Diama Dam. The Diama Dam was designed to prevent salinization by keeping out seawater. Before the construction of this dam the Senegal River had a rich flora thanks to the saltwater intrusion far inland. The two-dam system was created for purposes of irrigation, electricity generation and year-round drinking water provision. Electricity generation only started in 2002, but before that time many problems had already emerged (Malick N'Diaye 2007).

These problems entailed social conflict between different user and population groups as well as conflict between the planners and the traditional users: agriculturalists, animal breeders and the fishing population. The initial agricultural goal was to secure efficient and large-scale agricultural rice production involving massive irrigation operations. Until the building of the dams, land use rotated between the three groups mentioned above and was regulated by yearly flooding. Nature and not human intervention gave every group its turn, if they were lucky. But suddenly man rather than nature took charge which meant that a more sophisticated management system would need to be put in place and that was not so easy (Varis and Fraboulet-Jussila 2002). The usual agricultural procedure involved flood recession farming, which meant that the farmers would plant their crops in the flooded lands while the water was receding, thus taking advantage of the water supply as long as it lasted. With the arrival of the dry period cattle herders would move in. Irrigation, it was estimated, would make possible large-scale rice production for a national and international market. However, the needs of the local farmers were not really considered as they were not included in the planning and building process right from the very beginning (Adams 2000) so there were unexpected and undesired effects.

Although irrigation had come under human control, the water discharge from the Manantali Dam proved insufficient and irregular. This was not so much linked to water availability as to poor management (Rasmussen et al. 1999). Once the farmers had planted their rice, a sudden deluge would, for instance, provide too much water thus spoiling the harvest. Alternatively there was sometimes too little water, so that the quantity was altogether insufficient for rice cultivation. The planners prioritized keeping the level of water in the reservoir high and conducting technical experiments with it instead of allowing it to serve the farmers in the valley. Their status, they maintained, was higher and so they knew what was best. Only after several years of bad experiences and lost harvests did the government intervene and decide to keep the valley artificially flooded for a longer period of time and to a greater water depth. This, though, did not solve all the problems. The improved technology made irrigation farming attractive, not only to the local farmers, but even more so to upper-class city dwellers who had better access to the state authorities and to investment capital; they tended to be traditional leaders, traders, or men working abroad who made use of traditional family-based land rights to gain access to commercial agriculture. This gave rise to a massive influx of new farmers but the result was that the original independent farmers were virtually turned into mere land labourers due to pauperization. New problems arose during the 1980s following the introduction of policies of self-reliance and the withdrawal of subsidies. This was accompanied by other problems, for instance, regarding loan repayment which farmers had not been familiar with under the relatively amicable development regime of earlier times.

On the northern side of the Senegal River, in Mauritania, the situation was even worse. Irrigation farming attracted nomadic Arab tribes, the dominant class in the country, who came from the north. The government gave them access to all the land, not only expropriating Mauritanian landlords, but also the Senegalese black tribal groups living to the north of the river. The pressure of the drought of the 1970s had forced most of them to settle in the area as peasants. In supporting them the government ended traditional collective land tenure laws and introduced land tenure privatization which was designed to promote economic development and attract private investment. Conflict arose in April 1989 on the border between Senegal and Mauritania, thousands of people were massacred on both sides and about 130,000 black Africans were deported to Senegal. Because of this conflict between Senegal and Mauritania two parallel electricity lines were constructed on each side of the river, thus heavily pushing up costs (Malick N'Diaye 2007).

10.8 Case Analysis II

In this particular case one sees that different levels of government are involved: national and transnational and that there are conflicting groups and interests but, above all else, different values and value laden institutions.

1. The technical measures involving the building of huge dams, the introduction of agriculture regulated by irrigation and electricity production, were intended to improve life, to facilitate larger scale production and to stimulate international competition. Despite all the good intentions the system change ultimately bounced on a *reverse salient*, characterized by traditional *cultural patterns and value systems*. None of this was taken into account in the planned system transition. What was initially intended to provide a solution completely backfired and became a problem.
2. Traditional *hierarchical patterns* and the corresponding stratification of society caused specific classes and groups to take advantage of the new technologies. The priority given to electricity generation for the city at the expense of irrigation made things worse, both for the traditional poor farmers and for the new industrial farmers.
3. Competing *group identities*, national and tribal, turned the new burgeoning production capacity into a scarce commodity, precipitating competition and struggle.
4. Access to governmental bureaucracy turned out to be a decisive key to success, again in the absence of universalistic rules and neutral roles, and due to a lack, too, of an open civil society system. Because different collectivist groups took advantage of the state bureaucracy for their own collectivist purposes, neither a state characterized by *rule of law* nor a civil society characterized by multiple memberships and *shifting loyalties* was able to develop.

Here too, besides a system change, cultural transition would be desirable. As a rule, large-scale technology requires large-scale cooperation between different stakeholders. This, in turn, necessitates attitudes of wheeling and dealing with each other, which may very well be in place internally in one particular group (often a tribal group), but often does not exist between different groups. A negotiated approach and a corresponding change in values and communication patterns could have brought about the change required. If the initial plans had not been characterized by a technological fix but had taken into account the social context, and the cultural transition required, then proactive measures would have been possible, thus preventing the massive clash leading to so many deaths and to the stalemate controversy in the wake of the clash that persists to the present day.

10.9 Conclusions

In both cases it is clear that no change is not an option. The population explosion in Africa in general as well as the desire of its people to share in modern standards of health and wealth, make it impossible to continue traditional ways of life. That is not merely an evaluation from the point of view of Western "cultural imperialism". It is an unavoidable consequence of the introduction of large-scale technology. That said, it appears that several lessons can be learned in terms of what ought to be done and what should not be done to make such a transition possible and to innovate in a responsible manner.

1. The introduction of a *technological fix should be avoided.* Existing cultures cannot just be modernized by introducing large-scale technology. Mere technological modernization, though the intentions might well show respect for and non-interference in the autonomous development of the indigenous culture, easily becomes counterproductive. It catapults these cultures violently into the modern age. In Senegal the technical options might have worked, if at the same time a civil society had suddenly emerged based on an open association of individuals – if this could possibly have been a feasible option. But group identities and unequal access to hierarchies and bureaucracies stood in the way. In the Kenyan example it would not have helped much to put modern recycling technology in place without first carefully preparing for the required scale change and building the necessary institutions at all levels. Cultural change should therefore be part and parcel of the technological leap if the requirements of responsible innovation are to be met and explicit value trade-offs are to be made.
2. It is desirable to organize a conscious and deliberate trade-off and maintain respectful *dialogue between the old and the new.* The strong influence of group identities and the importance of relationships within government bureaucracies should not be viewed as evidence of backwardness and corruption but rather as the negative consequences of old collective value systems suddenly finding themselves in a new context. In former times, in agrarian or nomadic cultures, such a value system worked perfectly well but in Senegal a traditional society was exposed to modernity without respectful dialogue and mediation. In this clash of two sets of values the traditional system then appears to stand in the way of open cooperation and anonymous trust, universalistic rules and neutral roles, and so on. The Kenyan solid waste example conveys the same message.
3. A *carefully orchestrated sociocultural transition* should have accompanied the implementation of large-scale technology to manage the water flow in Senegal. The fact that the new water resources and arable land would become a source of strife and a scarce commodity to argue about could then have been foreseen. Though not identical, much the same can be said about the Kenyan case. Collectivist identities at the grassroots and top-down hierarchical attitudes (including status by position instead of by achievement) from the side of the government agencies stands in the way of universalistic rule, egalitarian exchange and civil

society-style coordinating mechanisms. This furthermore stands in the way of the modernization of the recycling process and waste treatment in general. One cannot change one without changing the other.

4. *Large-scale cooperation between different actors* is both difficult and unavoidable if large-scale technology is to occur. Whatever the implicit script of any technology might be, large-scale technology unavoidably demands large-scale cooperation between different people, interests and values, and without this it cannot function. The point is not to deny the historical value of African traditions nor is it expedient to do away with those traditions altogether. The point is simply that these values need to be integrated within the larger framework of values inherent to civil society or, if one prefers, the values of civil society have to be integrated into the traditional communal framework. The result envisaged is the same, it is to create a path-dependent step forward while preserving an equilibrium between the old and the new.
5. The concept of *an open civil society* is central to the process of change. The fact that this is the case may be corroborated by the knowledge that in Western history each step forward in the technological journey was also accompanied by progress in the self-organization reflected in a civil society (one need only think of medieval cities, German princes and English nobility) so that the emergence of a civil society, the large-scale implementation of technology and economies of a larger scale went hand in hand (Rosenstock-Huessy 1993; Ferguson 2011; Sassen 2006; Keane 2001). Step by step European society moved away from its original 'tribal' structure. Change is unavoidable, but it should be accompanied by respect so that a culture or societal group is not plunged into modernity by merely transplanting technology. Systematic renewal is what is required, responsible innovation, together with a conscious and deliberate integrating and reconciling of older layers of culture into new ways of life in a negotiated path forward.

This leads to our main conclusion to the effect that system transitions and value trade-offs, *cultural transitions*, or whatever one may wish to call it, should be part of *integral sustainable development*. In order to achieve optimal distributive justice different, often culture-dependent value priorities and interests, need to find a path-dependent equilibrium and/or trade-off. This way forward cannot be designed from the outside. It should be lived with and experimented with from within by the peoples and societies involved.

References

Adams, A. 2000. The Senegal River: Flood management and the future of the valley. IIED, http://scholar.google.nl/scholar?q=Flood+Management+and+the+Future+of+the+Valley+Adams&hl=nl&as_sdt=0&as_vis=1&oi=scholart. Consulted 19 June 2011.

Alberts, A., J. Van Huijstee, and M. Vogel. 2010. *Business plan AFRIWAC: Solid waste management in Nairobi*. Students internship report TU Delft.

Bressers, H., and S. Kuks (eds.). 2004. *Integrated governance and water basin management*. Dordrecht/Boston/London: Kluwer Academic Publishers.
Calderisi, R. 2006. *The trouble with Africa; why foreign aid isn't working*. New York: Palgrave McMillan.
Chang, H.J. 2007. *Bad Samaritans – The guilty secrets of rich nations and the threat to global prosperity*. London: Random House.
Collier, P. 2007. *The bottom billion: Why the poorest countries are failing and what can be done about it*. New York: Oxford University Press.
de Jong, M., K. Lalenis, and V.D. Mamadouh (eds.). 2003. *The theory and practice of institutional transplantation experiences with the transfer of policy institutions*, GeoJournal library, vol. 74. Dordrecht: Kluwer Academic Publishers.
Eade, D. 1997. *Capacity-building: An approach to people-centered development*. Oxford: Oxfam Development Guidelines.
Eberly, D. 2008. *The rise of global civil society: Building communities and nations from the bottom up*. New York/London: Encounter Books.
Ferguson, N. 2011. *Civilization, the west and the rest*. London: Penguin.
Fukuyama, F. 2011. *The origins of political order*. London: Exmouth House.
Gauchet, M. 1985. *Le Désenchantement du Monde, une Histoire Politique de la Religion*. Paris: Gallimard.
Gellner, E. 1997. *Nationalism*. New York: New York University Press.
Gupta, D. 2007. *Mistaken modernity, India between Worlds*. New Delhi: HarperCollins.
Hofstede, Geert. 1997. *Cultures and organizations; software of the mind*. New York: Mc Graw Hill.
Hughes, T. 1983. *Networks of power. Electrification in Western society 1880–1930*. Baltimore: The Johns Hopkins University Press.
Kasfir, N. 1998. *Civil society and democracy in Africa, critical perspectives*. New York: Routledge.
Kates, R.R., and I. Burton (eds.). 1986. *Geography, resources and environment, vol. 2: Themes from the work of Gilbert F. White*. Chicago: The University of Chicago Press.
Keane, J. 2001. Global civil society? In *Global civil society – 2001*, eds. H. Anheier, M. Glasius, and M. Kaldor, 23–47. Oxford: Oxford University Press.
Kissling-Näf, I., and S. Kuks (eds.). 2004. *The evolution of national water regimes in Europe*. Dordrecht/Boston/London: Kluwer Academic Publishers.
Landes, D.S. 1998. *The wealth and poverty of nations: Why are some so rich and others so poor*. New York: WW Norton.
Mahbubani, K. 2008. *The New Asian hemisphere*. New York: Perseus.
MalickN'Diaye. 2007. Construction in the Senegal River valley and the long-term socioeconomic effects. In *Knowledge, technology and policy*, ed. Martin de Jong, 44–60. New Brunswick: Rutgers University.
Moyo, D. 2009. *Dead aid, why aid is not working and how there is another way for Africa*. London: Penguin.
Nussbaum, M.C. 2006. *Frontiers of justice, disability, nationality, species membership*. Cambridge, MA: Harvard University Press.
OSSREA (Organization for Social Science Research in Eastern and Southern Africa). 2009. *Good governance and civil society participation in Africa*. Addis Ababa: OSSREA.
Pogge, P. 2008. *World poverty and human rights*. Cambridge/Malden: Cornwall Press.
Rasmussen, K., N. Larsen, F. Planchen, J. Andersen, I. Sandholt, and S. Christiansen. 1999. Agricultural systems and transnational water management in the Senegal River Basin. *Geografisk Tidsskrift – Danish Journal of Geography*, Bind 99, http://tidsskrift.dk/index.php/geografisktidsskrift/article/view/2447/4331. Viewed 29 Nov 2012.
Ravesteijn, W., and O. Kroesen. 2007. Tensions in water management: Dutch tradition and European policy. *Water Science & Technology* 56(4): 105–111.
Ravesteijn, W., L. Hermans, and E. Van der Vleuten. 2002. Water systems. Participation and globalisation in water system building. *Knowledge, Technology & Policy* 14(4): 4–12.
Rosenstock-Huessy, E. 1956. *Soziologie, Vol. I Die Übermacht der Räume*. Stuttgart: Kohlhammer.

Rosenstock-Huessy, E. 1958. *Soziologie, Vol. II Die Vollzahl der Zeiten*. Stuttgart: Kohlhammer.
Rosenstock-Huessy, E. 1981. *Origin of speech*. Vermont: Argo Books.
Rosenstock-Huessy, E. 1993. *Out of revolution – Autobiography of western man*. New York: Argo Books (original 1938).
Sachs, J.D. 2005. *The end of poverty*. New York: Penguin Press.
Samli, A.C. 2009. *International entrepreneurship, innovative solutions for a fragile planet*. New York: Springer.
Sassen, S. 2006. *Territory – Authority – Rights, from medieval to global assemblages*. Woodstock: Princeton University Press.
Sen, A. 1999. *Development as freedom*. New York: Anchor Books.
Stackhouse, M.L. 1984. *Creeds, society and human rights, study in three cultures*. Grand Rapids: Eerdmans.
Trompenaars, F., and C. Hampden-Turner. 1999. *Riding the waves of culture*. London: Brealey.
van der Pijl, P. 2007. *Nomads, empires, states, modes of foreign relations and political economy, volume I*. London: Pluto Press.
van Vleuten, E. 2004. Infrastructures and societal change. A view from the large technical systems field. *Technology Analysis & Strategic Management* 16(3): 395–414.
Varis, O., and S. Fraboulet-Jussila. 2002. Water resources development in the lower Senegal River basin: Conflicting interests, environmental concerns and policy options. *International Journal of Water Resources Development* 18(2): 245–260.
Wiarda, Howard J. 2003. *Civil society – The American model and third world development*. Amherst: University of Massachusetts.
Winkler, H.A. 2010. *Geschichte des Westens*. München: Beck.

Chapter 11
Sustainable Innovation, Learning and Responsibility

Udo Pesch

Abstract This paper takes as a starting point that it is a broad societal responsibility to stimulate the development and uptake of sustainable innovations. In order to pursue this societal responsibility, insights derived from systems' approaches to sustainable innovation will be connected to reflections on responsibility. A core understanding of systems' approaches is that actors from different institutional domains have to create a shared future orientation that directs innovation in a desirable way. This implies that boundaries between institutional domains have to be broken down, while these boundaries have been erected to hold individual actors accountable for their actions and decisions. The tension between these conflicting responsibility claims will be addressed here and described against the background of a number of societal developments that not only complicate the facilitation of sustainable innovation via the development of shared future visions, but also present new challenges for reflections on responsible innovation.

11.1 Introduction

One of the areas in which the need for responsible innovation is felt most strongly is sustainable development. Continuing the use of current technology will further contribute to environmental degradation, such as global climate change, the depletion of resources, and the pollution of our life world. It is estimated that in order to realize a society that can sustain itself, technologies that are a factor 20 more eco-efficient have to be in place (Weaver et al. 2000; Mulder 2006).

Although the field of environmental ethics still appears to lack a coherent framework (Carter 2007; Gardiner 2004), it does not seem out of line to say that

U. Pesch (✉)
Faculty of Technology, Management and Policy, Department of Values, Technology and Innovation, Delft University of Technology, Delft, The Netherlands
e-mail: U.Pesch@tudelft.nl

J. van den Hoven et al. (eds.), *Responsible Innovation 1: Innovative Solutions for Global Issues*, DOI 10.1007/978-94-017-8956-1_11,

it is an encompassing societal responsibility to contribute to the development and implementation of new, *sustainable technologies* (Jonas 1985; Giddens 2009; Beck 1992), which can be defined as technologies that fulfill the needs of mankind without the use of non-renewable resources, and without creating large scale, and/or irreversible damage (Mulder 2006). The question is how ethical reflection can contribute to the effectuation of this broad societal responsibility. How can we think in moral terms about the stimulation of sustainable innovation?

In the field of engineering ethics, the development of sustainable technologies is often framed either in terms of an individual moral duty of technology developers, or in terms of principles that can be integrated into the design of new technologies (Brumsen 2011; Michelfelder and Jones 2013; Basart and Serra 2013). This suggests a prevailing focus on the design phase of a technology. However, if we look at so-called systems' approaches to sustainable innovation, such a constrained focus appears to have some shortcomings, because systems' approaches claim that technology is not developed in isolation, but co-evolves with society, and as such we need to address the development of sustainable technologies as an integrative part of society – we have to approach technologies as part of so-called *socio-technical systems* (Geels 2002; Rip and Kemp 1998; Van De Poel 2000). In systems' approaches, the difficulty of developing sustainable technologies is not so much attributed to the lack of promising technologies or unwilling engineers, but to the presence of institutional and technological 'lock-in' (Unruh 2000; Rotmans and Loorbach 2009): the linkages between actors and technologies are reproduced in dominant practices, power positions, and new technologies – obstructing the proliferation of alternative, potentially more desirable technologies. The focus on the design phase of a technology therefore appears to be too one-sided, as it predominantly addresses the question how to embed societal values in new technologies, but not the question how to facilitate the societal uptake of desirable technologies.

This paper aims to frame the systems' perspective on sustainable innovation in terms of responsibility. It will do so by featuring *institutional domains* as social contexts that structure the decisions of individual actors, and as such have a distinct normative impact – in other words, they can be seen as 'accountability structures' (Van Gunsteren 1994; Bovens 1998; Pesch 2005). As we will see, using institutional domains in order to connect insights from systems' approaches to reflections on responsibility will bring about new moral and societal problems. These new problems involve a fundamental tension between two moral goods, namely sustainability and individual autonomy, because on the one hand, sustainability demands integration of separated societal spheres, while, on the other hand, the warrant of individual autonomy appears to be very much based on the compartmentalization of society in distinct institutional domains.

To address and question this moral tension, we will have to deal with a range of topics in this paper, which has the following outline. In Sect. 11.2, a number of insights will be derived from systems' theories on sustainable innovation. A core understanding that can be extracted from these theories is that boundaries between institutional contexts should be crossed somehow in order to facilitate sustainable innovation. Stakeholders and actors from different institutional domains, have to *learn* about the world views, problem perceptions, and so on, of other actors, and with that they can create a shared vision or a common idea about the future that directs innovation processes into a sustainable pathway. Hence, it can be inferred that the moral duty to facilitate sustainable innovation implies that boundaries between institutional domains have to be broken down by developing future orientations that are shared by actors from different institutional angles.

Removing institutional boundaries, however, brings about other issues that pertain to responsibility – especially because the institutional domains of market, politics, and science have a profound connection with the way responsibility and accountability is structured in modern society. In order to uncover these issues, Sect. 11.3 will give an ideal-typical description of the modern constellation of institutional domains, consisting of the market, state, and science. It will be contended that this constellation has been designed to warrant the autonomy of individuals, and to enable the capacity to hold individual actors accountable for their behavior. As accountability structures, institutional domains hand over the rules as well as the structure of these rules which individual actors have to follow. Hence, the idea of breaking down boundaries, as is promoted by systems' theories on sustainable innovation, appears to conflict with some of the most important accountability arrangements in modern society – were it not if the original goals of institutional domains have been subject to 'erosion' to start with. In other words, the tension between sustainability and autonomy knows more layers, adding even more complexity to the issue of responsibility and sustainable innovation.

Sections 11.4, 11.5, 11.6, and 11.7 will expand on developments that have contributed to this erosion of the original goals of institutional domains, which have led to a situation in which individuals are more and more confronted by demands posed by different institutional domains, and hence are faced by heterogeneous and conflictive sets of values. This situation makes it hard to assess the legitimacy of one's actions and decisions, but it also obstructs individuals to sharpen their moral intuitions. The capacity of overarching institutional domains to be responsive to the moral intuition of an individual has declined, which might be one of the biggest contributors to today's societal problems.

In the concluding section, the findings of the paper will be reflected upon. First, the implications of our analysis for methods connected to systems' approaches will be explored. Second, the ramifications for understanding responsible innovation, including sustainable innovation, will be addressed.

11.2 Learning for Sustainable Innovation

Systems' approaches to innovation include a range of different perspectives and theories, such as multi-level perspective and transition theory (Geels 2002, 2004; Kemp and Rotmans 2004; Smith et al. 2005), and innovation systems (Kamp 2002; Hekkert et al. 2007). The theoretical basis of these approaches is formed by so-called ‘quasi-evolutionary’ descriptions of technological change, which feature technological development as a result of a confrontation of different technological options developed by a variation environment (manufacturers, designers, producers, etc.) on the one hand, and the choice for a subset of these options by a selection environment (consumers, users, regulators, etc.) on the other hand. Essential here is that the variation and selection environments are not mutually independent, like in genuine evolutionary processes, but that there are feedback loops and linkages that have a profound effect on the development and societal implications of new technologies (Van den Bergh et al. 2006).

These linkages are for a great part constituted by rules, practices, expectations, routines, etc., that surround existing technologies. One may say that technologies are embedded in webs of significance which at the same time are reproduced by these technologies. Such webs of significance can be defined by the concept of ‘regime’:

> A technological regime is the rule-set or grammar embedded in a complex of engineering practices, production process technologies, product characteristics, skills and procedures, ways of handling relevant artefacts and persons, ways of defining problems; all of them embedded in institutions and infrastructures (Rip and Kemp 1998).

In terms of quasi-evolutionary theory, the presence of a regime establishes a repetition of the same connection between the variation- and the selection-environment; which could lead to a self-reinforcing pattern that becomes hard to avoid – a situation that is characterized as ‘lock-in’ (Unruh 2000). The most salient example of this situation concerns the use of fossil fuels as the basis of our energy provision, leading to all kinds of major problems, such as anthropogenic climate change and the depletion of natural resources. By all means it would be responsible to switch to renewable energy sources, however, this switch is hard to make due to the infrastructural and institutional dependencies that have been created in the fossil-fuel based energy system.

The definition of regimes given above is rather similar to the classic sociological version of ‘institutions’, which is used to describe social processes in which the practices of individuals and groups can be related to a shared body of understanding (Berger and Luckmann 1991; Geertz 1973; Fay 1975; Winch 2001). The notion of ‘regimes’ can be seen as an conceptual extension of institutional theory, as it emphasizes the relevance of socially constructed meanings and practices, which, in turn, are connected to technologies and technological systems. At the same time, the concept of ‘regimes’ stresses that the meanings and practices which enable and constrain technology development are not only embedded in *institutions*, but, as an expansion of the concept of institutions, also in *technological infrastructures*, which can be seen as an externalization and solidification of practices and bodies of meanings (Van De Poel 2000). In other words, ‘regimes’ accentuate the difference between ‘software’ and ‘hardware’, so to speak, in matters related to technology.

The challenge for scholars in the field of socio-technical systems is to construct methods that enable the ‘opening up’ of dominant regimes. In order to prevent the construction of undesirable future infrastructures, it is necessary that current infrastructures are transformed by the construction of new bodies of meanings that guide innovation processes. More concretely it means that long- and short-term orientations and expectations should be developed, which are shared and distributed by new networks and new institutions (Quist 2007; Van Lente 1993). An example of divergence of meanings and expectations that stalls the development of a new technology can be found in the case of biomass. From the 1990s onwards biomass has been put forward as an option for addressing the climate change problem. However, in recent years, the use of biomass for energy has been criticized increasingly. Discussions revolve mainly around the use of food crops for the production of biofuels, such as the use of palm oil and rapeseed oil for biodiesel and the use of corn and sugarcane for bioethanol. These types of biomass applications are controversial, because they may lead to an increase of food prices and a decrease of access to food for the poor. As a result of the controversies and the difficulties in determining sustainability of biofuel applications, the Dutch Minister of Environment decided to lower the compulsory use of biofuels in gasoline and diesel from 5.75 to 4 % in 2010 (Cuppen et al. 2010).

The construction of common future orientations and expectations is facilitated by the organization of *learning processes* among actors that represent these dominant regimes (Geels 2002; Verbong et al. 2008). In other words, learning is portrayed as a tool that can resolve complex social and ecological problems (Garmendia and Stagl 2010), by opening up the ingrained perceptions, values, practices, and mental routines, and allowing the establishment of a collective of body of meanings, including long-term orientations, to solve sustainability problems (Loorbach 2007; Quist 2007; Grin 2000; Cuppen 2009). It is important to observe that a collective future orientation does not automatically entail consensus; in order to be effective, these orientations should be broad enough to accommodate a certain amount of interpretative flexibility and diversity (Grin et al. 1997; Cuppen et al. 2010; Huitema et al. 2007).

Systems' approaches have come up with a wide array of methods that pursue the establishment of a shared body of meanings, especially by actively bringing together stakeholders from different societal angles (Te Kulve and Rip 2011; Rotmans et al. 2001; Schot and Rip 1997; Quist and Vergragt 2006). These participatory methods allow regime actors to be directly engaged in learning processes, so that these actors will be able to open-up their regime-induced patters of thinking and action; which should subsequently lead to a 'scaling up' of these new insights (Quist 2007; Brown et al. 2003). It has to be admitted that there are still quite some problems connected to the actual implementation of these methods, for instance, the question *who* will take the initiative to organize such participative projects remains quite an awkward issue (Pesch 2012; Meadowcroft 2009).

The account of learning presented above makes clear that systems' theories mainly focus on the 'software'-side of regimes: enhancing the learning of stakeholders is an attempt to integrate different bodies of meanings and problem definitions that are scattered over different institutions. The assumption it that by actively bringing actors together, barriers between institutions that produce different bodies of meanings are taken away. In other words, the development of collective future orientations is considered to be a prerequisite for the development of sustainable 'hardware'.

The institutions that are addressed by systems' approaches and the participative methods that are connected to these approaches are without exception formalized organizations, which, in turn, are very much structured by conditions conveyed to them by the broader societal contexts in which these organizations are found, namely the realms of market, politics, and science (Leydesdorff and Meyer 2003). These societal contexts that encompass a broad range of institutions and organizations that share a similar orientation will be featured here as *institutional domains*.

We may infer that the central premise that is given by systems' approaches to sustainable innovation is that the boundaries between institutional domains have to be overcome in order to develop a cognitive, institutional, and technological climate in which sustainable innovation is thriving. With that, the functioning of and interaction between institutional domains emerges as a relevant topic for further reflection, especially if one takes into account the moral ramifications that can be connected to institutional domains – as will be elaborated in the next section.

11.3 Accountability Structures

The normative role of the institutional domains of market, politics, and science, can be related to Max Weber's claim that modernity can be seen as an ever-expanding process of 'rationalization' (Weber 1972). This process refers to the eviction of otherworldly explanations of physical phenomena, as well as to the efforts to create a social structure in which personalized power relationships are replaced by universalistic, objective rules, so that no individual is bound by external authority.

The modern constellation of institutional domains is designed to serve this goal: the creation of different institutional settings has led to a society in which the legitimacy of an agent's actions and decisions of agents can be explicitly determined. Institutional domains can be seen as 'accountability structures', social contexts in which a reciprocal relation between the individual's moral intuition and a specific social context is warranted. Accountability structures allow individuals to be held accountable for their actions and decisions, by postulating conditions that enable a public to assess the validity those actions and decisions (Van Gunsteren 1994; Pesch 2005, 2008a, b).

In this section, the character of institutional domains as accountability structures will be presented in an *ideal-typical* way; our account involves an analytical reconstruction of the main elements and characteristics of the realms that constitute modern society, as well as it will present the mechanisms that create social order in these social realms, based on literature from philosophy, history, sociology, and political theory (Habermas 1999; Bobbio 1989; Poggi 1978; Gay 1973; Taylor 1989; Kunneman 1984). Later in this section, the empirical validity of these ideal-typical reconstructions will be qualified, but nevertheless the ideal-typical character of institutional domains given are recognizable in the discourses of both public and experts, and as such these reconstructions strongly influence the way we understand our world.

The modern accountability structures are the state, the market, and science. Each of these domains fulfills a specific role in society (Dahl and Lindblom 1963; Polanyi 2001). The realm of the state includes bodies such as government, parliament, and law. In liberal democracies, this realm is designed to establish and maintain collectively binding decisions that are aligned with the, admittedly enigmatic, 'public will' (Schubert 1960). Opposite to this public domain is the private domain of the market in which actors pursue the maximization of their economic self-interest. The dichotomous combination of these two domains is one of the basic ways to structure society, and in that sense the realm of science stands more or less separated from these realms. Instead of organizing society, science deals with the legitimacy of truth claims so that valid answers about the functioning of the world can be constructed.

In the realm of the state, actors are held accountable by several mechanisms, which relate to each other in a nested way (Montesquieu 2002). Government in liberal democracies, seen as the agency that executes the 'public will', is controlled by parliament. If government fails to live up to the job it has been designated to, parliament may decide to punish the government, for instance by sending away ministers. In turn, parliament has to represent the legitimate holder of the 'public will', which is the electorate. If parliamentarians fail to comply with the wishes of the electorate, they will be voted out of parliament during the next elections.

In the domain of the market, actors are held accountable mainly by the structure of the market itself. The presence of competition guarantees the functioning of the price mechanism, which, in turn, enforces suppliers of products to stick to prices that consumers are willing to pay (Smith 1998; Dumont 1977; Derksen et al. 1999).

The realm of science is predominantly characterized by the presence of explicitness. Both truth claims in themselves and the way these truth claims have been developed, have to be made object of external assessment, for instance in the form of the peer review system (Merton 1979).

In some important respects, institutional domains can be distinguished from mere institutions. The latter refers to any social context in which actors give meaning to their interaction, with which coordinated patterns of action may emerge (Berger and Luckmann 1991). This means that institutions bestow us with rules, values, and cognitive images. Institutional domains entail a more comprehensive scope; the rules provided by an institutional domain both encompass and constitute that whole domain. Moreover, these rules and the conditions with which values are enforced, are known by all actors involved in that domain, so that individuals are aware of what it takes to act responsibly.

With that, accountability structures provide a mechanism which allows 'passive' accountability and 'active' accountability to be matched (Bovens 1998). By internalizing the characteristics of behavior that is held to be accountable in a certain institutional context, individuals can develop intuitions about which kind of behavior is responsible or not, because they can actively and prospectively assess the consequences of their actions.

It is also important to realize that there are no encompassing accountability structures, instead they are all independent realms. According to Michael Walzer, the modern liberal society is based on the 'art of separation' (Walzer 1984), producing independent 'spheres of justice' (Walzer 1983). The conditions of legitimacy of different institutional domains can sometimes be incompatible or even contradictive. Contradictive sets of values emerge most clearly if the domains of the market and the state are merged (Jacobs 1992; Pesch 2005): these domains deal with contrastive goals, namely the pursuit of collective goals versus individual goals – mixing up these goals would lead to corruption and behavior that is controllable neither by the state nor by the market.

Given the complexity of today's problems, it is unthinkable to have political decisions without making use of science-based expertise. However, there is still a certain degree of awkwardness which emerges in cases where science and politics are mixed (Lindblom and Cohen 1979). The appeals of liberal democracy and of valid modes of truth-finding are sometimes contrastive, for instance, scientific results cannot simply trump political decisions that are decided by majority vote. At least the same level of discomfort arises if politics interferes with the domain of science. For instance, one may dispute whether the advocacy for intelligent design as an alternative approach to Darwin's evolution theory is motivated by intrinsically scientific considerations.

In Sect. 11.2, it has been concluded that institutional boundaries have to be taken down in order to facilitate sustainable innovation – and given the societal need for sustainable technology, it is nothing less than a broad societal responsibility to remove the boundaries between domains. This premise, however, appears to be at odds with the role of institutional domains as accountability structures that have to kept apart, as has been described above.

This does not mean that we can speak of an outright conflict between different responsibilities. As said, institutional domains have been presented in an ideal-typical manner; a closer look reveals that in empirical reality, the functioning of institutional domains is not as unproblematic as sketched above. To start with, the actual boundaries between the realm of science and the realm of politics are usually subject to contestation. There is no pre-given designation of tasks, duties, and responsibilities to either domain, but a role division has to be established in the interaction between experts and policy makers – the theoretical concept of 'boundary work' is used to describe this fluid and adaptive character of apparently rigid institutional boundaries (Gieryn 1995; Gieryn 1983; Shapin and Schaffer 1989; Jasanoff 1990; Halffman 2003). In other words, the delineations between the domain of science and politics are not cast in concrete – and the same can be said about the relation between the state and the market. In the words of Bruno Latour (1993), 'we have never been modern': the boundaries between institutional domains have always been ambiguous, permeable, applied in ad hoc and ex post fashion. Empirically, the boundaries between institutional domains are rather paradoxically constituted by the presence of effective boundary work. Concrete interactions between domains allow agreement about the actual character of the boundary between these domains (Pesch et al. 2012).

Another significant issue here concerns the relation between the modern institutional domains on the one hand and civil society on the other hand. In contrast with the institutional domains of market, politics, and science, civil society is fundamentally disorganized; rules and moral dispositions emerge spontaneously and are maintained by implicit agreement. From a political and ethical point of view, it is in civil society that the autonomy of individuals is to be found – which in turn has to be warranted by the appropriate functioning of the prevailing institutional domains. Especially the domains of the market and of politics can be conceived as derivations from two functionalities of civil society, namely the aggregation of individual citizens-as-consumers on the one hand, and the aggregation of citizens-as-electorate on the other hand (Beck 1992). In the ideal-typical description given above, the communication between civil society and institutional domains is portrayed as more or less automatic – empirical reality, however, shows something different. The relation between civil society and institutional domains is actively maintained, most notably by having organizations that represent certain stakes that are thought to be important by society. One may think here of labor unions, NGOs, churches, patient and consumer organizations, and so on. These civil society organizations play a pivotal role in the maintenance of the functioning of our institutional system. Not only do they allow the articulation of unvoiced values, opinions, and problem perceptions in society; to a large extent, they also provide the very fabric of society, contributing to a sphere a commonality that transgresses the realms of politics and the market (Putnam 1993, 2000).

A final empirical qualification is that there are a number of societal developments that have affected the functioning of institutional domains as accountability structures – so that their moral responsiveness has dramatically declined. In the following sections, this decline of responsiveness will be elaborated further. In Sect. 11.8, we will return to the issue of sustainable innovation, and discuss what our reflections on institutional domains imply for the premise that the boundaries between domains have to be broken down in order to facilitate the right climate for sustainable innovation.

11.4 The Decrease of Moral Responsiveness

The effectiveness of institutional domains as accountability structures has been decreasing, due to a number of historical developments that undermine their capacity to be responsive to the needs of individual actors. The first of these developments relates to one of the key insights that can be retrieved in systems' approaches to sustainable innovation, namely that the boundaries of institutional domains have become too stringent, which limits the capacity to address problems that transcend these boundaries. In other words, institutional domains have increasingly become autonomous entities, not responsive to individual needs, but concerned with the continuation of the rules that make up the domain. The second issue concerns a development that seems almost paradoxically opposed to this first development, which is the 'increased porosity' of institutional boundaries: institutional domains have a growing inclination to take over characteristics from other institutional domains. Possibly one of the most important causes of this problem involves the third issue discussed here, which is the historical emergence of organizations, most notably in the spheres of the state and the market, but also in the realm of civil society. As has been claimed, institutional domains seen as accountability structures allows a constellation in which there is a reciprocal relationship between the macro-level of an institutional domain and the micro-level of the individual. The meso-level that is represented by organizations has no place in the initial institutional constellation, and its appearance causes the effectiveness of accountability structures to be fundamentally affected. To a large extent, organizations function as accountability structures on their own, which not only affects the legitimacy of market and state organizations, but also of civil society organizations.

11.5 The Autonomy of Institutional Domains

The modern institutional constellation has been developed to allow individuals to act in autonomy, instead of being subjected to the caprice of political, economic, or religious powers. With the rise of modern institutional domains, individuals could be held accountable according to sets of rules that were – to a large extent – objective, impersonal and universal. What can be observed, however, is that institutional domains have acquired a 'life of their own', rules and practices that were non-personalized have become irresponsive to expressions of human need. Instead, the rules and practices of institutional domains have become normative for individual behavior instead of the other way around. Hence, institutional domains have lost a great share of their responsiveness to individual will.

What also can be observed is that institutional domains have bestowed people with difficulty to attend goals that transcend the short-term or particularistic level. Institutions have bestowed us with a serious degree of myopia with regards to the original goals of an institutional domain. The need to develop future orientations that are shared by various stakeholders, as was sketched in Sect. 11.2, emerges directly from this institutionally induced form of myopia.

For instance, the political domain has been set up to pursue the public will. However, the rules, procedures and practices that prevail in modern Western politics have given rise to a situation in which particularism and short-term gain are dominant. Political parties have become subject to Michels's iron rule of oligarchy; politicians are recruited from a narrow societal segment and stick to party discipline, which in turn is largely driven by the need to gain votes in upcoming elections. Political agendas are to a large extent set by media hypes and lobbyists pursuing a specific interest have more influence in politics than citizens (Lowi 1969). Another development that threatens the capacity of the electorate to discipline political actors, is the continuous role that polls and focus groups play in the formation of new political decisions. By adjusting political decisions to the desires expressed in these surveys, the effectiveness of elections as a tool to assess the past performance of political actors decreases.

Also in the market domain, a clear decline of responsiveness to individual consumers can be observed. Companies – especially the larger ones – are often more reactive to the wishes of shareholders than to the wishes directly expressed by consumers or groups of consumers (Galbraith 1998). Furthermore, the symmetry between the domains of the state and the market has disappeared. Where states are still, by definition, determined by national boundaries, the market has become an international structure. In this globalized market structure, companies have become entities that have no fixed locality. Production may be freely transferred to countries were wages are low and legislation is loose. Another aspect of how in the business domain the interests of individuals have eroded, is the way that to an increasing extent, people have adjusted their personal lives to the necessities of the flexible, almost fluid, globalized economy (Sennett 1998).

This decreased responsiveness of the domains of the state and the market shows itself most clearly in waves of widespread public dissatisfaction with the current political and economic system. Almost everywhere in the Western world there is a growing support for populist parties that claim to battle the traditional parties in power that are accused of only attending their own interest. The shared ideal that is stated by these new parties is that they want to restore the connection of politics with the 'people'. At the same time, the so-called anti-globalization movement fights the destructive forces of the market domain. Hirschman (1982) claims that resistance and enthusiasm for either the market or the state are alternating currents, each taking turns in promising what is the best way to solve social problems. Today however, both the public and the private domain are addressed by many as incapable of managing the problems of our time. Just as remarkable is that not the institutional domains and their original goals are contested; on the contrary, protesters aspire to restore or reinforce the effectiveness of these domains. Instead of replacing democracy, it is argued that we need a 'stronger democracy' (Barber 1984). This perpetuated support for liberal democracy contrasts strongly with the severe crisis of the 1930s when capitalism and democracy were seriously rivaled by political alternatives such as communism and fascism (Polanyi 2001; Hobsbawm 1995).

Although not to the same extent as distrust in the market and the state, one may still observe that science has lost a significant share of its authority. Scientific findings are now subjected to public distrust, as can be seen in numerous public controversies. One may for instance think of the so-called 'climategate'-controversy in which IPCC was accused of manipulating scientific findings, but one may also look at health issues where scientists cannot take away public fear for certain risks. In the UK, for instance, we have seen the BSE-scare and the widely held conviction that vaccination for measles might lead to autism (Jasanoff 1997). In the Netherlands, there has been a lot of public anxiety about the compulsory vaccination of Dutch children for swine flu. In these cases, experts are considered by a large section of society to be spokesmen of the elite, threatening the autonomy of individuals.

Besides the decrease of public trust, also the effectiveness of mechanisms that assess the quality of the work of scientists appears to be degrading. The stress on quality measures such as impact factors and past track record brings about all kinds of perverse effects, such as the hampering of groundbreaking scientific activities that do not belong to the dominant paradigms (Macdonald and Kam 2007).

Civil society organizations are not attached to the domains of market, state, or science. However, these organizations show the same kind of deficiencies in legitimacy as market, state, and scientific organizations, mainly because of problems to *represent* civil society. We may think here, for instance, about labor unions in the Netherlands, which have come to represent only a section of laborers, especially older, male workers (and even pensioners) are members (CBS 2011) – which may strongly bias the way these organizations defend the stakes of laborers in their negotiations with political and industrial representatives. Also other civil society organizations appear to have difficulty to clearly represent a societal stake, leading to the deterioration of their legitimacy (Breeman 2006).

11.6 Institutional Porosity

Next to the increasing autonomy of institutional domains, we can observe that institutional boundaries are increasingly crossed to the extent that institutional domains share constitutive characteristics. For instance, the state and the market have shown a growing amount of overlap (Pesch 2005, 2008b) and also the use of scientific information in political decision-making has been increasing (Ezrahi 1990; Lindblom and Cohen 1979). Such developments can be denoted as 'institutional porosity'.

Institutional porosity affects the possibility to assess the legitimacy of actions and decisions taken in institutional domains. If a market company assumes some of the monopolistic characteristics that are usually associated with the state, the functioning of the market system to allocate the appropriate prices to certain commodities is weakened. Also the replacement of democratic procedures by technocratic arrangements, or, contrastively, the application of political power in scientific discussions can be seen as boundary transgressions.

Examples of this development are quite omnipresent. The privatization of public services, for instance, may lead to a situation in which natural monopolies become private enterprises. One may think of the Dutch railways, which especially in 2002 were not responsive to both the desires of travelers and the government (Pesch 2005). As a company, the Dutch railways pursued to make money. Moving passengers in time and in a comfortable way appeared not to be the most profitable way to run the business. Commuters suffered bad service, while government remained impotent to steer the Dutch railways into a more consumer-friendly direction (Wessels 2003).

Another salient development is the entrepreneurship of civil servants, especially under influence of the so-called philosophy of New Public Management (cf. Osborne and Gaebler 2012). Public managers were told to be inspired by their commercial counterparts and be as daring and entrepreneurial as possible, in order to raise the benefits of public expenditure. This managerial philosophy however led to numerous financial debacles and cases of corruption, because the appropriate control mechanisms were simply lacking in the public sector. For instance, in 1999, when the Dutch province of South-Holland lost millions of euro's by the activities of one civil servant who, without any consultation, invested in a company that went bankrupt.

The recent banking crisis exhibited another issue related to institutional porosity. Even though the large financial companies are typically private enterprises, some of these banks received state support simply because the threat to their position also endangered the stability of socio-economic system as a whole. This implies that commerce has at least taken over some essential features of the public sphere, and vice versa.

One of the areas in which confusion of institutional domains has the most dramatic and far-reaching consequences is the health industry. The production of medicine is designated to the commercial sphere; the goal of involved enterprises is simply to make profit. This goal clearly goes against 'public' goals such as getting rid of most threatening diseases – which usually strike people who have least money to spend on medicine. The budget on development of new medicine is therefore mostly spent on 'luxury' diseases and furthermore lots of money is spent on convincing doctors to prescribe certain drugs instead of testing their efficacy (Pogge 2008).

Institutional porosity may also be observed in the realm of civil society. For instance, in many countries we see that labor unions and employers organizations are to a large extent 'captured' by the realm of the state; these types of civil society organizations have become fully embedded in the process of collective-decision making (Visser and Hemerijck 1997). One may also observe civil society organizations that have become embedded to a certain extent in the market. Many non-profit organizations, such as environmental NGOs, have private funding as their main source of income, which means that in order to survive as an organization, they have to apply a market logic in order to attract more funding (Lindenberg 2001).

11.7 Organizations

Conflicting sets of values not only occur in situations where different institutional contexts meet, they can also play a role within institutional contexts, because an actor's decisions might involve the choice between three sets of values: a first set that refers to the level of society as a whole; a second set which refers to organizational goals; and a third set of values that relate to personal morality (Moulton 2012; Wood 1991).

This confrontation of value sets relates to the emergence of organizations as entities that are positioned between the level of the individual and the level of the institutional domain. With that, organizations transfer, as well as mitigate, the conditions of an accountability structure to the individual, and vice versa (Pesch 2005). As has been discussed earlier, institutional domains have been set up so that they are in tune with the moral intuitions of individuals. The intermediary position of organizations however distorts that process of fine-tuning.

An organization can intercept the potential severity of an accountability structure by redistributing the accountability over 'many hands' (Thompson 1980), or by acting as an individual actor itself. With taking up such an intermediary role, organizations have come to fulfill the role of an accountability structure itself. Especially, the organizational form of bureaucracy advances a clear structure in which accountabilities are structured.

Historically, the institutional constellation which accommodates the level of the individual with the level of the institutional domain has been created before the establishment of modern organizations (Chandler 1977). The outcome is that instead of having companies run by individuals, and having a direct relationship between employer and employees, we now have commercial organizations that are literally faceless, in which the segments of the enterprise are run by managers, and which are owned by shareholders which have no direct relationship with the company.

In the domain of the state, the negative effects of bureaucracy are well-known to all of us. Merton's 'goal displacement' and the 'bureaucratic personality' (1952) have become infamous understandings. However, both in the domain of the state and the market, bureaucracies, with their specific pathologies, can be recognized. Moreover, also civil society organizations may show bureaucratic tendencies: once established, organizations will try to continue their existence, even if their representativeness withers away.

The introduction of bureaucratic organizations contributed greatly to the creation of conditions that allow the autonomy of institutional domains as well as to institutional porosity. The primacy of action has been taken away from individuals and has been transferred to organizations, which are not fully receptive to the conditions that are postulated by institutional domains. Organizations bring along their own specific social mechanisms that to a certain extent replace the mechanisms provided for by institutional domains.

11.8 Conclusion and Discussion

This paper started with the question how to align insights from systems' thinking on sustainable innovation with moral reflections on responsibility. This question is motivated by the idea that sustainable innovation is a broad societal responsibility, which appears to conflict with the rather particularistic focus on design practices and individual engineers in most work on responsible innovation.

The articulation of insights of systems' approaches to innovation into terms of responsibility has been done by focusing on institutional domains. In systems' approaches, the boundaries between institutional domains are considered to be obstructing elements in the establishment of shared future orientations, which, in turn, are seen as crucial prerequisites for having sustainable innovations. At the same time, institutional domains can be featured as accountability structures, as they convey the conditions that allow the assessment of the desirability of individual behavior in that domain. However, the emphasis on institutional domains has brought forward new complications. The analysis of institutional domains as accountability structures in contemporary society highlights important societal problems in relation to responsibility issues.

This final section will further discuss the findings of the paper. On the one hand, it will reflect on the implications of our analysis for systems' approaches; in other words, how can issues of responsibility be effectively incorporated in systems' approaches to sustainable innovation. On the other hand, some ramifications of our analysis for the wider issue of sustainable innovation and other forms of responsible innovation shall be explored.

To reiterate Sects. 11.3, 11.4, 11.5, 11.6, and 11.7, we see that the stringency of institutional domains has decreased due to the following developments. The *increasing autonomy of institutional domains* implies that the primacy of individuality is not guaranteed any longer. *Institutional porosity* causes confusion about the moral duties that are attached to different institutional domains. Finally, *organizations have acquired an intermediary function* between the level of the individual and the level of the institutional domain. The consequence is that the moral intuitions of individuals are not covered any longer by a singular institutional domain, but lay scattered over a world that has grown above the heads of individuals. Social problems that can be associated with the demise of institutional responsiveness are the neglect for both long-term interests and for genuinely public interests, but also the presence of wide-spread public distrust.

This analysis of institutional domains sheds a new light on the premise of systems' approaches to sustainable innovation that holds that institutional boundaries have to be crossed in order to develop shared future orientations for sustainable innovation. In first place, systems' approaches are very much aligned with the claim that the autonomy of institutional domains acts as a major problem for the uptake of sustainable innovation. The stringency of boundaries between domains prevents the development of a concerted future orientation towards sustainability. The motivations, problem definitions, agendas, and so on, of actors are, first and foremost, determined by the institutional domains in which these actors are found. Meaningful interaction transcending institutional boundaries has become utterly difficult due to institutional myopia.

However, our analysis reveals that the exclusive focus on crossing institutional boundaries may give rise to other problems. Especially if one looks at the actors who are involved in participatory methods related to systems' approaches, we can observe a strong, almost exclusive, emphasis on actors that represent formal organizations, that represent the market, politics, science, or civil society. As has been presented in Sects. 11.6 and 11.7, both the crossing of institutional boundaries and the strong role of organizations may have corruptive effects on the capacity of institutional domains to figure as accountability structures. Hence, emphasizing these two elements in badly designed participatory methods may lead to a decreased capacity to hold actors accountable for their actions and decisions. Following this decrease, there may be difficulties to support the legitimacy of sustainable future orientations that are developed. Parties that are not involved, most notably individual citizens which have not (yet) found the opportunity to organize themselves, might resist these future orientations as invasive assaults to their life world.

With that, the task of designing and organizing participatory projects aimed at facilitating sustainable innovation becomes even more complicated than it is already. Responsible innovation does not only mean the effective crossing of institutional boundaries to create shared future orientations, it also means that the functioning of institutional domains as accountability structures should be guarded in order to maintain (or restore) the societal legitimacy that is embedded in the modern institutional constellation.

How to actually achieve this double assignment is yet to be found out. A balance has to be reached between contrastive moral forces, but the exact position of this equilibrium can only be retrieved from actual practice. Empirical research is needed to gain more insight into these dynamics and their implications for striking a balance between crossing and maintaining boundaries between institutional domains. In sum, research should be developed into practices of establishing shared future orientations that takes explicit account of the relationship between the actors involved and the institutional domains from which these actors originate.

Systems' approaches to sustainable innovation may benefit from incorporating considerations about responsibility, because this may increase the efficacy and legitimacy of the results of participatory projects aimed at developing shared future orientations. At the same time, our thinking about responsibility issues connected to sustainable innovation is also enriched by including insights from systems' approaches. To start with, these insights show that responsibility in relation to sustainable innovation means more than installing the right codes of conduct or design procedures for engineers. Without any doubt, such initiatives are essential to further sustainable innovation, but it should not be overlooked that sustainable innovations are a broad societal responsibility that can only be effectuated by a similarly broad societal engagement. The emphasis on the design phase of new technologies is mainly driven by the aspiration to embed societal values in to engineering practice, but systems' approaches urges us to also look at the other side of the coin: how can we responsibly stimulate and implement new technologies? We may think here, for instance, of the development of methods that enable stakeholders *and* the general public to learn about new, desirable technologies.

Next to establishing broad societal engagement in technology development, our analysis also shows that the societal demand for effective responsibility arrangements requires the restoration of institutional domains. The pathologies observed here – which are the autonomy of institutional domains, the porosity of institutional boundaries, and the role of organizations – should be addressed in order to even consider responsible innovation. It has to be emphasized that the attention for these institutional issues does not contrast with the focus on responsibilities in the design phase of technologies. On the contrary, engineers and designers are, to a large extent, 'agnostic' in relation to the prevalent institutional domains. Engineers and designers can be found in science, in industry, and in policy-making, and as such their work always relates to a given institutional context. In turn, this implies that the actions and decisions of engineers and designers are embedded in an already existing accountability structure. Hence, responsible innovation, even explicitly aimed at an individual technology developer, cannot evade the conditions of responsibility that are already in force. In sum, the relation between engineering, society, and institutional domains has to be subjected to further theoretical and empirical exploration.

References

Barber, Benjamin. 1984. *Strong democracy. Participatory politics for a new age*. Berkeley: University of California Press.

Basart, Josep, and Montse Serra. 2013. Engineering ethics beyond engineers' ethics. *Science and Engineering Ethics* 19(1): 179–187.

Beck, Ulrich. 1992. *Risk society. Towards a new modernity*. London: Sage.

Berger, Peter, and Thomas Luckmann. 1991. *The social construction of reality. A treatise in the sociology of knowledge*. London: Penguin.

Bobbio, Norberto. 1989. *Democracy and dictatorship: The nature and limits of state power*. Minneapolis: University of Minnesota Press.

Bovens, Mark A.P. 1998. *The quest for responsibility: Accountability and citizenship in complex organisations*. Cambridge: Cambridge University Press.

Breeman, Gerard. 2006. *Cultivating trust. How do public policies become trusted?* Rotterdam: Optima Grafische Communicatie.

Brown, Halina Szejnwald, Philip Vergragt, Ken Green, and Luca Berchicci. 2003. Learning for sustainability transition through bounded socio-technical experiments in personal mobility. *Technology Analysis & Strategic Management* 15(3): 291–315.

Brumsen, Michael. 2011. Sustainability, ethics, and technology. In *Ethics, technology, and engineering. An introduction*, ed. I. Van de Poel and L. Royakkers. Chichester: Wiley.

Carter, Neil. 2007. *The politics of the environment ideas, activism, policy*. Cambridge/New York: Cambridge University Press.

CBS. 2011. http://www.cbs.nl/nl-NL/menu/themas/arbeid-sociale-zekerheid/publicaties/artikelen/archief/2011/2011-3490-wm.htm. Centraal Bureau voor de Statistiek 2011. Cited 13 Sept 2012.

Chandler, A.D. 1977. *The visible hand. The managerial revolution in American business*. Cambridge/London: Belknap Press.

Cuppen, Eefje. 2009. *Putting perspectives into participation. Constructive conflict methodology for problem structuring in stakeholder dialogues*. Oisterwijk: Box Press.

Cuppen, E., S. Breukers, M. Hisschemöller, and E. Bergsma. 2010. Q methodology to select participants for a stakeholder dialogue on energy options from biomass in the Netherlands. *Ecological Economics* 69(3): 579–591.

Dahl, R.A., and C.E. Lindblom. 1963. *Politics, economics, and welfare. Planning and politico-economic systems resolved into basic social processes*. New York: Harper & Row.

Derksen, W., M. Ekelenkamp, F.J.P.M. Hoefnagel, and M. Scheltema (eds.) 1999. *Over publieke en private verantwoordelijkheden*. Edited by W. R. v. h. Regeringsbeleid. The Hague: WRR.

Dumont, L. 1977. *From Mandeville to Marx. The genesis and triumph of economic ideology*. Chicago/London: Chicago University Press.

Ezrahi, Y. 1990. *The descent of Icarus. Science and the transformation of contemporary democracy*. Harvard: Harvard University Press.
Fay, Brian. 1975. *Social theory and political practice*. New York: Holmes & Meier Publishers.
Galbraith, John Kenneth. 1998. *The affluent society*. New York: Houghton Mifflin Company.
Gardiner, Stephen M. 2004. Ethics and global climate change. *Ethics* 114(3): 555–600.
Garmendia, Eneko, and Sigrid Stagl. 2010. Public participation for sustainability and social learning: Concepts and lessons from three case studies in Europe. *Ecological Economics* 69(8): 1712–1722.
Gay, Peter. 1973. *The enlightenment: An interpretation. The science of freedom*. London: Wildwood House.
Geels, Frank W. 2002. Technological transitions as evolutionary reconfiguration processes: A multi-level perspective and a case-study. *Research Policy* 31(8–9): 1257–1274.
Geels, Frank W. 2004. Understanding system innovations: A critical literature review and a conceptual synthesis. In *System innovation and the transition to sustainability: Theory, evidence and policy*, ed. B. Elzen, F.W. Geels, and K. Green. Cheltenham/Northampton: Edward Elgar.
Geertz, Clifford. 1973. *The interpretation of cultures*. New York: Basic Books.
Giddens, Anthony. 2009. *The politics of climate change*. Cambridge/Malden: Polity.
Gieryn, Thomas F. 1983. Boundary-work and the demarcation of science from non-science: Strains and interests in professional ideologies of scientists. *American Sociological Review* 48(6): 781–795.
Gieryn, T.F. 1995. Boundaries of science. In *Boundaries of science*, ed. S. Jasanoff, G.E. Markle, J.C. Petersen, and T.J. Pinch. Thousand Oaks: Sage.
Grin, John. 2000. Vision assessment to support shaping 21st century society? Technology assessment as a tool for political judgement. In *Vision assessment: Shaping technology in 21st century society*, ed. J. Grin and A. Grunwald. Heidelberg: Springer.
Grin, John, Henk Van de Graaf, and Rob Hoppe. 1997. *Technology assessment through interaction*. The Hague: Rathenau Institute.
Habermas, Jürgen. 1999. *The structural transformation of the public sphere. An inquiry into a category of Bourgeois society*. Cambridge: MIT Press.
Halffman, Willem. 2003. *Boundaries of regulatory science. Eco/toxicology and aquatic hazards of chemicals in the US, England, and the Netherlands, 1970–1995*. Amsterdam: Amsterdam University Press.
Hekkert, M.P., R.A.A. Suurs, S.O. Negro, S. Kuhlmann, and R.E.H.M. Smits. 2007. Functions of innovation systems: A new approach for analysing technological change. *Technological Forecasting and Social Change* 74(4): 413–432.
Hirschman, A.O. 1982. *Shifting involvements. Private interest and public action*. Princeton: Princeton University Press.
Hobsbawm, Eric J. 1995. *The age of extremes: The short twentieth century, 1914–1991*. London: Abacus.
Huitema, Dave, Marleen van de Kerkhof, and Udo Pesch. 2007. The nature of the beast: Are citizens' juries deliberative or pluralist? *Policy Sciences* 40(4): 287–311.
Jacobs, Jane. 1992. *Systems of survival. A dialogue on the moral foundations of commerce and politics*. New York: Random House.
Jasanoff, Sheila. 1990. *The fifth branch. Science advisers as policymakers*. Boston: Harvard University Press.
Jasanoff, Sheila. 1997. Civilization and madness: The great BSE scare of 1996. *Public Understanding of Science* 6(3): 221–232.
Jonas, Hans. 1985. *The imperative of responsibility: In search of an ethics for the technological age*. Chicago: University of Chicago Press.
Kamp, Linda M. 2002. *Learning in wind turbine development. A comparison between the Netherlands and Denmark*. Utrecht: Utrecht University.
Kemp, Rene, and Jan Rotmans. 2004. Managing the transition to sustainable mobility. In *System innovation and the transition to sustainability: Theory, evidence and policy*, ed. B. Elzen, F.W. Geels, and K. Green. Cheltenham/Northampton: Edward Elgar.

Kunneman, Harry. 1984. *Habermas' theorie van het communicatieve handelen*. Meppel/Amsterdam: Boom.

Latour, Bruno. 1993. *We have never been modern*. Cambridge: Harvard University Press.

Leydesdorff, Loet, and Martin Meyer. 2003. The Triple Helix of university-industry-government relations. *Scientometrics* 58(2): 191–203.

Lindblom, C.E., and D.K. Cohen. 1979. *Usable knowledge: Social science and social problem solving*. New Haven: Yale University Press.

Lindenberg, Marc. 2001. Are we at the cutting edge or the blunt edge?: Improving NGO organizational performance with private and public sector strategic management frameworks. *Nonprofit Management and Leadership* 11(3): 247–270.

Loorbach, Derk. 2007. *Transition management new mode of governance for sustainable development*. Rotterdam: Erasmus University.

Lowi, Th.J. 1969. *The end of liberalism: Ideology, policy, and the crisis of public authority*. New York: Norton.

Macdonald, Stuart, and Jacqueline Kam. 2007. Ring a Ring o' Roses: Quality journals and gamesmanship in management studies. *Journal of Management Studies* 44(4): 640–655.

Meadowcroft, James. 2009. What about the politics? Sustainable development, transition management, and long term energy transitions. *Policy Sciences* 42(4): 323–340.

Merton, R.K. 1952. Bureaucratic structure and personality. In *Reader in bureaucracy*, ed. R.K. Merton. Glencoe: The Free Press.

Merton, Robert King. 1979. *The sociology of science. Theoretical and empirical investigations*. Chicago/London: The University of Chicago Press.

Michelfelder, Diane, and Sharon Jones. 2013. Sustaining engineering codes of ethics for the twenty-first century. *Science and Engineering Ethics* 19(1): 237–258.

Montesquieu. 2002. *The spirit of the laws*. Cambridge: Cambridge University Press.

Moulton, Stephanie. 2012. The authority to do good: Constraining and enabling socially responsible lending behavior in a public mortgage program. *Public Administration Review* 72(3): 430–439.

Mulder, Karel F. 2006. *Sustainable development for engineers: A handbook and resource guide*. Sheffield: Greenlead.

Osborne, D, and T. Gaebler. 2012. *Reinventing government. How the entrepreneurial spirit is transforming the public sector*. Reading: Addison-Wesley Publishing.

Pesch, Udo. 2005. *The predicaments of publicness. An inquiry into the conceptual ambiguity of public administration*. Delft: Eburon.

Pesch, Udo. 2008a. Administrators and accountability: The plurality of value systems in the public domain. *Public Integrity* 10(4): 335–344.

Pesch, Udo. 2008b. The publicness of public administration. *Administration & Society* 40(2): 170–193.

Pesch, Udo. 2012. Overcoming regimes by regimes. In: *Proceedings of IST 2012. 3rd international conference on sustainability transitions*, Copenhagen.

Pesch, Udo, Dave Huitema, and Matthijs Hisschemöller. 2012. A boundary organization and its changing environment: The Netherlands environmental assessment agency MNP. *Environment and Planning C* 30(3): 487–503.

Pogge, Thomas. 2008. Access to medicines. *Public Health Ethics* 1(2): 73–82.

Poggi, Gianfranco. 1978. *The development of the modern state: A sociological introduction*. Stanford: Stanford University Press.

Polanyi, K. 2001. *The great transformation. The political and economic origins of our time*. Boston: Beacon Press.

Putnam, R.D. 1993. *Making democracy work: Civic traditions in modern Italy*. Princeton: Princeton University Press.

Putnam, R.D. 2000. *Bowling alone: The collapse and revival of American community*. New York: Simon & Schuster.

Quist, Jaco. 2007. *Backcasting for a sustainable future: The impact after 10 years*. Delft: Eburon.

Quist, Jaco, and Philip Vergragt. 2006. Past and future of backcasting: The shift to stakeholder participation and a proposal for a methodological framework. *Futures* 38(9): 1027–1045.
Rip, Arie, and R. Kemp. 1998. Technological change. In *Human choice and climate change*, ed. S. Rayner and E.L. Malone. Columbus: Battelle Press.
Rotmans, Jan, and Derk Loorbach. 2009. Complexity and transition management. *Journal of Industrial Ecology* 13(2): 184–196.
Rotmans, Jan, Rene Kemp, and Marjolein Van Asselt. 2001. More evolution than revolution: Transition management in public policy. *Foresight* 3(1): 15–31.
Schot, Johan, and Arie Rip. 1997. The past and future of constructive technology assessment. *Technological Forecasting and Social Change* 54(2–3): 251–268.
Schubert, G. 1960. *The public interest. A critique of the theory of a political concept*. Glencoe: The Free Press.
Sennett, Richard. 1998. *The corrosion of character. The personal consequences of work in the new capitalism*. New York: Norton.
Shapin, S., and S. Schaffer. 1989. *Leviathan and the air-pump: Hobbes, Boyle, and the experimental life*. Princeton: Princeton University Press.
Smith, Adam. 1998. *An inquiry into the nature and causes of the wealth of nations. A selected edition*. Oxford/New York: Oxford University Press.
Smith, Adrian, Andy Stirling, and Frans Berkhout. 2005. The governance of sustainable socio-technical transitions. *Research Policy* 34(10): 1491–1510.
Taylor, Charles. 1989. *Sources of the self. The making of the modern identity*. Cambridge/London: Harvard University Press.
Te Kulve, Haico, and Arie Rip. 2011. Constructing productive engagement: Pre-engagement tools for emerging technologies. *Science and Engineering Ethics* 17(4): 699–714.
Thompson, Dennis F. 1980. Moral responsibility of public officials: The problem of many hands. *The American Political Science Review* 74(4): 905–916.
Unruh, Gregory C. 2000. Understanding carbon lock-in. *Energy Policy* 28(12): 817–830.
Van De Poel, Ibo. 2000. On the role of outsiders in technical development. *Technology Analysis & Strategic Management* 12(3): 383–397.
van den Bergh, Jeroen C.J.M., Albert Faber, Annemarth M. Idenburg, and Frans H. Oosterhuis. 2006. Survival of the greenest: Evolutionary economics and policies for energy innovation. *Environmental Sciences* 3(1): 57–71.
Van Gunsteren, Herman. 1994. *Culturen van besturen*. Amsterdam/Meppel: Boom.
Van Lente, Harro. 1993. *Promising technology: The dynamics of expectations in technological developments*. Enschede: University of Twente.
Verbong, Geert, Frank W. Geels, and Rob Raven. 2008. Multi-niche analysis of dynamics and policies in Dutch renewable energy innovation journeys (1970–2006): Hype-cycles, closed networks and technology-focused learning. *Technology Analysis & Strategic Management* 20(5): 555–573.
Visser, Jelle, and Anton Hemerijck. 1997. *A Dutch miracle: Job growth, welfare reform and corporatism in the Netherlands*. Amsterdam: Amsterdam University Press.
Walzer, Michael. 1983. *Spheres of justice. A defense of pluralism and equality*. Philadelphia: Basic books.
Walzer, Michael. 1984. Liberalism and the art of separation. *Political Theory* 12(3): 315–330.
Weaver, P., Leo Jansen, G. Van Grootveld, E. Van Spiegel, and Philip Vergragt. 2000. *Sustainable technology development*. Sheffield: Greenleaf Publishers.
Weber, Max. 1972. *Wirtschaft und Gesellschaft. Grundriss der verstehenden Soziologie*. Tübingen: J.C.B. Mohr.
Wessels, Kees. 2003. *Verkeerd spoor. De crisis bij de NS*. Amsterdam/Antwerpen: L.J. Veen.
Winch, Peter. 2001. *The idea of a social science and its relation to philosophy*, 2nd ed. London: Routledge.
Wood, Donna J. 1991. Corporate social performance revisited. *The Academy of Management Review* 16(4): 691–718.

Chapter 12
The Family of the Future: How Technologies Can Lead to Moral Change

Katinka Waelbers and Tsjalling Swierstra

Abstract We increasingly rely on technological artefacts for supporting or replacing personal interactions. But such delegation is not always unproblematic: ever so often unexpected alterations in our relationships occur. More particularly: new technologies tend to destabilize established norms and values. What does this techno-moral change imply for the normative project that is responsible innovation? Is it possible to anticipate these alterations, at least to some extent? In this article we develop a heuristic matrix that identifies patterns and mechanisms of techno-moral change. We then present as a case study the ambient intelligence systems that are currently developed to coordinate the domestic lives of family members to explain how understanding these patterns and mechanisms can help to discuss future techno-moral change.

12.1 Introduction

Since decades, baby monitors are helping parents to take care of their newborns. In the last 15 years, it has become quite common not to walk to your colleague next door, but to delegate the delivery of your messages to an email program. At present, several companies are developing high-tech robots to take care of the elderly of tomorrow, and to help raise our future children. These examples show that we increasingly rely on technological artifacts for supporting or replacing personal interactions. We delegate human interactions (fully or partially) to technologies (Waelbers 2009). But such delegation is not always unproblematic: ever so often unexpected and less desirable alterations in our relationships occur.

K. Waelbers • T. Swierstra (✉)
Faculty of Arts and Social Sciences, University of Maastricht, Maastricht, The Netherlands
e-mail: t.swierstra@maastrichtuniversity.nl

J. van den Hoven et al. (eds.), *Responsible Innovation 1: Innovative Solutions for Global Issues*, DOI 10.1007/978-94-017-8956-1_12,

Much work on how technologies influence human agency is inspired by Latour's groundbreaking work (e.g. Latour 1992, 2005). Technologies can "authorize, allow, afford, encourage, permit, suggest, influence, block, render possible, forbid, and so on" human action (Latour 2005, p. 72). In his view, not only human agents have a program of action, but non-human (technological) agents possess one too. When both types of agents interact, the resulting human-nonhuman association obtains a new program of action that differs from both original action programs. Agency does not determine this interaction but arises from it (Latour 2005). This implies with regard to *technologically mediated* social relationships, that it would be naïve to expect technologies just to perform their delegated tasks without altering these relationships.

Latour provides a perspective on human-technology interactions that stimulates the analyst to adopt a third person's or observer's perspective in which the difference between people influencing people and technologies influencing people becomes irrelevant. It does not make much difference whether a police officer or speed bump persuades you to slow down. This third person perspective enables scholars to observe the mutual shaping of things and humans, of technology and society. In this spirit, a host of sociological, historical, or post-phenomenological case studies of technologically mediated relations and behaviors have by now been carried out.

But Latour has said little about the way his work could be utilized for creating better technologies. If responsible innovation is our aim, what advice does he have to offer? Although we to a considerable degree build on his work, we find his work, when seen from this practical, first person's, normative perspective, unhelpful in two respects.

First, the whole notion of 'responsibility' is hard to combine with the complete symmetry between human and non-human actors (e.g. Swierstra 1999). Latour describes the social role of technologies as if we were dealing with the impact of a technical object on a human object. But objects cannot take moral responsibility: they only perform actions in the sense of reactions. Humans (and some animals; see De Waal 2006) are distinct from objects because they have *reasons* for their action and can reflect on these reasons. As a result, technologies affect our actions not just by altering the course of action (like billiard balls do to each other) but also by mediating *our reasons* or motives to act in a particular way (Waelbers 2011). And this means that conscious reflection on how our reasons are affected by technologies, can also determine our actions to some extent. This ability to reflect on their reasons, is what practical reason is all about. And because humans have this practical reason, they are the ones that can act in a responsible or irresponsible fashion.

Secondly, when we grant agency to technologies, responsible innovation becomes much more demanding than only assessing 'risks', i.e. technology's capacity to harm us or other stakeholders. After Latour, responsible innovation has also to rely on the anticipatory exploration of future human-technology interactions, of the *social impacts* of emerging technologies. But such an exploration is hardly imaginable, if the past does not teach us some recurring patterns, some common mechanisms of technological mediation of our actions. Such patterns and mechanisms would allow for 'controlled speculation' (Arie Rip) about future social impacts. Unfortunately, there is nothing in Latour's work that resembles such patterns or mechanisms. We think it is fair to say that describing and classifying mechanisms would go against the grain of his theoretical work, which stresses contingency and complexity. In his view, social impacts just emerge from unpredictable power interactions. But there is a price attached to that intellectual ethos: he has little to offer in terms of a method or heuristics for the anticipatory exploration of technology's social impacts. But how can we take responsibility for future technologies if we have no starting point to anticipate future techno-social change? We therefore think that even though we side with Latour on the fundamental unpredictability of the socio-technical future, it is still essential to invest in what John Dewey has called 'dramatic rehearsals' (Dewey 1957) of the consequences of our actions. Even if these rehearsals never materialize, the exercise of rehearsing will leave us better prepared for, because more attentive to, the interactions and consequences that do materialize.

In this article we try to remedy these two – from our normative perspective – shortcomings, so as to make Latour's insights fruitful for responsible innovation. We believe it is necessary to study how technologies can affect the reasons of human agents, and that it is possible to identify patterns and mechanisms of techno-social change. More specifically, we want to concentrate on a particular form of techno-social change, that is: techno-*moral* change (Swierstra et al. 2009; Swierstra and Waelbers 2010; Boenink et al. 2010; Stemerding et al. 2010). Of course morality is part of society, but it is interesting that until recently very little attention was devoted to exploring how technological change influences moral controversy, and sometimes – when this controversy results in a novel closure – moral change. But since 2007, several authors have argued that imagining techno-moral change is important, e.g. to show how certain technological promises are implausible, or to organize a productive public debate (e.g. Swierstra and Rip 2007; van Asselt et al. 2010; Verbeek 2011).

This article presents a matrix that hopes to be useful in the context of responsible innovation as it enables technology actors, policy actors, and other stakeholders to imaginatively explore possible future techno-moral changes in advance. The two axes of the matrix are constituted by the answers to the two fundamental questions with regard to techno-moral change:

1. How does the new technology mediate our relations to three key elements of any moral judgment: (a) the parties that are affected by our actions; (b) the consequences of our actions; and (c) to the beliefs and practices that constitute our conceptions of the good life.
2. How does the new technology mediate our beliefs about (a) how the world is, (b) how we can act in that world, and (c) how we should act in that world?

These two axes of the matrix are explained in Sect. 12.2 before presenting the matrix itself. In Sect. 12.3 we then present as a case study the ambient intelligence systems that are currently developed to coordinate the domestic lives of family members to explain how understanding these mechanisms can help to discuss future techno-moral change, see Sect. 12.3. In Sect. 12.4, finally, we provide some means for further academic exploration of plausible techno-moral change in the future and we return to the question how exploring technomoral change is not merely descriptive, but also important from the normative point of view implied in the concept ‘responsible innovation’.

12.2 The Matrix

Morality, as the pragmatist philosopher John Dewey was the first to point out, primarily exists in the form of practical routines that seem so self-evident that their impact on how we act, think and feel usually goes unnoticed (Gouinlock 1994, pp. 21–22). And this is how it should be. Explicitness and reflexivity are not things to be valued under normal circumstances. For instance, if you first deliberate whether or not to kill your obnoxious colleague, only to decide after careful reflection that you cannot do this because it would be immoral, then there is already something deeply disconcerting about you (Williams 1985). The taboo on killing should be so self-evident that under normal circumstances one obeys it unthinkingly: the more self-evident a norm, the less visible; the less visible, the more effective. *Ethics,*[1] by contrast, refers to the conscious reflection and discussion on morality. One does not do ethics just for fun. That effort needs an occasion.

Moral self-evidences thrive best in a stable environment where they find constant confirmation. But modern societies are defined by their dynamism, fuelled by scientific and technological development, with endemic moral uncertainty and controversy as a consequence.

This section discusses some mechanisms of techno-moral change by presenting the two axes of the matrix introduced above.

12.2.1 The Technological Remediation of Stakeholders, Consequences, and the Good Life

Before asking how technologies mediate our moral judgments, we have to identify the key elements of such judgments. Building on earlier work (Swierstra and Waelbers 2010) we want to differentiate three of such elements. Broadly speaking, the domain of ethics can be divided into two chambers: rule ethics and good life ethics. Rule ethics aims at governing the relations between parties with different, and sometimes conflicting, interests. The leading question of rule ethics is: what do these parties owe each other? According to rule ethics, acting morally involves taking the legitimate interests and rights of our fellow-beings seriously when deliberating how to act. In other words: actors are under a general moral obligation to make sure

[1]Or, in Dewey’s terminology: ‘reflective morality’.

that the consequences of their actions don't conflict with the legitimate interests of others. We can therefore identify two key variables in rule ethics: the *consequences* of our actions, and the *stakeholders* who are affected by these consequences. If our perception of either of those variables changes, so does our moral judgment.

First, the concept 'stakeholder' is used to mark out those parties affected by an actor's practical choices. Of course, as the actors have rights and interests – a 'stake' – too, they will often be stakeholders themselves as well, but this is not necessarily the case. Stakeholders have a 'stake' in our (in)actions, and a moral claim on us, e.g. to be treated fairly, to be helped, or to be given an explanation for why we chose to do what we did. When deciding how to act morally, it is therefore always necessary to identify such stakeholders and their interests and rights. And if our perception of who the stakeholders are was to change, so would our moral judgment. For instance, when parents see on television that a toy is cheap because it is made in a factory that employs 8-year-olds who work 12 hours a day, 7 days a week, they may be less inclined to buy it for the amusement of their own 8-year-old.

Secondly, acting morally implies trying to anticipate the consequences of our (non)actions, and to establish whether these are morally desirable (obligatory) or not. In everyday life we commonly justify our norms, values, or practical choices by pointing at (intended) consequences of our choices. Realizing that our choice does not have the intended consequences, commonly leads to changing our moral assessment of that action. Now that – thanks to technological instruments – more and more people become convinced that CO2 emissions are causing climate change, the pressure to decrease the emissions increases.

Finally, morality not only deals with the question what actors owe each others, but also to the question of how to live a good life. This is the case even if in contemporary, pluralistic, liberal societies, this question has to a considerable extent been banned from the public domain (Swierstra 2002; Swierstra et al. 2009; Waelbers and Briggle 2010). The good life thus constitutes our third key element of moral judgments. Insofar as our aims central to what we consider essential to human flourishing change, our conception of the good life does too (Swierstra 2010). This implies that just as stakeholders and consequences, our conceptions of the good life can change too under the influence of technology. Technologies typically promise to help realize our goals more efficiently, to satisfy our desires, to diminish suffering and pain, and so forth. But they also help define those goals, they create new desires, new forms of pain and suffering, and so forth.

12.2.2 How Can New Technology Remediate Our Relations to the World?

So, the first axis of our model is made-up by the three key variables of moral judgments: stakeholders, consequences, and the good life. This axis informs us about *what* can be technologically mediated. The second axis now addresses the question *how* this technological mediation takes place by identifying three different practical perspectives on these variables.

Latour explains that technologies are interfering with our actions and vice versa: by such interaction people may act differently when they employ certain technologies (Latour 1992, 2005): they bring back the hotel key when a large fob is attached and they slow down to pass a speed bump. But when it comes to imagining future human-technology interactions, it is not enough to show that people may act differently: we need to know why they may do so. People leave the hotel key at the lobby because it is too large, and so it is uncomfortable to put it in their pocket. Drivers slow down for speed bumps because otherwise they might damage their car and their backs. Thus, the presence of the technologies changes the reasons for their actions. In other words, technologies do not just simply interfere with our actions, but they influence our reasons for actions.

If we acknowledge that technologies mediate our actions by influencing our reasons for actions, we can proceed with our search for the mechanisms, which we can then try to "apply" to future cases. To this end, three types of beliefs that inform our actions can be distinguished (Waelbers 2011): our factual beliefs (what we *know*), our prudential beliefs (what we *could* do), and our moral beliefs (what we *should* do).

In the context of practical reasoning, these beliefs are offered as reasons. For example, when responding to the question: Why don't you pay back the money you borrowed from me?, one may answer: because I never borrowed money from you in the first place (factual); because so far I haven't been able to (prudential); or, because you are so much richer than I am that giving you back your money wouldn't maximize collective utility (moral).

Of course, these three types of reasons are interrelated to each other: sometimes we ought to do something because we can, and often we can only do something because we understand some factual aspects. But to be able to analyze how technologies affect the reasons for people's actions, it is useful to introduce these three types of reasons for action separately.

12.2.2.1 What We Believe to Be the Case

As Don Ihde explains, technologies can alter our factual beliefs or perceptions (1993). He argues that we observe the world through technologies that transform our observations or microperceptions (e.g. the baby monitor). In addition to microperceptions, Ihde identifies macroperceptions, which consist of our worldviews, or our understanding of the world. These macroperceptions are informed by microperceptions, or in other words, macroperceptions are interpretations of microperceptions. Since microperceptions are mediated by technologies, macroperceptions are also technologically mediated. In the context of practical judgment, the mediation of our perceptions, or the technological influence on our factual beliefs, provides reasons for action on which we can reflect.

12.2.2.2 What We Believe We Can Do

Our reasons for actions are also related to what we believe what we can do. In other words, we act in certain manners because we recognize certain options that are available to us. It is commonly argued that technologies provide new options for actions. But often, previous options become less recognizable due to new technologies or they disappear all together. With an example of Latour (1992): an automatic door groom makes it harder to pass through the door with a wheelchair.

Often new technologies are offered with the promise that people are free to adopt the technologies or not. But, as Albert Borgmann (1984) already pointed out, in real life, there is always technological and social pressure to use the new technologies and the choices that are left are rather superficial. For future technologies it is important not to overlook such mechanisms. If for instance many future parents will provide their children with psycho-pharmaceuticals to improve their learning abilities, it may become an issue whether it is responsible parenthood not to give your children these drugs: they will suffer a huge disadvantage for the rest of their lives if they belong to the minority who did not get these enhancers (Gezondheidsraad 2002).

12.2.2.3 What We Believe We Ought to Do

We have multiple beliefs about how the world ought to be, and we use these beliefs for the evaluation of our actions. Moral beliefs are convictions about what is good to be and do in relation to the flourishing of oneself, other humans, animals, and nature. Our moral beliefs are for instance beliefs that address moral values (i.e. that what we believe to be essential for human and environmental flourishing), beliefs that focus on moral norms such as duties, or beliefs that deal with the notion of virtue. Technologies mediate our moral beliefs since they alter our factual and practical beliefs. Old norms, values and virtues disappear and new norms, values and virtues arise because we understand the world in a different manner and because we have other options for our actions.

12.2.3 The Matrix

Now the question is: how can these distinctions be used to explore instances of techno-moral change that may be induced by new technologies? We have constructed the following matrix (see Fig. 12.1) to help people enquire what the possible morally relevant, social role of the technologies might be. On the horizontal axis, we distinguished the three basic types of beliefs (reasons) that play a role in practical judgment, and on the vertical axis we distinguished the variables of moral judgment that constitute the subject matter of those beliefs. Our claim is that a practical judgment rests on combining both axes. When a moral actor reasons

	a. Is	b. Can	c. Ought
1. Stakeholders	Presence	Empowerment	Rights
2. Consequences	Anticipatory knowledge	Practical affordances	Responsibilities
3. Good life	Contingency	Freedom	Flourishing

Fig. 12.1 Matrix for exploring techno-moral change. The horizontal axis represents the three types of reasons for action and the vertical axis represents the points of moral focus

about what to do, s/he bases her/himself on what she beliefs to be the case (who are the stakeholders, what are the consequences, where lies the borderline between what has to be accepted as given and what can be changed); on what she beliefs s/he can do (what can I do for these stakeholders, what can I do about the consequences; to what extent am I free to chose my life); and finally on beliefs on what s/he should do (what do I owe to these stakeholders, which consequences are my responsibility; what should I do to flourish).

The upcoming subsections illustrate each box of this table. Note that for each point, technologies can simultaneously work to increase or decrease, expand or limit, frustrate or support the aspects under investigation. Furthermore, as a *stakeholder* is defined as someone who suffers or enjoys the *consequences* of our (non)actions, or vice versa, morally relevant consequences are defined in terms of whether they affect stakeholders or not.

12.3 The Family of the Future

To illustrate how the matrix may support the imagination needed to anticipate future techno-moral change, a new technology for realizing the home of the future is discussed. Scientists and engineers work to convert current IT devices, domotics, ambient intelligence, and care systems for domestic use to alter our houses into smart surroundings that are more comfortable, sustainable, entertaining and safer, while easily connecting with the homes of family and friends. To enable this convergence, the European Commission funded[2] a group of fifteen European companies to develop an open-source software architecture (called middleware) that would enable most devices to communicate with each other and to actually realize the intelligent home of the future. This software is called ISTAmigo. In a promotion movie, the full potential of this middleware is presented by telling the story of an average day of an average family living in an intelligent and networked home.[3] In Textbox 12.1, an almost literal account of the short movie is given.

[2]The project received funding from 2004 to 2008 for further development of the middleware.

[3]http://www.youtube.com/watch?v=wey94w-pNVI
http://www.hitech-projects.com/euprojects/amigo/

Textbox 12.1. The Promotion Film of ISTAmigo

Maria is waking up with her favorite music. We are told that the volume of the music is adapted to the situation in the bedroom. It stops when she leaves and her husband turns around to sleep some more. Maria starts her day with a working out on the home trainer: the system adapts the exercise program to the recent training history of her profile. When she is done, she goes back to the bedroom to get dressed. The system follows her to deliver a message from her office. She removes it by touching a screen on the wall.

In the bathroom, Maria checks her health data, and the mirror provides her with advice about her diet and exercise. In the mean time, her husband Jerry has prepared breakfast, and the food guide of the system creates some dinner plans. This guide not only takes into account the food requirements of the individual family members, but also those of the regular guests. It generates menus and recipes to choose from. When the choice is made, it checks what is in stock, and it generates a shopping list which is automatically sent to the grocery shop.

After breakfast, all family members prepare to leave the house. But someone left the refrigerator open, and the system warns that it first has to be closed. The system also warns Maria about forgetting her identification card: it senses that her card has not been removed from the bedroom yet, while on other days she has taken it with her much earlier. The card is a contact source for the Amigo system, just like the family's cell phones and PDA's.

Eventually, the family leaves the house. The entrance management turns on automatically so that after some hours, when Maria comes home, the system recognizes her voice and opens the door. The board shows her that her husband is at the office and that Roberto (her son) is at school. She leaves a message for Roberto about the food in the kitchen, and she tells the systems that she does not want to be disturbed while she is working at her home office. So, when later on the day Roberto (about 10 years of age) is not feeling well, her husband receives the phone call from school telling him that Roberto is sent home. Jerry goes to the privacy bubble that is installed in the office building to talk shortly to his son to see how he is doing.

In the next scene, Roberto, who is by now home alone and lying on bed, sees that his grandpa John is online and contacts him. John also has a networked home system, which is connected with Maria and Jerry's system. They share experiences, photos and activities and, according to the movie, they "maintain a feeling of social presence".

A little later that day, John becomes unwell while he is watching TV. The system detects the sudden movement of John when he passes out. And it detects an irregularity in his heart beat. Since the food is still on, the system immediately turns of the burner, and it tries to wake up John but fails.

(continued)

Textbox 12.1. (continued)

The system contacts Maria and Jerry, but they both have the "do not disturb" setting on. The system then warns John's neighbor. When he arrives at the house, the system recognizes him and opens the door. A few moments later, John wakes up and leaves with the neighbor to an ambulance. The system closes everything down, forwards all messages of John to Maria and sends a message to Maria's entrance board that John has left the house with the neighbor after getting unwell.

Peter, a friend of Roberto's, is at the door. The system recognizes Peter as a friend of the family and lets the boy in. The notification that Peter is in the house is sent to the board in Maria's home-office and it is sent to Peter in the living room. Peter and Roberto select a play from the playlist that is derived from Roberto's profile and the parental protection filter. When they start the game, the ambience of the living room adapts to the game, changing light in synchrony with the action of the game.

Maria comes out of her home office. She reads the message that John has become ill, sees that Roberto and his friend are playing a game and that Gerry is with them.

Peter's father is at the door. He is a member of, but not a friend within, Maria's and Jerry's community. So, the system recognizes him, but does not let him in. So, they have to manually open the door: he collects Peter and goes home.

Jerry starts the food guide to see the recipe that was selected that morning: he makes dinner and they eat together. After having watched some television, it is bedtime for Roberto and Jerry and Maria stay in the living room to watch a movie together: the ambience light changes color from green to red. After a while, Maria activates the surveillance mode: thermostats and doors are set, lights are turned off. The day has ended.

As the description makes clear, this movie nicely presents us with a view of the home of the future. So the question is: what is there to add? What can the matrix add to this prospective view? In numerous sociological and philosophical contributions, the issue of privacy has been brought up: if all that intimate information is collected by a system that is connected to the internet, then how to protect that data? This is a very legitimate worry, but it is important to take the discussion beyond the current focus on protecting the value of privacy, for this is only one of the many changes an intelligent home of the future is likely to bring.

Here, the aim is to identify some moral changes in personal relationships that might be brought about by the home of the future. We expect these changes to be rather substantial: consider how many social changes computers, internet, email, blogs and tweeds have brought and are likely to bring about (for a large collection of observations, see Brockman 2011). The aim is to help to move the debate beyond the obvious moral problems and to identify possible future techno-moral change.

We did not develop an exhaustive techno-moral scenario as for such a scenario further steps are required. The aim is solely to show how the matrix can be used for imagining future change in family relationships.

12.3.1 The Stakeholders

12.3.1.1 Ad 1a. Presence

We start by asking how technology can affect beliefs about the presence or absence of stakeholders. Technologies can make actors more or less aware of other stakeholders. The awareness of stakeholders' presence is morally relevant, as it is a precondition for taking their interests and rights into account. Middleware such as ISTAmigo increases the awareness of stakeholders. The intelligent home of the future requires middleware, whether it is ISTAmigo or any other system. Middleware is designed also to increase the awareness of the users about the other family members and friends. But as we said in the introduction: technologies never simply serve our intentions, but also – more or less subtly – modify them. Which means in this case: it changes the meaning of presence. The parent can for instance check via the screen where the child or the other parent is, but s/he cannot see how they are feeling, how they are doing. Family members become more aware of who is in what room performing which activity, but since this information is not derived from real contact but is mediated by technology, the kind of information they receive is different. They can see that their children are doing their homework, but they cannot see how seriously they are involved, whether they really understand it and whether they are enjoying it or whether they are frustrated. So, while on the one hand the system provides an increase in information about who is doing what in the house, on the other hand the system hides certain information about others from sight.

12.3.1.2 Ad 1b. Empowerment

Change in presence can empower or depower the stakeholders. In the promotion film, empowerment is presented as an important feature of the intelligent home. It enables for instance the son to enter the house without help from his parents when he feels a little ill. Also when grandpa is not with him, the son can choose to contact him and invite his grandpa to play with him. This empowers the child for it makes living in the house and sharing experiences with other people less dependent on constraints like distance and grown-ups being available. But there is also a decrease in empowerment. Since the house can perform some monitoring tasks the son for instance cannot so easily pretend that he is ill or that he has made his homework. The same is true for other members of the household: hiding (for instance for preparing Christmas presents), truancy, and occasional laziness become less of an option.

12.3.1.3 Ad 1c. Rights

In this respect, we can wonder how the intelligent home of the future may change the rights (and corresponding duties) of family members: people simply need an occasional slack and they need privacy. If the home of the future becomes a reality, we need new privacy rules not only to protect it from hackers or nosy governmental institutions, but also to teach us anew how to respect the privacy of other family members. For instance, when grandpa fell ill, all his messages were automatically forwarded to his daughter. When is she morally allowed reading those? And at what age should a parent stop to check the screen to see what the kids are doing?

Another example is the intelligent door: if it becomes the standard, we need moral rules on how to use it. Who do you, as a house owner, give full access, who gets partial access to the house and who has to wait at the door? As to the visitors, 24/7 access should not imply that you should make use of it at all times. But what are the rules? Could you access someone's house if they are not there? You are allowed technically, but also morally? When does something count as trespassing and when not?

12.3.2 The Consequences

12.3.2.1 Ad 2a. Anticipatory Knowledge

The introduction of a new technology can change the factual beliefs of the users with regard to the consequences of their actions, as these may become illuminated or blurred from view by the employment of the technology. An intelligent home, such as enabled by the ISTAmigo system, can fundamentally change our beliefs about our bodies and lifestyles, as it provides us non-stop with a variety of real-time biomedical data (such as nutrition weight, sports and movement, heartbeat, temperature) that tell us how our actions are affecting our health.

Often, what can be measured and controlled by computers is believed to be more objective and therefore more accurate then what we feel. Therefore, such a system can prevent laziness, overindulgence, or asking too much of oneself during exercise and diet. But it also changes how we evaluate our and our family members' subjective comfort. Do we listen to our bodies when we feel tired, or do we let our intelligent home convince us that we are just lazy? "Objective" biomedical data do not only influence how doctors think about their patients health, but how also people feel and think about their own body, as has been demonstrated for the case of diabetics meters and insulin intake (Mol 2008). What the system thinks may become more important than how you actually feel, for better and for worse. Individual biomedical differences and preferences get ignored, although one can question the legitimacy of such an interaction. If the intelligent home of the future is to provide us with constant feedback on our biomedical condition, we need to learn how to deal with this knowledge and how to prudently combine it with other information (such as our bodily response).

12.3.2.2 Ad 2b. Practical Affordances

The promise to create new practical affordances underlies almost all technological expectations, and often for good reasons. However, when new options surface, existing options may become diminished or cease to exist. For instance, the home of the future enables people to follow a diet quite easily: next to calculating your calories and exercises, it presents menus from which people can choose. From the selected menu, it prepares a grocery list, which is sent directly to the shop. The family only has to collect the groceries or make sure that the shop delivers them at the door. But the question is: what selection criteria are used by making the list? Obviously, such systems provide people with options to select regarding diet and perhaps selecting biological products. But how about the broad range of preferences such as wanting to eat GMO-free, vegetarian, minimum salt, and so forth? In any selection a certain system makes, certain values are adopted and others are neglected: how can we wisely program and make our choices?

12.3.2.3 Ad 2c. Responsibilities

Technologies can increase or decrease both our knowledge of our actions' consequences as well as our ability to influence those consequences, which directly translate into our moral responsibilities (de Vries 1989). The intelligent home relieves people from certain responsibilities regarding life style choices since the technology takes over some of the management of one's life on the micro-level. Now, people are responsible for their own planning of diet and exercise: guidance they receive are (if any) rather general and non-restricting. A physician or dietician can give people advice, but this advice will be presented to them, say, once a week, month or year. A system like ISTAmigo offers people real time advice: "this morning, you should bike for 10 more minutes", or "you have not eaten enough today, take some fruit before you go to bed". However, the responsibility to remain slim, fit and healthy also increases. The more options someone has to reach certain goals, the greater his or her responsibility to reach those goals becomes: there are less excuses to fail since you had all the information and help to succeed. Hence, adopting an unhealthy lifestyle in the intelligent house of the future might be considered to be even more irresponsible (and thus blamable) then it is now.

12.3.3 Good Life

12.3.3.1 Ad 3a. Contingency

New technologies affect the dividing line between on the one hand what has to be accepted as given, as being determined by outside forces, as simply happening, and on the other hand what can be altered by choosing alternative options. For instance, being available, or not, for your family members is nowadays usually to some extent a matter of choice (one can close the door and say one doesn't want to be disturbed) and somewhat a matter of fate (one's son ignores the prohibition and just barges into the room). More importantly, in most situations it will be not so clear whether one is available to other family members. We give of more or less subtle messages (Pfff, I am tired, so I am going to read my newspaper for a while.) that aim to inform the rest of the family how open we are to contact, and these signals get taken up or not. This balance between what happens to you and what you make happen, subtly shifts when ISTAmigo is introduced. It is likely to become more normal to delegate to the technology the task of communicating whether one is available or not. The result will be that availability becomes much more binary and explicit. When grandfather suffered from a heart attack, his daughter had just enabled the do-not-disturb function in her home office. Consequently, she was only much later notified of what happened. In some cases this may well be considered progress, but it is not hard to imagine that if the heart attack was serious, she would have liked to be informed immediately and not hours later via an automatically generated text message.

Another example of how the middleware for developing the home of the future alters the contingency of family lives is given by the constant mediation of the communication between the family members. Mediated communication could be more efficient and it may enable people to spend time with their loved ones even though they are very busy, far away or not feeling well. These are all important advantages. But if people can just touch the screen to tell their family members that dinner is ready, they would probably feel less inclined to take the stairs to tell them in person. Consequently, they will not see what the others are doing, and perhaps miss a chance to compliment the children on their new Lego-project or to quickly kiss their partners. When you increase control over reality, reality protests by withholding its surprise gifts.

12.3.3.2 Ad 3b. Freedom

Technologies create and limit our options to live what we believe to be a good life. With the increased opportunities to shape your life rather than simply obey the role that is connected to your given status in society, the dominant conception of the good life has moved away from 'obeisance' towards 'autonomy' and an activist stance. Freedom or autonomy is increased in the home of the future since people are empowered to be more self-supportive: remember grandfather living on his own while having a bad hart. But the freedom is also limited for a system like ISTAmigo only works when it is connected to multiple other homes that run middleware. It is like having a fax machine: it is only a useful device when sufficient people in your network happen to have one too. So for grandpa to be safe in his intelligent house, the neighbors and his daughter have to adopt the system too. When the majority has adopted such a technology (as is for instance the case with email and mobile phones) you might become a social outcast when you refuse to go with the flow.

Such network based technologies can severely limit people's freedom, especially when its use becomes socially controlled. For instance, if the technology is available, then it would not be a bad idea for employers to require people to install it in their home-offices or studies, and make the switched-off-option obligatory during working hours: after all, is it not your duty to be unavailable for private matters and to spend your time for your boss?

12.3.3.3 Ad 3c. Flourishing

By altering our perceptions and practical options, technologies co-shape what we believe to be a good life. We need to explore and evaluate how the technology might influence what we believe to be virtuous. What will it mean to be a good family member or friend when the intelligent home of the future becomes a reality? Already SMSs and Tweets made it normal to constantly share, in a few lines, experiences, feelings, activities and thoughts with your "connections". In other words, on the one hand being extrovert, while on the other hand being brief in your communication is promoted. The home of the future takes this trend a step further since in all rooms multiple devices are present to enable such sharing. In a future in which the intelligent home is the standard, one can wonder whether people can still be good grandparents or best friends if they do not constantly share their daily experiences, feelings and thoughts. And what if they are "not present" and do not (immediately) respond to the things the grandchildren or friends share?

Further, contact with your friends and family via the intelligent system feels like real social presence, the movie tells us. You can hear and see each other as if you were actually there. But you cannot smell, feel or touch each other through the screens. And you only see a part of the person, and not the full person and the room. Children do not just want to "share their experiences" with their grandparents: they want to experience their grandparents. They want to hug them, wrestle with them. The question is how to learn to technologically (e.g. adaptations in design) and socially combine the comfort and efficiency of the intelligent home of the future with the need for real life attention.

12.4 Conclusion

The study of the home of the future illustrates that understanding technological mediation of human relationships in terms of influencing reasons for actions can help support anticipatory reflection on techno-moral change. The home of the future is likely to influence our factual ideas, options for actions and moral beliefs with regard to the three key elements of any moral judgment: stakeholders, consequences, and conceptions of our place in the world.

Of course, our analysis of the home and family of the future must be read as a proof of principle, showing how the matrix might be applied in practice. We do not claim that family relations will exactly evolve like this if we delegate all kinds of tasks to ISTAmigo. Humbled by many failed attempts in the past, people have learned that the future is impossible to predict. Not only do we lack the necessary knowledge, but the future is essentially open and contingent on our choices, as is clear from phenomena like self-denying or self-fulfilling prophecies. Still, we are bound to prepare for it, since we need to take important development and design decisions now. Purposeful action by definition assumes some degree of speculation about future impacts. (Swierstra et al. 2009, p. 120). Therefore, people need tools to imagine how technologies might influence our reasons for action (perceptions, options, and moral beliefs) before deciding what tasks to delegate to which technologies. In the words of Hans Jonas" the "[f]irst duty of ethics of the future is visualizing the future" (Jonas 1984, p. 27).

Responsible innovation too often concentrates on so-called risks, that is: the chance that the new technology will adversely affect our safety or health, or the environment. By choosing this focus, it 'risks' to completely miss out on the consequences of technology that affect people most, that is: the ways their daily lives are affected. Even if technologies don't explode or pollute, they do something more and different than simply serving us. They are not passive instruments, but active forces shaping our lives. Responsible innovation is not truly responsible if it avoids these less tangible, but very real, impacts of emerging technologies. This means that responsible innovation cannot avoid developing rich imaginations, thick descriptions, of how technology might affect our practices, our values, our aims, our aspirations, our morals. Our matrix aims to provide a valuable tool for responsible innovation by stimulating anticipatory explorations of plausible techno-moral change.

Of course, placing techno-moral change in the centre of the analysis cannot avoid introducing a certain type of relativism, as it accepts that morals coevolve with technology. This relativism seems to condemn us to a form of neutral, impotent descriptivism, that would be incompatible with the explicitly normative stance at the heart of responsible innovation. For is, the answer to this puzzle doesn't lie in unearthing deep moral foundations, but in a more dialogue-based approach exemplified by philosophical hermeneutics (Gadamer 2004[1975]; Ricoeur and Thompson 1981) This theoretical approach justifies our contention that the imaginative presentation of techno-moral futures is an important element in TA exercises.

Hermeneutics explains how we can learn from works of art without succumbing to moral relativism, even in the absence of fixed moral standards that can function as an Archimedean point of reference. Works of art, or in our case: of the informed techno-moral imagination, can reveal our common preconceptions about ourselves and our world; disturb the taken-for-granted nature of morality; and thus help expand our horizons-of-understanding (Gadamer). Moral learning can occur where and when people are confronted with 'strange', new, conflicting morals. Even when, as in art or in technomoral scenarios, we ourself have to devise these conflicting perspectives in our imagination. Moral plurality invites reflection, (self)criticism, dialogue and the open exchange of ideas. By developing techno-moral scenarios, we travel to future worlds where slightly different technologies and morals prevail. It is by seeking this confrontation between present and imagined morality, that we learn to guide technological change in a manner both reflective and flexible, without reifying either the present or the possible future.

Again, one cannot predict the future. The results of our deliberations may never materialize. But still, anticipatory exercises are important to explore what we regard as *desirable* and what we regard as *undesirable* techno-moral change, and they allow us to develop a keen eye for the instances of techno-moral change that do happen. Only by making well-informed decisions can we try to work towards responsible innovation.

References

Boenink, M., T. Swierstra, and D. Stemerding. 2010. Anticipating the interaction between technology and morality: A scenario study of experimenting with humans in bionanotechnology. *Studies in Ethics, Law, and Technology* 4(2): 1–38.

Borgmann, A. 1984. *Technology and the character of contemporary life*. Chicago: University of Chicago Press.

Brockman, J. (ed.). 2011. *Is the internet changing the way you think? The net's impact on our minds and future*. New York: HarperCollins Publishers.

De Vries, G. 1989. Ethische theorieën en de ontwikkeling van medische technologie. *Kennis en Methode* 13: 278–294.

de Waal, F. 2006. *Primates and philosophers. How morality evolved*. Princeton: Princeton University Press.

Dewey, J. 1957. *Human nature and conduct: An introduction to social psychology*. New York: The Modern Library, Inc.

Gadamer, H.G. 2004 [1974]. *Truth and Method*. Trans. J. Weinsheimer and D.G. Marshall). London: Continuum.

Gezondheidsraad. 2002. *De toekomst van onszelf*. Den Haag: Gezondheidsraad. Available on-line: http://www.gr.nl/pdf.php?ID=517.

Gouinlock, J. 1994. *The moral writings of John Dewey* (Revised edition). Amherst: Prometheus.

Idhe, D. 1993. *Postphenomenology*. Evanston: Northwestern University Press.

Jonas, H. 1984. *The imperative of responsibility: In search of an ethics for the technological age*. Chicago: University of Chicago Press.

Latour, B. 1992. *Where are the missing masses? The sociology of the new mundane artefacts shaping technology, building society*. Cambridge: MIT Press.

Latour, B. 2005. *Reassembling the social: An introduction to actor-network-theory*. Oxford: Oxford University Press.
Mol, A. 2008. *The logic of care: Health and the problem of patient choice*. New York: Routledge.
Ricoeur, P., and J.B. Thompson. 1981. *Hermeneutics and the human sciences: Essays on language, action and interpretation*. Cambridge: Cambridge University Press.
Stemerding, D., T. Swierstra, and M. Boenink. 2010. Exploring the dynamic mutual interaction of technology and morality in the field of genetic susceptibility testing: A scenario study. *Futures* 42: 1133–1145.
Swierstra, T. 1999. Moeten artefacten moreel gerehabiliteerd? *Kennis en methode* 23(4): 323–334.
Swierstra, T. 2002. Moral vocabularies and public debate: The cases of cloning and new reproductive technologies. In *Pragmatist ethics for a technological culture*, ed. T.E. Swierstra, J. Keulartz, J.M. Korthals, and M. Schermer, 223–240. Deventer: Kluwer Academic Publishers.
Swierstra, T. 2010. Het huwelijk tussen techniek en moraal. In *Moralicide. Mens, techniek en symbolische orde. (Jaarboek Civis Mundi i.s.m. Rathenau Instituut)*, ed. M. Huijer and M. Smits, 17–35. Rotterdam: Lemniscaat.
Swierstra, T., and A. Rip. 2007. Nano-ethics as NEST-ethics: Patterns of moral argumentation about new and emerging science and technology. *Journal for Nanoethics* 1(1): 3–20.
Swierstra, T., and K. Waelbers. 2010. Designing a good life: The matrix for the technological mediation of morality. *Engineering Ethics* 18(1): 157–172.
Swierstra, T., D. Stemerding, and M. Boenink. 2009. Exploring technologically induced moral change. The case of the obesity pill. In *Evaluating new technologies*, ed. P. Sollie and M. Düwell, 119–138. Dordrecht: Springer.
van Asselt, M., A. Faas, F. van der Molen, and S. Veenman. 2010. *Uitzicht: toekomstverkenningen met beleid*, Wetenschappelijke Raad voor het Regeringsbeleid. Amsterdam: Amsterdam University Press.
Verbeek, P.-P. 2011. *Moralizing technology. Understanding and designing the morality of things*. Chicago: University of Chicago Press.
Waelbers, K. 2009. Technological delegation: Responsibility for the unintended. *Journal for Science and Engineering Ethics* 15: 51–68.
Waelbers, K. 2011. *Doing good with technology: Taking responsibility for the social role of technologies*. Dordrecht: Springer.
Waelbers, K., and A. Briggle. 2010. Technology, the good life, and liberalism: Some reflections on two principles of neutrality. *Techné: Research in Philosophy and Technology* 14(3): 176–193.
Williams, B. 1985. *Ethics and the limits of philosophy*. London: Fontana Press/Collins.

Part IV
Ethical and Societal Aspects of Concrete Technological Developments

Chapter 13
Quandaries of Responsible Innovation: The Case of Alzheimer's Disease

Yvonne M. Cuijpers, Harro van Lente, Marianne Boenink, and Ellen H.M. Moors

Abstract The interest in responsible innovation has led to various activities to include social, economic and moral concerns in the process of innovation. This ambition, however, brings along several fundamental questions. We encountered these in a project on responsible innovation in the case of new molecular early diagnostics for Alzheimer's disease (AD). Currently, a number of novel technologies are being developed for in vivo early diagnosis of AD, by identifying and testing new molecular biomarkers. In our project, we study scientific and clinical uncertainties in technology development, analyze the social and cultural as well as the moral implications of existing and alternative ways to deal with them. In this chapter we summarize the fundamental questions about responsible innovation in terms of six 'quandaries': problematic, difficult and ambiguous conditions that somehow require fundamental and practical decisions.

13.1 Introduction

In recent years, the notion of 'responsible innovation' has become fashionable amongst policy makers, firms and researchers. Based on the insight that technologies are not neutral and that innovation may have serious side effects, the ambition is proposed to include concerns about the social, economic and moral consequences of new technologies and their embedding in society. The European Commission, for

Y.M. Cuijpers (✉) • H. van Lente • E.H.M. Moors
Innovation Studies, Copernicus Institute of Sustainable Development, Utrecht University, P.O. Box 80115, 3508TC Utrecht, The Netherlands
e-mail: y.m.cuijpers@uu.nl; h.vanlente@uu.nl; e.h.m.moors@uu.nl

M. Boenink
Department of Philosophy, University of Twente, P.O. Box 217, 7500AE Enschede, The Netherlands
e-mail: m.boenink@utwente.nl

J. van den Hoven et al. (eds.), *Responsible Innovation 1: Innovative Solutions for Global Issues*, DOI 10.1007/978-94-017-8956-1_13,

instance, urges researchers to investigate the possibilities of responsible innovation, defined as "[. . .] a transparent, interactive process in which societal actors and innovators become mutually responsive to each other with a view on the (ethical) acceptability, sustainability and societal desirability of the innovation process and its marketable products (in order to allow a proper embedding of scientific and technological advances in our society)" (Von Schomberg 2011).

Likewise, the Dutch research foundation NWO has launched a program to explore and support 'responsible innovation', which, in their definition "[. . .] concerns research, development and design and reflects social values, interests, needs, rights and welfare" (Nederlandse Organisatie voor Wetenschappelijk Onderzoek 2008a).

Unsurprisingly, the ambition of responsible innovation is not straightforward. It entails important challenges for policy makers, technology developers and social science researchers that seek to unravel the possibilities and limits of responsible innovation. We are involved in a project on responsible innovation in the case of new molecular early diagnostics for Alzheimer's disease (AD). We collaborate with Leiden Alzheimer Research Netherlands (LeARN; van Buchem 2007), a public-private partnership of several Dutch academic medical centers, universities and companies (a.o. Organon and Philips), funded by the Dutch Centre for Translational Molecular Medicine (CTMM). LeARN develops a number of novel technologies for in vivo early diagnosis of AD, by identifying and testing new molecular biomarkers made visible by PET-, MRI scans and/or Cerebro Spinal Fluid (CSF)-analysis. Such biomarkers are promising tools to enable earlier and more reliable diagnosis of AD, to identify leads for drug development, and to enable monitoring of disease development and/or drug response. In our project, we study scientific and clinical uncertainties in technology development, analyze the social and cultural as well as the moral implications of existing and alternative ways to deal with them. Eventually, we hope to design strategies for responsible uncertainty reduction in innovation of AD diagnostics.

When we started our study on the possibilities of responsible innovation in the case of new molecular early diagnostics for AD 2 years ago, we came across various basic questions concerning the ambition, assumptions and approaches of responsible innovation. In this chapter we summarize our findings and struggles in terms of what we have labeled as 'quandaries': problematic, difficult and ambiguous conditions that somehow require fundamental and practical decisions. We think that this reflection is of general interest for researchers, technology developers and policy makers.

13.2 Quandary One: Technocentric or Multi-actor Views on Innovation

Responsible innovation, in a basic sense, points to the integration of viewpoints. By explicitly coupling research, development and design, and social values, interests, needs, rights and welfare, responsible innovation stresses the alignment of the social landscape, and research and innovation within this landscape. This, however, raises questions about *where to start* any thinking about responsible innovation. One could, for example, start in the context of ongoing technological and scientific developments, including potential controversial ones, because: "Considering the solutions that technological and scientific know-how is capable to offer to societal issues and problems, it is important to examine their ethical and societal aspects" (Nederlandse Organisatie voor Wetenschappelijk Onderzoek 2008b). Another starting point is the articulation of societal needs and 'grand' challenges, because: "When it comes to solving global problems (....), people have great expectations from technology and science" (Nederlandse Organisatie voor Wetenschappelijk Onderzoek 2008b).

Clearly, the development of early diagnostics of Alzheimer's disease is highly intertwined with the societal challenges posed by an aging society. The fact that the population is aging confronts public health systems, social care as well as the economic system as a whole with tough questions, requiring innovators and policy makers to rethink current practices. Against this backdrop, research programs aim to develop a more reliable and earlier diagnosis of AD based on biomarkers, working towards a future in which, hopefully, prevention and personalized treatment of AD will be available. Scientific and clinical efforts, as well as public funding are being invested in this type of research. Where should thinking about responsible innovation start, in the first place?

A technocentric perspective on responsible innovation would focus on the promises of early diagnostics and investigate questions like: How to responsibly embed this technology in society? What will be the social, cultural, ethical consequences of such techniques and how can we deal with them? In that case thinking about responsible innovation starts with the innovative development itself.

This, however, is not the only option. One may also start with for example the aging population and the care for the elderly, which concerns many actors and their viewpoints. Such a starting point would employ a multi-actor perspective on responsible innovation. It would focus on a societal problem or need, in which many actors are involved. In this case, different technological and non-technological options may be expected to provide some sort of solution, a means to deal with the problem or to fulfill the need.

Both the technocentric and the multi-actor perspectives have a history and their drawbacks have been reported in various ways. The technocentric perspective has been accused of a deterministic bias. It puts the expectations and promises of technology developers centre stage, while other stakeholders only enter the scene when they react to these expectations. The focus is on reducing negative side effects of an innovation in order to improve the acceptance of technology. Moreover, by closely collaborating with persons who have a strong interest in a particular technology, there is a risk of being co-opted and becoming less critical (Johnson 2007). Being co-opted brings along the risk of neglecting questions concerning the *need* for the development of these technologies in the first place. It thus tends to ignore questions such as: How will this technology solve social problems? How will research address the social problems? Is this technology a response to these needs or issues? Who will benefit from this development? Should we invest our scarce sources in this development? What are alternative ways to deal with a specific societal issue?

A multi-actor perspective, on the other hand, does not take the promising development as a starting point, but starts with a social problem and the various ways in which this is voiced. Hence, it does not privilege the perspective of technology developers but emphasizes that technological developments are social developments. In this view "emerging technologies are emerging social arrangements, social relationships and meanings" (Johnson 2007). And since sociotechnical developments embody values, the multi-actor view highlights how values are infused in social practices, social arrangements, systems of meaning, as well as in the technological artifacts themselves (Johnson 2007). Likewise, the need for and the development of an early diagnosis (including the social institutions, mindsets and values), are being constructed, ignored, or destructed in multiple places simultaneously. According to the multi-actor perspective, it is relevant that these processes are all constitutive for early diagnosis and they can all be useful starting points. So, the fear of getting demented, ideas of successful aging, social workers wanting to prevent crisis situations, visits to a doctor when there is a suspicion of dementia, support and care for elderly when getting the diagnosis AD, changing diagnostic criteria and protocols, TV programmes for elderly practicing memorizing shopping lists – all these developments may be seen as parts of the distributed construction of early diagnostics for AD. In this perspective, responsible innovation appears as a task to acknowledge this richness and to shape innovation accordingly. Yet, the same richness and multi-directionality of the perspective may paralyze the whole endeavor. Where to start? With the current instruments? With the patients? The clinical practices? The public perception of AD? Arguably, all these starting points are justified, yet they cannot be followed at the same time.

13.3 Quandary Two: Singular or Multiple Futures?

Any inquiry for responsible innovation will entail sketches of a future, or futures. The question, then, is whether one should assume a sketch of a singular future, or prefer the ambivalence of multiple futures. This, then, is the second quandary for responsible innovation, which relates to the *goal* of the exercise: singular or multiple futures?

In research conducive to early diagnostic instruments for AD a strong, singular, future is being sketched. Research on biomarkers and advances in imaging techniques, as the dominant argument goes, will enable an earlier and more reliable diagnosis of AD, which will have two advantages. An early diagnosis is valuable for patients because it reduces uncertainty about their health status and it enables them to prepare for dementia and to organize care and support. Second, the diagnosis of AD at an early stage enables biomedical research to study the early development of

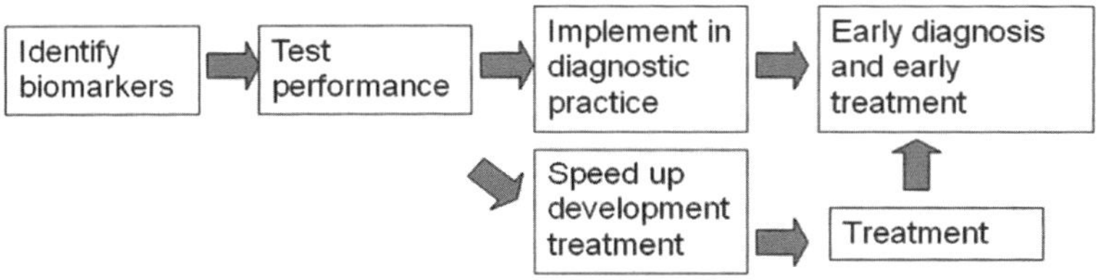

Fig. 13.1 The singular future of early diagnosis of AD

the disease and to monitor the treatment through biomarkers at an early stage of the disease, at which it is expected to have most effect. Within this future image, AD will be diagnosed early and treated with disease modifying drugs. And while disease modifying drugs are not yet available, an early diagnosis will provide support and care for patients and informal caregivers.

This future of early diagnostic instruments entails a chain of research stages, starting with hypotheses about the most important mechanisms in the brain causing AD and moving onwards to the identification of biomarkers which allow to signal (or mark) these processes. Then, these biomarkers will be visualized through dedicated MRI or PET scans, or measured with chemical analysis of the cerebrospinal fluid. If these tests offer proof of sufficient sensitivity and specificity they can be implemented in the diagnostic process, providing more certainty to patients and the possibility to organize care and support. These tests could then be used to speed up research into drug development. The final promise is that this leads to an earlier diagnosis and treatment of AD. See Fig. 13.1.

These expectations are articulated in the Dutch research program LeARN (LeARN), which is working on these developments, and are embedded in broader expectations about molecular medicine that guide the research center CTMM which co-funds the research of LeARN (Center for Translational Molecular Medicine 2006). The vision of CTMM is as follows:

> The practice of medicine in the 21st century will be very different from how it is today. We are on the brink of a paradigm shift both in medical technology and in its therapeutic applications and effects. New technologies will enable clinicians to take great strides forward in addressing the main obstacles to effective healthcare: (too) late diagnosis of disease, medication that is ineffective or has serious side effects, and delays in translating therapeutic innovations from the lab to clinical practice. The impact of the most common lethal and debilitating diseases, such as cancer, cardiovascular diseases, and neurodegenerative diseases like Alzheimer's will be significantly reduced, and people who must live with disease will enjoy an improved quality of life. Mere stargazing? It need not be. Molecular Medicine holds the promise to realize this paradigm shift

These promises are indeed held to be more than mere stargazing and have already led to proposals for new diagnostic guidelines for AD (e.g. Alzheimer Association 2011, Dubois et al. 2007), which include molecular imaging techniques and chemical analysis of biomarkers, both in the research and the clinical context. "Expansion of the conceptual framework for thinking about Alzheimer's disease to include a "preclinical" stage characterized by signature biological changes [i.e. biomarkers] that occur years before any disruptions in memory, thinking or behaviour can be detected. The new guidelines [. . .] propose a research agenda that builds on promising preliminary data emerging from recent studies" (Clifford et al. 2011).

This stipulated future of AD, thus, is underpinned by results, but is contested as well. It is uncertain whether such research eventually will lead to these particular futures. The uncertainties are also fueled by disputes about definitions (What is the distinction between normal and pathological aging?); about limits of the current knowledge on AD (What is the relation between specific changes in the brain and the symptoms of dementia?); about moral questions (What is the value of early diagnosis when treatment is lacking?); about strategic issues (Should we not spend the money and effort on better care?); about the innovation trajectory (Will the research trajectories of PET, MRI, CSF succeed in developing diagnostic instruments?); about the future implementation/embedding (Who will be offered early molecular diagnosis for AD?) and about visions on 'good diagnosis' (How early do we want to diagnose AD?).

The promises of early diagnostic instruments measuring biomarkers are based on the expected future availability of disease modifying treatments. Yet, if one considers the development, or the possible effectiveness of disease modifying treatment under development, as uncertain, and when an earlier identification of the disease merely serves to organize the best care and support available, there are more possible routes to provide early care and support and to achieve an earlier diagnosis, besides imaging or measuring biomarkers. And when there are many possible routes to innovate the diagnosis of AD, the question of responsible innovation also multiplies.

Instead of investigating the singular future and its particular uncertainties, we decided to explore the multiple futures at stake. As an example we will describe alternative futures that were prominent during our observations in so-called Alzheimer Cafés (Cuijpers and Van Lente forthcoming). Alzheimer Cafés are monthly events in The Netherlands where patients, family, and local professionals in the field of dementia meet to exchange experiences, ask informally for advice and discuss a specific theme.

The futures of AD that circulated here were diverse. For instance, the problem of dementia was not so much considered as a medical problem, but as a care problem. Also in this future image the identification of dementia at an early stage is important. It refers to 'early signaling' of dementia by care professionals, general practitioners, as well as the general public. This may avoid crisis situations, misunderstandings and may provide persons timely with needed care and help. This is perceived as better than the present situation in which often persons go to see a general practitioner very late in the development of the disease, when they are already running into a lot of problems. Signaling problems at an early stage and receiving a diagnosis, thus, is not seen as a stepping stone to 'cure', but provides the possibility of timely organizing the care, support and guidance a person needs.

To provide good care for persons with dementia, the disease modifying treatments were not put central, but the main concern was the development of customized, patient centered care arrangements, and the 'tinkering' needed to achieve the best care in that specific situation (Mol et al. 2010). In this future image, the differences between the development of the disease in individuals, as well as the coping strategies of patients and their partners is acknowledged. Since the problem is not singular, there will never be one solution to strive for, but always a careful balancing of options.

Other future images we encountered concerned the development of Alzheimer's disease from a societal perspective. For instance, our society is facing a growing aging population with a growing number of persons with AD. Ageing baby boomers will increasingly put pressure on the health care system and the economic system. Another desired future development concerned the social status of persons with Alzheimer's disease. The Alzheimer Cafés aim to improve the position of patients and their relatives by reducing the stigma and taboo on AD, and emancipate AD patients and their families in order to better deal with the condition. To conclude, efforts for responsible innovation may be predicated on a particular future, or may embrace the plurality of futures. Depending of the problem definition and the perspective, responsible innovation of early diagnosis of Alzheimer's disease is likely to take a different shape.

13.4 Quandary Three: Identifying with Whom?

Questions about what constitutes responsible innovation are often triggered by new technological developments, like the rise of genomics, nanotechnology or synthetic biology. The funding for research on responsible innovation may even be closely linked to the funding of technological development itself, as in the case of genomics and nanotechnology in both the USA and Europe. Moreover, it is now quite common for researchers in the field of responsible innovation to use methods in which they collaborate or engage with technology developers (Guston and Sarewitz 2002; Fisher and Mahajan 2006). As a result, it is easy for researchers in responsible innovation to identify with the scientists and engineers working on a specific innovation. As noted above, a close collaboration with actors who have a strong interest in bringing about a particular technology, brings the risk of 'going native' and thus to become less critical (Johnson 2007).

To avoid such lock-in, one should go beyond the perspective of the technology developers, as was already stipulated in the first quandary. The third quandary points to another difficulty of identifying with the ideas of the developers: the moral question of whose interest to pursue. In general, one may argue that one of the conditions that makes innovation responsible is that it is aligned with important social needs and moral values. Some work in the field of Science & Technology Studies seems to be implicitly driven by the desire to support groups or views that tend to be marginalized in political, public or professional debates. However, it does not suffice simply to side with the perspective of more marginalized stakeholders either. Yes, highlighting what is less visible or not taken seriously is a valuable contribution to making innovation more responsible. However, an ethical interpretation of responsible innovation requires that *all* relevant stakeholders and their views and interests are taken into account, including the dominant ones.

The question with whom to identify relates to the issue of users in the innovation process. Users often develop new functions for technologies, solve unforeseen problems and propose or even develop innovative solutions. Therefore, users are recognized as important sources or even co-developers of innovations, and can have an impact on the direction of technological developments and innovations, especially in early stages of technology development (e.g. Von Hippel 1976; Oudshoorn and Pinch 2003; Lüthje et al. 2005). Smits and Boon (2008) summarized the reasons for user involvement as follows: (1) users can address market failures and suggest ways to overcome them; (2) they contribute to adoption of innovations by articulating their creative potential in the form of wishes and experiential knowledge; (3) they can support the boundary conditions of innovation processes and by this are instrumental to processes; (4) they can 'champion' innovations and by this form a counterforce to potential (ethical) objections; (5) and they have the moral and democratic right to co-decide on and co-produce innovations that have a great impact on their lives.

Likewise, multi-actor involvement can contribute to more responsible innovations. Research in responsible innovation thus should investigate how this inclusive form of deliberation can be facilitated (Gutman and Thompson 1996). Ultimately, this means that research in responsible innovation should engage with all stakeholders but identify with no one in particular. This aim does not presuppose a view from nowhere (Nagel 1989), a detached moral point of view. It does, however, require the researcher to continuously compare and mutually assess all possible viewpoints and considerations.

This is easier said than done. In the case of innovating technologies for diagnosing AD, for example, many actors may be potentially affected by this development. An earlier diagnosis addresses governments and all citizens by promises to reduce public health care costs, by providing timely home care allowing persons to live at home longer. It influences the future prospects of persons suffering from AD. And there actors who for several reasons do not use, or are against the use of these innovations (Henwood et al. 2003; Katz et al. 2002). In the case of AD, for example, often patients do not want to get diagnosed due to a fear of the prognosis of AD itself (denial), or a self-chosen and conscious 'blissed ignorance'. For insurance companies early diagnosis might be a way to assess the risks of a person to develop AD. For researchers it provides new possibilities for research on the causes of AD and interventions. Other stakeholders involved are municipalities, nursing homes, home care institutions, welfare organizations, all elderly people (or even all healthy people who may be at risk – which means everyone), neighbors, industry, housing corporations, and more. To include all these stakeholders in deliberation on the desirability of emerging diagnostic technologies for AD is an immense task. In practice, then, one has to focus on some stakeholders and leave others aside, due to limitations of time and funding. How to make a well considered selection?

To identify with all stakeholders, thus, is a complicated route, to say the least. An additional complicating factor is that different stakeholders will have different meanings of 'Alzheimer's disease'. AD can be an existential problem for patients and caregivers, a process in the brain for biomedical researchers, and policy makers may approach it as a socio-economic issue. While one may consider all these meanings as valid, it is not easy to acknowledge them at the same time. Any practical effort of deliberation will imply a choice. The quandary, thus, is: identifying with whom?

13.5 Quandary Four: Process or Outcome?

The ambition of responsible innovation, in principle, entails two possible questions: 'How to innovate in a responsible way?', and 'What kind of innovation (as a result of an innovation process) is responsible?' In other words, does responsible innovation refer to the process or the outcome of a process? This basic distinction leads to very different kinds of questions and activities.

When responsible innovation refers to the *outcome* – the innovative product and the societal embedding of this product – a researcher on responsible innovation should assess the products and systems as envisioned and might advise on conditions in which this innovation may be responsible. In the case of early diagnostics for AD there are many different kinds of outcomes envisioned. Generally three scenarios are mentioned by the researchers in the field: (1) the use of these instruments as an add-on in current diagnostic practice; (2) the use of these instruments to distinguish between patients with mild memory complaints (Mild Cognitive Impairment) who will develop Alzheimer's Disease, and those who will not; or (3) a pre-symptomatic diagnosis of Alzheimer's Disease, even before any symptoms are present (which is then positioned far in the future). We could try to analyze possible and plausible outcomes of this innovation and the conditions in which early diagnostics for AD would be responsible.

Mattson et al. (2010) and Gertz and Kurz (2011) pursued this approach. Mattson reviews possible clinical consequences of early diagnosis of AD. The issues that should be anticipated include (a) the risks of erroneous tests, misdiagnosis and wrong treatment; (b) the consequences of an early diagnosis for a patient and for the relatives, including the role of stigmatization, feelings of despair and hopelessness; (c) the attitude of doctors bringing the bad news. A big advantage of an early diagnosis is that patients can prepare at an early stage of the disease, and get the help they need at a later stage, when they will be too demented to decide on this. An ethical problem in this case is whether a patient at the early stages of the disease might misjudge his or her future self's best interest. There is a problem in making decisions about a future self when developing such a thoroughly life changing disease, such as AD. All these issues could already be discussed or decided upon.

Gertz and Kurz (2011) discuss the improvements of diagnostic methods to enable a very early diagnosis of AD, while there is no such progress in the development of disease-modifying treatments. They emphasize the need to change the current practice of diagnosing AD, to more actively include the patient in the decision to undergo an early diagnosis, and to make very clear to this patient that there will be a lack of therapeutic options when the diagnosis is positive.

These two articles discuss conditions under which such an early diagnosis could be responsible and the measures that should be taken, or discussed in order to decrease the undesirable consequences of this development for the patients involved. By focusing on the outcome of the innovation process, it 'black boxes' the decisions taken during the innovation process.

The other approach would be to open the black box, and to try to make the innovation *process* more responsible. Hence, process criteria become more important. A researcher of responsible innovation could try to broaden the issues taken into account *within* the innovation process, by informing stakeholders on different possible perspectives, facilitate the sharing of perspectives, values and interests between stakeholders, and stimulate social learning. Scenario- or multi-stakeholder workshops or organizing public dialogues could be examples of this. In the case of early diagnostics for AD, this might involve additional activities from the side of researchers on responsible innovation, to broaden the current Health Technology Assessment (HTA) undertaken in the LeARN research program. The HTA currently involves scientists, clinicians and health economists only and focuses solely on financial costs and quality of life. This HTA could include contextual factors, pre-conditions and broader considerations. De-contextualized early diagnostics euphoria can create constraints with regard to aligning disease management, integrated care, or life-course perspectives on AD.

So, the basic ambiguity in the term 'innovation', which may refer to either outcome or process, resonates in the ambition of responsible innovation. The two are not automatically aligned: a responsible outcome of an innovation process does not need to be the result of a responsible innovation process. And vice versa, holding to process criteria in an innovation process does not need to result in a responsible outcome.

13.6 Quandary Five: Speculation or Plausibility

Innovation (in particular in emerging technology) is a rather elusive subject: it is, by definition, about entities that do not exist. Technological developments, which aim at innovations in the future, largely consist of promises and expectations that cannot directly be assessed in terms of veracity. They may even be highly speculative. At the same time, such claims are grounded in currently (perceived) problems and in current ideas on what the world is like.

Futures, moreover, are not innocent. From the sociology of expectations we learn that promises and expectations are 'performative', meaning that expectations 'do' something. Innovations, as they tend to go with many expectations, already have consequences before they are embedded in society, or even developed, through these expectations. Through their content, expectations are able to coordinate action, by allocating roles, creating linkages and obligations between actors and by defining agendas. In this way they shape technological developments. Expectations can also be used by actors to legitimize actions, mobilize funding and attention of other actors. They are used in decision making processes to reduce the uncertainty inherent in technological development (Van Lente 1993; Van Lente and Bakker 2010).

Research in responsible innovation (and its funding) is also often triggered by the same visions of the future, asking whether the envisioned future is desirable. As Nordmann and Rip have pointed out for the case of ethics of nanotechnology, this type of 'parallel research' runs the risk of uncritically assuming that these expectations are plausible (Nordmann 2007; Nordmann and Rip 2009). Similar warnings could be issued for social and legal (ELSA) research into emerging technologies more generally.

Nordmann and Rip warn that in the case of nanotechnology, and other emergent technologies, ethicist have the tendency to go along too easily with speculative visions and expectations concerning technological development (or even describe speculative future scenarios themselves). Ethicists then continue to ask attention for the ethical concerns these (expected) technologies raise, "*as if such technologies were upon us already*". Moreover, when ethicists discuss the ethical aspects of an expected outcome of technological developments they contribute to the credibility and the power of these expectations, even if they stress the negative consequences these developments might have. It is thus problematic that the ethicist presents remote possibilities as plausible technological developments. When these expectations fail to come true, research in responsible innovation may be futile, irrelevant, and squander the scarce resources for this type of research. Another drawback of such speculative ethics is that one misses out on (often more mundane) ethical issues occurring *during* the technology development process. The development process itself is black boxed. Nordmann and Rip suggest two strategies to deal with these issues. The first is to increase discussion about the quality of promises and representations of emergent technologies: some sort of reality check. The second is to focus on more specific technologies (in our case, say, a specific biomarker test for AD), rather than on general ideas of technological developments (for example the tendency towards molecularization in medicine).

Grunwald, on the other hand, stresses the value of speculating about the future, especially when considering the societal issues of new technologies. The purpose of a more speculative form of ethical reflection is (1) to provide a preliminary conceptual and substantive structure for a future field of ethics; (2) to point out critical questions that require increased examination in the future; (3) to contribute to identifying gaps of knowledge; (4) to learn something about and for us today (e.g. what is their implicit criticism about the present, how do they suggest us to change?). Rather than a 'reality check' Grunwald emphasizes vision assessment, to uncover the cognitive and normative content of the visions, to evaluate their validity and plausibility, and to confront diverging images of the future with each other, analytically, or with different stakeholders (Grunwald 2004, 2007, 2010).

The development of molecular diagnostic instruments for Alzheimer's disease is definitely liable to speculation, and the question is how to deal with that. The Nordmann & Rip strategy would be to focus on a specific technology, like the combination of biomarker tests developed in LeARN, together with a reality check of the claims being made. Lucivero (Lucivero et al. 2011) elaborates what such a reality check (or rather plausibility assessment) would entail. She proposes to distinguish claims about the technology in the lab, about the use of the technology, and about its desirability. A careful check is needed of, for example, claims about the 'early' in early diagnostics. Are we still talking about patients with subjective complaints, or about testing a-symptomatic individuals? This has immediate implications for the context of use. But even if molecular diagnostics only concerns patients with complaints, the role of the biomarker tests may be envisioned as a complete diagnostic tool in itself, or as an addition to a complex set of tests. Also the reason why different stakeholders are interested in these diagnostics may differ, from getting knowledge about one's health state, receiving clues how to arrange care and treatment, gaining knowledge about the pathological disease mechanisms underlying the disease, or searching for reassurance that everything is all right. Desirability claims cannot be assessed on the basis of invariable norms and values; morality itself may shift partly because of technical developments. So, careful reflection on interaction of technology and morality is necessary. For example: how will norms about cognitive functioning change as a result of developments in AD diagnostics? And how does this affect the experience of AD?

Grunwald's proposal, on the other hand, would entail that we explicate the visions implicit in the LeARN project and more generally in molecular diagnostics. The problem definitions and the presuppositions of these visions should be assessed, and alternative scenarios should be developed to create a broad public debate on what kind of future vision is desirable.

13.7 Quandary Six: Responsibility for the Future or Responsibility for the Present

A final quandary that we encountered in the aim to contribute to a responsible diagnostic practice of AD, is whether we should focus on a responsible *future* practice, or on a responsible *current* practice. This issue is related to some of the ambiguities discussed above, in particular the issue of process or outcome and the issue of speculation versus plausibility. Again, we adopt promises and expectations of the Alzheimer researchers, and try to formulate conditions any practice of early AD diagnostics should satisfy to be responsible. Or we could take a more skeptical stance and question how current innovations should proceed to ensure a responsible research practice. What is at stake here is not just the object of the responsibility claim, but also its time-frame.

There may be a difference here between social and ethical approaches of responsible innovation. From a social perspective, responsible innovation is usually about acceptability: an innovation can be considered responsible if it is actually accepted by all actors involved. This means that the product of innovation can be judged on its own, regardless of the innovation process. From an ethical perspective, however, it is possible to say that an innovation that is accepted by all involved is nonetheless not responsible, because either some stakeholders or specific considerations were neglected- or both. From an ethical point of view, then, the process is more important, implying that responsible innovation encompasses both the present and the future.

For our case, this means that contributing to a responsible practice of (early) diagnostics of AD should start right *now*, by facilitating the translation of research into clinical practice in such a way that the views of all relevant stakeholders are taken into account. Considerations of patients, informal caregivers, elderly people in general, and medical professionals should receive attention already in the research phases. After all, their views on what constitutes the potential benefit (and drawbacks) of the aimed for innovations may differ from what researchers perceive as its benefit (or drawbacks).

In our first explorations of the field, such discrepancies became already visible. After introduction our research one doctor responded with the remark "It is only the persons holding test tubes who are interested in this." And clinicians, for example, asked: What is the value of these biomarkers for the diagnosis in clinical practice? What is *really* in it for the patient? Patients who go to a hospital to get diagnosed are send there by the general practitioner, a nurse said. This means that they already have complaints. If you want to have an earlier diagnosis, you don't need novel diagnostic tools, but need to go to the general practitioners. Now, they often do not recognize signs of early dementia and do not refer patients to the hospital. Furthermore, the clinical diagnosis AD is not equal to interpreting images from MRI scans, which are mainly used for additional information or research purposes. Basically, some clinicians do not have high expectations about this type of research on the short term in clinical practice, and they suggest other routes to diagnose persons at an earlier stage, for example the education of general practitioners in early signs of dementia. If such considerations are left aside, the result of the innovation process risks rejection and contestation.

The quandary is not solved, however, by rendering the *now* responsible, because even in facilitating a responsible process here and now, we anticipate the future. Such anticipation itself can be more or less responsible. We indicated already that it is fraught with the risk of speculation. Nordmann's criticism of what he calls 'if and then ethics' (Nordmann 2007) implies that researchers on emerging technologies should take responsibility for the images of the future they use. After all, images of the future do have repercussion in the present. If we go along too easily with the expectations of the research on early diagnostics, for example, we may reproduce an irresponsible bias towards biomedical definitions of the problem as well as technical solutions for this problem (George and Whitehouse 2009). Taking responsibility for the present then also means that we should take a critical stance towards the problem definitions and assumptions underlying current attempts at innovation. Finally, working on the present process of innovation is inevitably directed towards the future in another sense as well. Responsible innovation, whatever its form, aims at a better technology for a better world. So even if we decide to focus on the process of innovation only, we inevitably claim to contribute to a *future* world as well, in which the innovations will be embedded in practice.

Yet, we should avoid the pitfalls of simplistic thinking about shaping the future. After all, the interaction of technology and society is replete with complexity and contingencies. Does it make sense at all, then, to claim that attending to the present innovation process will guarantee a responsible outcome in the future? Of course not. What we can do, however, is try to define minimum conditions for a future practice of AD diagnostics to be responsible. In addition, and perhaps even more important, we had better think about ways to ensure that innovation processes can be redirected once it becomes clear that the most recent outcome does not satisfy such minimum criteria. Responsibility for the future then takes the form of permanent and flexible guiding.

13.8 Conclusion

Responsible innovation is not an oxymoron but not a straightforward task either. Our basic finding concerns the tension between simplicity and complexity. Any practical translation of the notion of responsible innovation has to find a path through the intrinsic and intricate complexities of socio-technical change – a path that has to avoid overly simplistic assumptions regarding innovation and responsibility, as well as a surrender to the full complexity of social and technical life.

In this paper we delineated six basic tensions that we encountered in our research into the early diagnostics of AD. The six quandaries refer to basic questions about responsible innovation. See Table 13.1. The quandaries echo the ambiguity of the term responsible innovation itself: is it to safeguard innovation by making it acceptable, or is it to enhance responsibility through innovation or other means?

Does this set of quandaries imply that responsible innovation is an evasive concept? Yes and no. It is impossible to certify innovations as responsible, because innovations are never finished and they are part of bigger social, technical and moral changes. That is, innovations will continue to raise questions about responsibility. Yet, the concept seems to be helpful as it points to the capability to choose. The

Table 13.1 Six quandaries of responsible innovation

	Basic question about responsible innovation	Quandary of responsible innovation
1	Where to start?	Technocentric or multi-actor perspectives?
2	Where to end?	Singular or multiple futures?
3	With whom?	Developers or stakeholders?
4	What's the goal?	Process or outcome?
5	What to question?	Speculation or plausibility?
6	Responsible for whom?	Responsibility for the future or the present?

identification of the six quandaries could help both researchers and policy makers, not only to make their choices more explicit, but also to be aware of choices that could be made.

References

Alzheimer's Association. New diagnostic criteria and guidelines for Alzheimer's disease. http://www.alz.org/research/diagnostic_criteria/. Accessed 11 July 2011.

Center for Translational Molecular Medicine. 2006. Business plan. http://www.ctmm.nl/pro1/general/start.asp?i=2&j=0&k=0&p=0&itemid=52&folder=About%20CTMM&title=Business%20Plan. Accessed 11 July 2011.

Clifford, R.J. Jr., M.S. Albert, D.S. Knopman, F.M. McKahnn, R.A. Sperling, M.C. Carillo, B. Thies, and C.H. Phelps. 2011. Introduction to the recommendations from the national institute on aging and the Alzheimer's Association Workgroup on Diagnostic Guidelines for Alzheimer's Disease. *Alzheimer's & Dementia* (16 April 2011): 1–6.

Cuijpers, Y., and H. van Lente. (forthcoming) Early diagnostics and Alzheimer's disease: Beyond 'cure' and 'care'. *Technological Forecasting and Social Change*. doi: 10.1016/j.techfore.2014.03.006.

Dubois, Bruno, Howard H. Feldman, Jacova Claudia, Steven T. DeKosky, Barberger-Gateau Pascale, Cummings Jeffrey, Delacourte André, et al. 2007. Research criteria for the diagnosis of Alzheimer's disease: Revising the NINCDS–ADRDA criteria. *The Lancet Neurology* 6(8): 734–746.

Fisher, E., and R.L. Mahajan. 2006. Midstream modulation of nanotechnology research in an academic laboratory. In *Proceedings of ICEME2006 ASME international mechanical engineering congress and exposition*. http://csid.unt.edu/files/Fisher_MM_IMECE-06%20_.pdf. Accessed 6 July 2011.

George, D., and Peter J. Whitehouse. 2009. The classification of Alzheimer's disease and mild cognitive impairment: Enriching therapeutic models through moral imagination. In *Treating dementia, do we have a pill for it?* ed. Jesse F. Ballenger, Peter J. Whitehouse, Constantine G. Lyketsos, Peter V. Rabins, and Jason H.T. Karlawish, 5–25. Baltimore: The John Hopkins University Press.

Gertz, H.-J, and A. Kurz. 2011. Diagnosis without therapy – Early diagnosis of Alzheimer's disease in the stage of mild cognitive impairment. *Nervenarzt* 82(9): 1151–1159.

Grunwald, Armin. 2004. Paper 5: Vision assessment as a new element of the FTA toolbox. Paper presented at EU-US seminar: New technology foresight, forecasting & assessment methods, Seville, http://forera.jrc.ec.europa.eu/fta/papers/Session%204%20What\T1\textquoterights%20the%20Use/Vsion%20Assessment%20as%20a%20new%20element%20of%20the%20FTA%20toolbox.pdf. Accessed 11 July 2011.

Grunwald, A. 2007. Converging technologies: Visions, increased contingencies of the conditio humana, and search for orientation. *Futures* 39(4): 380–392.

Grunwald, A. 2010. From speculative nanoethics to explorative philosophy of nanotechnology. *NanoEthics* 4(2): 91–101.

Guston, D.H., and D. Sarewitz. 2002. Real-time technology assessment. *Technology in Society* 24(1–2): 93–109.

Gutman, A., and D. Thompson. 1996. *Democracy and disagreement*. Cambridge: Harvard University Press.

Henwood, F., S. Wyatt, A. Hart, and J. Smith. 2003. 'Ignorance is bliss sometimes': Constraints on the emergence of the 'informed patient' in the changing landscapes of health information. *Sociology of Health and Illness* 25(6): 589–607.

Johnson, Deborah G. 2007. Ethics and technology 'in the making': An essay on the challenge of nanoethics. *NanoEthics* 1(1): 21–30.
Katz, J., M. Aakhus, H.D. Kim, and M. Turner. 2002. Young user's attitudes toward ICTS: A comparative semantic differential study of the mobile telephone. *Annales Des Telecommunications/Annals of Telecommunications* 57(3–4): 225–237.
LeARN. Public summary, in vivo molecular diagnostics in Alzheimer's disease. http://www.ctmm.nl/pro1/general/start.asp?i=0&j=0&k=0&p=0&itemid=78. Accessed 4 Aug 2010.
Lucivero, F., T. Swierstra, and M. Boenink. 2011. Assessing expectations: Towards a toolbox for an ethics of emerging technologies. *NanoEthics* 5: 129–141.
Lüthje, C., C. Herstatt, and E. Von Hippel. 2005. User-innovators and "local" information: The case of mountain biking. *Research Policy* 34(6): 951–965.
Mattson, N., D. Brax, and H. Zetterberg. 2010. To know or not to know: Ethical issues related to early diagnosis of Alzheimer's disease. *International Journal of Alzheimer's Disease* 2010: 1–4.
Mol, A., I. Moser, and J. Pols. 2010. *Care in practice, on tinkering in clinics, homes and farms*. Bielefelt: Transcript Verlag.
Nagel, T. 1989. *The view from nowhere*. Oxford: Oxford University Press.
Nederlandse Organisatie voor Wetenschappelijk Onderzoek. 2008a. Maatschappelijk verantwoord innoveren, ethische en maatschappelijke verkenning van wetenschap en technologie, MVI programma notitie april 2008.
Nederlandse Organisatie voor Wetenschappelijk Onderzoek. 2008b. Maatschappelijk verantwoord innoveren, ethische verkenning van wetenschap en technologie, beschrijving themaprogramma.
Nordmann, A. 2007. If and then: A critique of speculative nanoethics. *NanoEthics* 1(1): 31–46.
Nordmann, A., and A. Rip. 2009. Mind the gap revisited. *Nature Nanotechnology* 4(5): 273–274.
Oudshoorn, N., and T. Pinch. 2003. *How users matter: The co-construction of users and technologies*. Cambridge, MA: MIT Press.
Smits, R.E.H.M., and W.P.C. Boon. 2008. The role of users in innovation in the pharmaceutical industry. *Drug Discovery Today* 13(7–8): 353–359.
van Buchem, M.A., B.N.A. van Berckel, et al. 2007. *Project description Leiden Alzheimer Research Netherlands*. Eindhoven: CTMM.
Van Lente, H. 1993. Promising technology, the dynamics of expectations in technological development. Ph.D., University of Twente.
Van Lente, H., and S. Bakker. 2010. Competing expectations: The case of hydrogen storage technologies. *Technology Analysis and Strategic Management* 22(6): 693–709.
Von Hippel, E. 1976. The dominant role of users in the scientific instrument innovation process. *Research Policy* 5(3): 212–239.
Von Schomberg, Rene. 2011. Prospects for technology assessment in a framework of responsible research and innovation. In *Technikfolgen abschätzen lehren: Bildungspotenziale transdisziplinärer Methode*, ed. M. Dusseldorp and R. Beecroft, 39–61. Wiesbaden: Vs Verlag.

Chapter 14
Towards Responsible Neuroimaging Applications in Health Care: Guiding Visions of Scientists and Technology Developers

Marlous E. Arentshorst, Jacqueline E.W. Broerse, Anneloes Roelofsen, and Tjard de Cock Buning

Abstract To develop responsible innovations, the potential impacts on society, both positive and negative, should be identified and incorporated into research, development and design of new technologies. In this research, neuroimaging applications in health care are subject to a constructive technology assessment (CTA) process which is combined with vision assessment that acknowledges the mechanisms and dynamics surrounding innovations. The 'guiding visions' of scientists and technology developers which are currently shaping the future of neuroimaging are presented. Results show that these experts expect that future advances in neuroimaging technologies will make it possible to obtain more insight into both the healthy brain and brain disorders. They consider that these advances will lead to improved prevention, diagnosis and treatment options. The barriers that need to be overcome to realize these guiding visions are identified. In addition, findings show which aspects need further exploration and follow-up activities in order to ensure that medical neuroimaging develops in a more responsible direction.

14.1 Neuroimaging Technologies

Innovations in neuroimaging technologies, for example functional Magnetic Resonance Imaging (fMRI), Positron Emission Tomography (PET), Electro Encephalogram (EEG) and Magneto-encephalography (MEG), make non invasive study of the human brain possible in an increasingly profound way. These technologies

M.E. Arentshorst (✉) • J.E.W. Broerse • T. de Cock Buning
Athena Institute for Research on Innovation and Communication in Health and Life Sciences, VU University Amsterdam, Amsterdam, The Netherlands
e-mail: m.e.arentshorst@vu.nl

A. Roelofsen
Dutch Cancer Society, Amsterdam, The Netherlands

J. van den Hoven et al. (eds.), *Responsible Innovation 1: Innovative Solutions for Global Issues*, DOI 10.1007/978-94-017-8956-1_14,

facilitate study of brain structure, brain function, connectivity and biochemistry. Future developments in these technologies are expected to contribute to solutions for some of the health challenges facing high-income countries. The challenges are the result of an ageing population, rising trends in the number of chronically ill patients and increasing demands for adequate evidence-based care. To date, neuroimaging technologies have contributed to insights into neural processes associated with three major types of disorders: psychiatric (e.g. Malhi and Lagopoulus 2007), behavioral (e.g. Dickstein et al. 2006) and degenerative (e.g. Rosas et al. 2004) brain disorders. Moreover, these technologies have contributed to improved diagnosis and therapies for some of these disorders.

The technological advances to image the function, connectivity and biochemistry of the brain, such as increased resolution and improved options for data-analysis, will probably lead to more understanding of the brain and its disorders. This is expected to result in the development of improved diagnosis and treatment options. Above all, neuroimaging technologies could contribute to novel options for prevention. For example, Alzheimer's disease is a degenerative brain disorder causing a major burden on society, families and the individual. In the near future, neuroimaging technologies are expected to provide more accurate tools to determine preclinical or early-stage Alzheimer's disease, making it possible to prescribe targeted drugs to delay the onset of the disease at an earlier stage (e.g. Petrella et al. 2003). Another example concerns common psychiatric disorders. Diagnosis and treatment of psychiatric disorders is currently based on external symptoms rather than on biological insights. Treatment is therefore often a case of trial-and-error, taking some time before adequate, symptom mitigating medication or other therapy is found (Glahn 2008). With neuroimaging technologies, possibilities open up to diagnose psychiatric disorders in a functional, accurate and timely way and to develop novel pharmacological approaches (Willmann et al. 2008; McGuire et al. 2008). In this way, neuroimaging technologies might potentially improve the quality of life of patients and decrease the burden of these disorders on society (Glahn 2008).

However, besides opportunities, concerns are raised. For example, should people who do not display any symptoms know that they have a subclinical disorder? What is the individual and societal impact of receiving such a diagnosis before the onset of symptoms? Will a person at risk of developing a certain brain disorder endure stigmatization and discrimination when seeking medical insurance or employment? Will the growing knowledge on the brain further increase disease mongering and thereby raise the demand for medical services, medicines and other products (Fuchs 2006; Glannon 2006; Illes and Racine 2005)? Without due attention, these concerns may impede successful realization of the intended benefits.[1]

[1]i.e. the failure to anticipate on controversies surrounding genetically modified crops led to the failure of some industrial innovations (Chilvers and Macnaghten 2011).

14.1.1 Hype-Horror and Promise-Disappointment Cycles

Innovations in science and technology typically follow a pattern known as hype-horror cycles (Swierstra and Rip 2007) and promise-disappointment cycles (Brown et al. 2003). Since the consequences of an innovation are uncertain because the innovation does not yet exist in practice, the expected desirable consequences are speculative and therefore communicated as expectation by its proponents, often scientists and technologists (Swierstra and Rip 2007). These cycles show a pattern in which proponents of the innovation inflate expectations by making promises that are mostly unrealistic (known as 'hypes' or 'breakthroughs') but are desired by many people, in order to gain attention and hence attract financial and political support for further research and development. However, these inflated promises also raise questions and doubts and can result in fear. In this situation, the hype turns into a horror story. This phenomenon is also referred to as 'hopeful monstrosities' by Mokyr (1990, 291): hopeful because proponents believe in the promising future of the innovation and monstrous because the images are not in proportion to the technical possibilities. In response to this created horror story or monstrous image, proponents frequently begin to trivialize the novelty of the innovation and deflate expectations by toning down the inflated promise (Swierstra and Rip 2007). Until, at some point, the innovation has not lived up to its initial promise, resulting in disappointment and extinguishment of the hype. In some cases, this results in severe damage of the reputation and credibility of professional groups, industry, institutions and investors (Brown and Michael 2003; Brown et al. 2003). Retrospectively, promises seem rather naive and unrealistic by those who once held promises regarding an innovation (Brown and Michael 2003). However, to get attention and support for another innovation, the cycle starts all over again (Brown et al. 2003; Brown and Michael 2003) (Fig. 14.1).

Many promising innovations followed the pattern described above and successful realization of the intended benefits to society failed at least partially and in the short-term. Sometimes there are even unexpected negative effects[2] (Brown and Michael 2003; Swierstra and Rip 2007). In addition, the concept behind the inflated promise has the potential to benefit certain stakeholders and is still hawked by its proponents.

In recent decades, several approaches have been developed and applied which aim to manage scientific and technological innovations in a more realistic, desirable way, aiming for more responsible applications and societal support. Examples of

[2]For example biofuels did not live up (to now) to their 'promise' of a clean, sustainable and environmentally friendly way to produce fuel that would enable energy independence and reduction of greenhouse gas emissions. In addition, it is questioned whether the possible benefits outweigh the possible damage (for example, using up food resources) the production of biofuels may cause (e.g. Kleiner 2008; Laney 2006).

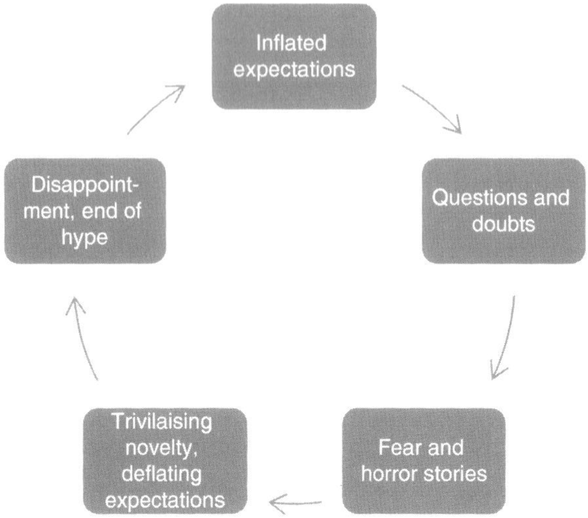

Fig. 14.1 Pattern of hype-horror and promise-disappointment cycles. Expectations are inflated to gain attention and attract financial and political support. These inflated expectations also raise questions and doubts and can result in fear. In this situation the hypes turns into a horror story. In response to this, proponent frequently trivialize the novelty of the innovation and deflate expectations until, at some point, the innovation has not lived up to the initial promise, resulting in disappointment and end of the hype. However, to get attention and support for another innovation, the cycle starts all over again

these approaches are forms of constructive technology assessment (CTA) (e.g. Rip et al. 1995), upstream engagement (e.g. Wilsdon and Willis 2004), and midstream modulation (e.g. Fisher et al. 2006). All these approaches aim to facilitate innovations in science and technology with desired positive impacts and with few (or at least manageable) negative impacts.

14.1.2 Manage Innovations in an Early Phase Towards Responsible Innovations

Approaches to develop more responsible innovations aim at maximizing desired positive impacts and minimizing potential negative impacts of an innovation. These approaches aim to facilitate the embedding of innovations in society. For this reason, the potential positive and negative impacts of an innovation on society must be identified and incorporated into research, technology development and design. There is a growing awareness that all relevant stakeholders should be involved, including the public, in order to avoid bias by the researchers (e.g. Broerse et al. 2009; Hagendijk and Irwin 2006; Rip et al. 1995). The challenge here is to address these implications in an early phase of development of an innovation, when options are still open for exploration and there is still opportunity to steer the development in a particular direction. At the same time, this early phase is characterized by uncertainty about which technological developments will be realized, what scientific knowledge will be generated, what applications will be developed and the societal impacts of these developments. Given this uncertainty of this early phase, the motivation of scientists and societal stakeholders to engage into a joint reflection about potential applications and implications of a technology is rather low. In later phases of technology development, the situation is reversed: applications and implications are apparent but possibilities to steer the developments are limited. This has been described as the Collingridge dilemma of control (Collingridge 1981).

The research described in this chapter is part of a project which aims to direct medical neuroimaging innovations in an early phase towards shared desirable applications with few, or at least manageable, negative impacts.[3] In our research, we aim to bypass hype-horror and promise-disappointment cycles In other words, during the research and development phase of innovations, the focus should not be on unrealistic expectations and hypes but, instead, the focus should be directed towards what is possible and desirable, in order to steer towards responsible directions and to minimize negative impacts, both direct and indirect, as much as possible.

In this chapter, we present the 'guiding visions' of experts, namely scientists and technology developers, involved in medical neuroimaging. These scientists and technology developers are, respectively, developing and working with neuroimaging technologies in a research setting. They are currently shaping the future of neuroimaging with their passion and ideas. In this way, a script of expected user behavior is materialized in the applications (Akrich 1992). It is therefore highly relevant to identify and analyze their visions of the future of medical neuroimaging. Because, at the moment, we are only aware of the hypes in the neurosciences. These are undifferentiated and not detailed about the context of application. However, this approach does not imply that these stakeholders are central to our research. Instead, it is their guiding visions which are translated into past and current developments that are the focal point of this chapter. In later phases of our research, this approach will facilitate the development of more widely shared desirable futures.

In addition, we aim to present a responsible innovation agenda, raising issues that need to be addressed if the guiding visions are to be realized and identifying other aspects that need further exploration and follow-up if neuroimaging is to be developed in more responsible directions. Finally, we discuss undesirable trends identified by scientists and technology developers.

---

[3]This research project is funded by the thematic programme Responsible Innovation. Ethical and societal exploration of science and technology (MVI) of the Netherlands Organisation for Scientific Research (NWO).

14.2 Methodology

In this research, a specific operationalization of a CTA approach, namely the Interactive Learning and Action (ILA) approach (Broerse and Bunders 2000), is used to challenge the Collingridge dilemma. The ILA approach has been developed to steer scientific and technological innovations in an early phase in more desirable directions defined by the stakeholders involved. This involves active participation of stakeholders from science and society using interactive methods, such as focus groups and dialogue meetings. Key features of the ILA approach are: active participation of relevant stakeholders on an equal footing early in the innovation process, explicit use of experiential knowledge, development of a shared vision, knowledge creation through mutual learning (via dialogue), enhancement of trust relationships, coalition building, and independent and competent process facilitation. The approach is characterized by an emergent and flexible design and can be roughly divided into five phases: (1) initiation and exploration, (2) in-depth study of problems, needs and visions of involved stakeholder groups separately, (3) integration of different perspectives of stakeholder groups through mutual learning, (4) prioritization and agenda setting, and (5) implementation through reflexive learning cycles of planning, action, observation, reflection and re-planning. During the entire process, the output of one group of participating stakeholders is used as input for another group, in order to redefine and deliberate on outcomes. How this basic design takes shape varies between contexts.

To develop responsible medical neuroimaging innovations, implicit long-term directions of the research and development process need to be identified. Moreover, these long-term directions are the driving force behind an inflated promise and may need to be re-shaped towards more commonly shared desirable directions. There are different approaches to analyze long-term directions of new innovations. Although closely related and often used interchangeably, broadly three types of futures can be distinguished: probable, possible and desirable futures. Probable futures are analyzed with forecasting approaches and extrapolate current trends into the future. This future is the one which participants feel is most likely to happen. Possible futures are central in most scenario approaches and take a broader view of expectations. Desirable futures are the focus of vision assessment and aim to provide long-term directions that guide developments in scientific and technological innovations (Roelofsen et al. 2008).

In this study, we combine the ILA approach with vision assessment (Grin and Grunwald 2000). Roelofsen et al. (2008) show that this combined ILA approach is applicable for assessing emerging technologies. In this process, the ‘guiding visions’ made explicit in vision assessment are continuously assessed and reflected upon, parallel to research and development, in a societal learning process. This is consistent with the ILA approach in which stakeholders continuously reflect on the outcomes, and create knowledge through mutual learning in order to develop shared desirable visions.

14.2.1 Vision Assessment

With vision assessment (Grin and Grunwald 2000), shared future visions can be shaped which guide the directions of innovations. Central to the approach is the identification of so-called 'guiding visions', which are mental images of an attainable future shared among stakeholders. These visions are neither restricted to an extrapolation of knowledge what the future probably or possibly will look like, nor are they science fiction images of the future. Instead, guiding visions describe neither facts nor pure fictions, but are a mixture of both (Grunwald 2004). It is considered that these mental images function as a 'common language' that guide the actions in concrete practices of technology development, and guide the interaction between groups of stakeholders. Therefore, these mental images are called guiding visions and are, as a rule, held by scientists and science managers (Grunwald 2004). Guiding visions are relatively stable and open to steering. It is therefore assumed that by actively collecting and critically reflecting upon one's own and others' visions, shared desirable visions can be shaped (Mambrey and Tepper 2000).

14.2.2 Towards Guiding Visions

Our research started with an exploration of the literature to make an inventory of the scientific state-of-the-art concerning technological and scientific developments of neuroimaging in health care, as well as an exploration of the potential relevant societal issues (phase 1 of the ILA-approach). Next, guiding visions of medical neuroimaging were collected via semi-structured interviews and focus groups with scientists and technology developers who currently shape the future directions of neuroimaging with their passion and ideas (phase 2) (Akrich 1992). Nine semi-structured interviews were conducted with department leaders from different scientific research institutes in the Netherlands and two with representatives of two large imaging equipment companies in the Netherlands. The interviewees were selected on the basis on their expertise with the use of neuroimaging, their working field (for example, neurodegenerative disorders) and their position within the institute they work (including those who will write research proposals in the coming years, such as coordinators of research themes) (see Table 14.1). The interviewees were questioned about the research executed in their institute in order to gain insight into the current application of neuroimaging in their specific field of research and about their future expectations, 20–40 years from now, of scientific developments with respect to neuroimaging. The interviewees were asked what impacts the developments could have in a societal context, both in a positive and negative way, and what their desirable futures of neuroimaging would look like. After obtaining permission, all interviews were recorded and transcribed verbatim for further analysis. Summaries of the interviews were sent back to the interviewees for member check.

Table 14.1 Expertise and position of interviewees

Expertise	Position
Anatomy, psychopharmacology	Section head
Development of attention and memory	Coordinator research group & professor
Language disorders and cognition	General director research institute & professor
Neurobiology of psychiatric disorders	General director research institute & professor
Neuroimaging, functional and structural brain connectivity, pharmacologic fMRI, aging and dementia	Project leader & professor
Neuroimaging in psychiatric, anxiety and mood disorders	Project leader & professor
Neurointerventional radiology	Radiologist
Neurosciences, psychiatric disorders	Section coordinator & professor
Radiotherapy, RF radiation modelling, brachytherapy biological modeling	Project leader
Development MR applications	Project coordinator
Development imaging applications	Product manager

The results from the literature study (phase 1) and the interviews served as input for the design of the focus groups. Focus groups with a relatively homogeneous composition (such as only people working in the field of psychiatric brain disorders) are considered to be an effective tool to stimulate experts in articulating their implicit guiding visions in detail (Roelofsen et al. 2008). For this reason, four focus groups were organized in April 2011 involving total of 19 scientists and technology developers (see Table 14.2). The scientists and technology developers who participated in the focus groups were selected based on the same criteria as was used to select people for the interviews. Three of the focus groups participants had also been interviewed. The participants were divided in four groups of four to five participants based on their specific field of research: neurodegenerative disorders, psychiatric disorders, behavioral and learning disorders, and technology and biomarker developers.

Each focus group was led by a facilitator who guided the process, a monitor who observed the group process, and an assistant who took notes. With permission of the participants under the condition of anonymity, all focus groups were audio recorded and transcribed verbatim for further analysis. All focus groups were conducted according to the same design which built on the approach developed by Roelofsen et al. (2008, 2010) to inventory guiding visions (see for methodological details, validation and discussion Roelofsen et al. 2008, 2010). The focus groups lasted four hours with a break halfway and comprised two assignments. During the first assignment, the guiding visions of neuroimaging of the participants in their specific field of expertise were made explicit. This was established by asking the participants to formulate their desirable future medical neuroimaging application, 20–40 years from now, assuming there are no barriers, neither technical nor societal. In this

Table 14.2 Expertise and position of focus group participants

Focus group	Expertise	Position
Behavioral and learning disorders	Attention and information processing, mental fatigue and gender differences	Section coordinator
	Functional, diffusion and perfusion magnetic resonance imaging	Program leader & professor
	Neurosurgery, preclinical epilepsy	Post-doc
	Cognitive neurosciences, strokes	Post-doc
Neurodegenerative disorders	Brain computer interface	Post-doc
	Clinical neuroimaging	Head of research
	Early diagnosis of dementia	Post-doc
	Neuroimaging, functional and structural brain connectivity, pharmacologic fMRI, aging and dementia	Project leader & professor
	Sleep, psychiatric and neurological disorders	Senior scientist
Technology and biomarker developers	Biomarkers prognosis and diagnosis neurological disorders	Head department
	Clinical neurosciences	Section head
	Cognitive neurosciences, TMS	PhD candidate
	Development molecular imaging applications	Product manager
	Development MR applications	Project coordinator
	Neonatal neurology, biomarkers	Post-doc
Psychiatric disorders	Brain changes in developmental disorders	Project leader
	Neuroimaging and neurodevelopment	Post-doc
	Psychopharmacology, psychophysiology, learning and memory	Post-doc
	Simultaneous EEG/EMG/fMRI, multichannel EEG	Project leader & professor

way, the participants were not restricted in their imagination of desirable futures. In addition, the participants were asked which developments of neuroimaging they considered undesirable. To gain insight into the assumptions underlying these guiding visions, the participants were subsequently asked which stakeholders would benefit from the introduction of the identified applications and which stakeholders might be disadvantaged by these applications. The second assignment was a joint backcasting analysis to formulate the developments and potential interdisciplinary collaborations that need to take place, and barriers that need to be overcome in order to realize desirable applications and explore potential win-win options.

In order to ensure that all key aspects were addressed, six scientists, who had been unable to join the focus groups but whose input was considered relevant and necessary to explore the field of medical neuroimaging developments, were invited in separate feed-back interviews to reflect upon the results of the focus groups (see Table 14.3).

Table 14.3 Expertise and position of feedback interviewees

Expertise	Position
Biomedical MR imaging and spectroscopy, with special interest in stroke patients	Project leader
Clinical neurophysiology, with special interest in epilepsy patients and ICU patients	Coordinator & professor
Elementary reactions in proteins and enzymes, nonlinear microscopy	Project leader & professor
Experimental molecular imaging, with special interest in neuropsychiatric disorders, addiction, neurotoxicity	Project leader & professor
PET pharmacokinetic modeling, neuroreceptor imaging	Project leader & professor
Cognitive neuroscience, psychiatry and child psychiatry	Project leader & professor

14.2.3 Analyzing Guiding Visions

In analyzing guiding visions the following six elements, based on research of Grin and Grunwald (2000) and Roelofsen et al. (2008), are distinguished:

- **Current state of knowledge and technology**: This element concerns the knowledge and technological developments that currently take place, generally formulated as objectives and technological challenges (for example, increasing the resolution of images).
- **Problem definition**: Different visions can entail various problem definitions and ways to assess solutions. Assessing the assumptions underlying a problem definition uncovers empirical and normative assumptions.
- **Purposes to be fulfilled**: This element refers to the objective the technological and knowledge developments are aiming at, namely the problem definition.
- **Relevant contextual aspects**: This element explores the relation between the technological artefact and contextual aspects, such as in which context will the artefact be used, how, by who, who will benefit and who will possibly experience disadvantages.
- **Barriers**: This element entails factors that may hamper the realization of the envisaged application, namely barriers that need to be overcome. Although these barriers are part of the contextual aspects, these were identified and analyzed (via backcasting) separately.
- **Basic features of the desirable state**: This element refers to basic assumptions around which visions develop: the preferred state of affairs the vision entails and ideas about what the world should look like (for example, the future world will be engaged in cognitive quality of life).

The guiding visions described on the following pages are based on the framework above. As the aim of this article is to present the guiding visions with respect to medical neuroimaging of scientists and technology developers, the current state of knowledge and technology is described concisely.

14.3 Guiding Visions

During the interviews and the focus groups, the participating scientists and technology developers considered a range of future directions of neuroimaging as desirable. All articulated directions are envisioned to be applied in the field of prevention, diagnosis or treatment. Below, the desirable applications are illustrated per field of application. First, the problem definition is described, based on the current state of knowledge and technology. Next, the desirable application is outlined as well as the purposes it is intended to fulfill. Subsequently, the contextual aspects are described. Sections 14.3.4 and 14.3.5 describe, respectively, the articulated barriers and the basic features of the desirable states.

14.3.1 Prevention

At present, the diagnosis of brain disorders is made when people consult a health professional as soon as the first symptoms of disease appear. The current clinical focus is therefore on the treatment of people who have developed a brain disorder and less on the prevention of these disorders. Applications of neuroimaging technologies in the field of prevention are considered to be of added value in shifting the focus from treatment towards prevention.

Desirable applications of neuroimaging technologies in the field of prevention focus on the use of these technologies as a tool for (1) detecting very early stage, sub-clinical brain disorders, namely early diagnosis, or for (2) determining predispositions for brain disorders. Both applications are envisioned as desirable by a majority of the scientists and technology developers with the condition that therapy is available. Without therapeutic options, these applications are often regarded as less desirable.

14.3.1.1 Neuroimaging as a Tool for Early Diagnosis

This application is concerned with the detection of brain disorders in a very early stage of disorder development when symptoms are not yet perceived. Many of the participants mentioned as desirable neuroimaging technologies which can detect (changing) trends and patterns in brain activity and the development of coping mechanisms in an early stage of the onset of a disorder. One participant stated:

> Especially in the context of behavioral and psychiatric disorders and disease processes that you do not yet see anatomically, those functions show a certain trend already. Within the brain, a certain coping mechanism is developing to delay the symptoms of the disorder as long as possible. In the context of early diagnosis, this [these developments] is a step forward.

Some scientists and technology developers compared this application with the current population screening methods in the Netherlands for cervical and breast cancer. This involves the screening of women who meet certain indication criteria at an interval of approximately 5 years. Some other participants envisioned neuroimaging to become part of an annual health check for the whole population in order to detect brain disorders as soon as possible. The execution of both the screening and the annual health check would take place at the general practitioner's office or a mobile screening centre. This is possible because it is expected that neuroimaging technologies in the future will be much cheaper, mobile, compact and able to visualize disorder specific biomarkers at an individual level. One of the participants envisioned this application as follows:

> So, we will have imaging through which you can see [scan] many aspects of the brain, the entire spectrum of the brain; development, disorders, defects, etcetera, in five minutes (. . .) Then, [when mobile imaging technologies become available] you can just put them in the waiting room of the general practitioner.

After detection of an early stage of a brain disorder, secondary prevention strategies can be applied.

14.3.1.2 Neuroimaging in Combination with Biomarkers as a Tool to Determine a Predisposition

The other application of neuroimaging in the field of prevention focuses on neuroimaging technologies as a tool to determine whether an individual has a predisposition for a brain disorder. Brain disorders are considered to be expressed through interaction with certain environmental factors (exposure to chemical substances, smoking, and alcohol use) and some people genetically have a higher risk to develop a brain disorder than others. With the help of neuroimaging technologies that visualize biomarkers for specific characteristics of a brain disorder, these predispositions could be determined. As a result, people can be advised on healthy behavioral patterns in order to delay the onset of the disorder. One of the participants explained:

> You will have a screening and based on this you receive a profile in which it is clearly stated what to do to conserve your health.

It is envisioned that the determination of predisposition will be undertaken at the general practitioner's office or mobile screening centre. This will screen individuals who meet certain criteria, such as children who differ significantly in school performance when compared to their siblings. In this way, individuals will be able to anticipate on their future health instead of addressing a disorder when it manifests itself. This application is envisioned as follows:

> To use the scan as a preventive tool. So, before manifestation of the disorder, to communicate in advance the chance someone has to develop a psychiatric disorder and that we can treat that 'disorder' [via participation in a preventive program] immediately.

Besides the above mentioned benefits for individuals, the previously mentioned applications in the field of prevention are also considered to be an addition to the toolkit of health professionals because the applications will allow them to give accurate and more efficient options for advising their (future) patients on healthy ageing. Moreover, these applications will make it possible for professionals to guide their patients quickly towards efficient therapeutic options. Pharmaceutical companies can benefit from the preventive applications by responding to the demand for the development of preventive medication and by developing disorder specific biomarkers. Commercial companies are expected to offer health checks which will not be included in the standard health insurance packages.

In addition to opportunities, concerns are also being raised with respect to the burden of knowing a predisposition. This burden is considered to be heavy, especially when no treatment is available. Furthermore, predisposition represents a chance of developing a disorder, not a certainty. Another concern mentioned is that health insurance companies might use the outcomes of health profiles and annual check-ups to determine the health insurance premium. Indeed, the health profile and annual check-ups could become compulsory elements for insurance acceptance. This raises questions of determinism and stigmatization, as formulated by one participant:

> That they [insurance companies] say, we insure you, but we want to have a whole brain scan of you, otherwise we will not insure anymore.

Furthermore, many participants indicated that the use of neuroimaging technologies on a large scale and an increase in preventive (pharmaceutical) options may jeopardize the overall affordability and accessibility of health care in general through a major increase of health care costs.

14.3.2 Diagnosis

Diagnoses of brain disorders are currently made in comparison with control groups, and are not possible at an individual level. Consequently, there is always an overlap with the control group, resulting in false positive and false negative diagnoses. Furthermore, the cause of many brain disorders is (partially) unknown. Diagnosis is therefore often based on external symptoms, rather than on biological insights on changing functions, patterns or coping mechanisms in the brain. In the case of neurological brain disorders, such as Alzheimer's disease and Parkinson disease, external symptoms are explored with structural neuroimaging technologies, resulting in a diagnosis. Usually diagnosis takes place, 10–15 years after the onset of the disorder when the first symptoms are perceived. In addition, diagnostic tools for many neurological brain disorders have limitations. Although generally sufficient at a structural level, the technologies lack the ability to diagnose at a functional level, or are not yet employed in this way. Diagnosis of psychiatric disorders is often based on external symptoms because most of these disorders are not visible at a structural level within the brain. Moreover, there is a major overlap in manifestations which complicates the ability to differentiate between various psychiatric disorders, for example the difference between a depression and an early phase of dementia. Furthermore, many subtypes of psychiatric disorders cannot be diagnosed which complicates the search for an effective treatment.

Desirable applications of neuroimaging technologies in the field of diagnosis focus on diagnostic tools to make an efficient and effective diagnosis of a disorder. In the future, it is expected that these tools will be fast, accurate and tailor-made, able to diagnose a disorder in all its aspects: which subtype, which stage of progression, etcetera. The neuroimaging technologies will be cheap, mobile and compact, and are not only used in hospitals but also at the general practitioner's office, at centers for monitoring health of the under five's (*consultatiebureau*) and at primary schools. Options for fast, accurate and tailor-made diagnostic neuroimaging applications are seen as highly desirable by most participants. With these applications, the prognosis of the development of the brain disorder and options for treatment can be communicated to the patient. Moreover, the treatment can be adjusted to this accurate and tailor-made diagnosis. As one participant noted:

> I am thinking of sub-typing! When we know, this subtype can be treated this way and that subtype can be treated that way, well, then you would like to see on one scan whether these networks are parietal Alzheimer, or temporal Alzheimer... or are these networks a depression?

During the interviews and focus groups not all participants articulated all aspects of these kinds of diagnostic methods. Some stressed the need for more efficient diagnostic options, while others stressed the need for tools to make an individual diagnosis. However, all mentioned desirable directions in the field of diagnosis aimed at improving the quality of life of patients by making the diagnosis fast, accurate and tailor-made in order to be able to intervene as soon as possible when a disorder develops. Applications in this field are seen as desirable by some scientists working in the field of neurodegenerative disorders because knowing the cause of symptoms and complaints is satisfying for patients, despite the absence of effective treatment. However, scientists working in the field of psychiatric and behavioral disorders consider these applications only desirable when options for treatment are available.

In addition, some participants envisioned a new standard for diagnosis using neuroimaging technologies. One of the participants explained:

> And we will then [when knowledge concerning the onset and progression of brain disorders is better known] be able to put all results together and obtain the idea that there are patterns of brain activity. Dementia looks like this, depression like that (...) You can put these patterns into the scanner, and if you scan a new patient, where it is not clear what is wrong, you can use the obtained patterns to make a diagnosis, so what disorder fits this patient.

In addition to the benefits for the patients, health professionals are, in this case, also considered to be beneficiaries. In particular, psychologists and psychiatrists are seen as potential beneficiaries because the applications will allow them to make a specific diagnosis and adjust treatment accordingly which is, at the moment, sometimes very difficult. The participants expressed their concerns that health professionals may experience ethical objections to making diagnosis of psychiatric disorder for which no treatment is available.

14.3.3 *Treatment*

Current therapies are based on their effectiveness at group level. As a result, particularly in the case of psychiatric disorders, many therapies are not effective at an individual level. Patients have to endure a long period of trial-and-error before the appropriate therapy and/or medication and the right dose are found. Furthermore, for certain treatments resistance may occur at some point, such as with treatments for epilepsy and depression. The following applications were mentioned as desirable by the participants: tailor-made treatment, on-demand treatment, image-guided interventions and enhancement of brain functions.

14.3.3.1 Tailor-Made Treatment

Most of the desirable applications in the field of treatment are expected to involve the use of neuroimaging technologies for personalized medicine: the ability to provide effective and tailor-made treatment. Testing the efficacy of therapeutic options (being medication, neurostimulation or cognitive therapy), adjustment of these options towards the specific deficiency in the brain or region of the brain that is not functioning well (based on the individual diagnosis, see above), and monitoring the progress of disorders and therapeutic options, are seen as desirable by all participating scientists and technology developers. This is especially the case in treating psychiatric disorders, as explained by one of the participants:

> There are good medication options available for depression. However, these only have an effect for 50 percent of the patients. And those patients, in whom the medication is effective, have a chance of relapse. It is hard to predict for whom a medication is working, therefore it is important that this can be predicted better through neuroimaging.

Dose-specific medication is a frequently mentioned part of this desirable direction. Instead of using trial-and-error, the extent to which a given treatment may be effective in an individual patient could be determined individually by using neuroimaging as a (monitoring) instrument. Moreover, new medications can be developed. To quote one of the participants:

> We hope, we think, that by using these technologies, we, together with pharmaceutical companies, will be able to produce better medication. Cheaper, faster and better, through which you can actually see, with those scans, does the medication do what we thought it would do. Now, there is more an indirect proof needed and there are many assumptions what a drug more or less does (...) Being able to see it on a scan, that's better.

Furthermore, tailor-made options to predict the recovery of patients are articulated by some scientists working in the field of psychiatric disorders and by those working in the field of acquired brain injuries. As one participant explained:

> When you have a stroke or something like that, then it is currently not clear which functions have a chance to recover and which not. I think that if we better understand the system, we can determine, based on the original damage, when it is appropriate to start a therapy and when this is not useful anymore. Because you can predict how the brain [functions] (...) what the plasticity of the brain is, and what the developments in the brain are.

Health professionals will be able to give patients detailed information about which therapy they receive, the progress of the therapy and disorder, and chances for recovery. Moreover, providing this information is expected to result in more dedication of patients towards their therapy.

The neuroimaging technologies for these applications are envisioned to be compact, so that they can be applied to every individual including, for example, those suffering from claustrophobia. Furthermore, they will be mobile enough to cover the influence environmental aspects have on the brain, for example the temptation perceived by an alcoholic when seeing a bottle of wine in a restaurant can be visualized and measured.

14.3.3.2 On-Demand Treatment

The treatment of chronic brain disorders, such as epilepsy, depression and bipolar mood disorders, currently takes place via continuous use of medication to prevent attacks. It would be desirable if these patients are able to only use the medication when an attack is likely to occur. In other words, to treat these disorders on-demand. This is especially seen as desirable when these applications can be used by patients themselves so that they can manage their own disorder. If patients have their own portable neuroimaging device they would be able to check their health status and the device would notify them when an attack is likely to occur, regardless of their physical location (e.g. home, work).

> It is actually like carrying an insulin pump when you have diabetes, that you manage your own treatment when the device tells you: now it is necessary

14.3.3.3 Image-Guided Interventions

In addition to using neuroimaging technologies as a tool, neuroimaging technologies could, according to some participants, also be used as an intervention. Envisioned are technologies through which, for example, patients with depression, learn to control their brain activity by operant training with feedback of specific EEG or MRI components, called neurofeedback. Furthermore, technologies were mentioned that enable health professionals to remove only that brain tissue that needs to be removed. For example, the possibility to distinguish between tumor tissue and healthy tissue during surgery in a precise way. These applications were not elaborated upon in detail by the participants.

14.3.3.4 Enhancement of Brain Functions

Enhancement is mentioned in the context of improved cognitive functions of dementia patients. Dementia patients can probably better cope with their disorder when their cognitive functions are enhanced via medication. The application of options for enhancement to people who do not have impaired cognitive functions (i.e. who are healthy) is usually seen as an undesirable direction. However, some participants formulated a future in which happiness becomes a choice. Using neuroimaging technologies, happiness may be regulated in the brain. For patients with depression, this is seen as a desirable treatment. Whether the use of this application by healthy people is also desirable raised discussion. As stated by one of the participants:

> It is not an undesirable direction, but I think this [enhancement] will be an important issue in the future of applying neuroimaging technologies. Are we allowed to boost a brain and who decides on this? Is it a financial matter? When I am rich, I am allowed to boost the brains of my children?

With respect to the abovementioned desirable applications for treatment, participants mentioned many advantages for both patients and health professionals. Furthermore, consultancies are seen as beneficiaries through an increasing demand for cost-effectiveness analyses of the applications. Commercial companies are expected to offer treatments which are not included in the health insurance package. Concerns that were raised related to the enhancement of healthy people that potentially results in questioning what is normal and healthy. In other words, when is something an impairment or disorder that should be treated and when is something enhancement. With this, the participants expressed their fear of medicalisation of relatively healthy people.

14.3.4 Basic Features Desirable State

The preferred state of affairs of the articulated visions, namely ideas about what the world should look like, were only implicitly articulated by interviewees and focus group participants. Participants mentioned that the overall purpose of their desirable applications is to improve the quality of health care and to improve quality of life.

14.3.5 Overcoming Barriers

In order to realize the guiding visions, barriers were frequently mentioned. According to the participants, barriers need to be overcome on a technological, knowledge, research, health care organizational and societal level. To frame it differently, there are challenges that need to be addressed in order to achieve the desirable applications of neuroimaging in health care.

14.3.5.1 Knowledge and Research Barriers

Currently, the knowledge of the healthy brain and the knowledge concerning the causes, onset, development and progression of brain disorders is not sufficient to realize the desirable applications. Improvement and development of neuroimaging technologies is mentioned as a potential solution to overcome this barrier, a frequently mentioned example is the combination of imaging technologies, for example a MRI-PET-CT scan. However, participants indicated that this runs into another barrier: this type of research is expensive and time consuming. Furthermore, to develop the knowledge that is needed, most of the participants stressed the necessity for interdisciplinary research teams which, they argued, is also an expensive and time consuming type of research. This type of research is time consuming because of the professional jargon of each individual discipline which makes it difficult to understand and communicate with scientists from other disciplines Moreover, interdisciplinary research is considered not to be 'rewarding'. This is related to the few suitable journals that publish the results from interdisciplinary research and the lower impact factors these journals have.

14.3.5.2 Technical Barriers

There are currently no options to make diagnoses at group level applicable to diagnosis at an individual level, as explained by one of the participants:

> We can do a lot with imaging. However, this is at group level (. . .) you also want an articulation for individual patients, what is the most likely diagnosis. This is the challenge.

In addition, the interpretation of signals measured with neuroimaging technologies are based on hypothesis of the functioning of the brain that are not known to be correct. For both barriers, options to improve the temporal and spatial resolution is considered to be part of the solution.

Subsequently, data-analysis is a time consuming activity based on statistical assumptions. Methods and software to draw quick conclusions about the scan made, envisioned via tools that enable automatic online data analysis with a direct standard norm, such as is the case for the IQ test, are considered to be potential solutions for this barrier. Furthermore, many disease-specific mechanisms in the brain cannot currently be visualized. Biomarkers are expected to provide solutions for this, together with improved neuroimaging technologies.

Another technical challenge is to have enough space for data storage. A scan of one patient currently generates one terabyte of data. Moreover, neuroimaging technologies are only applicable in a hospital setting. Technological developments to minimize and make the technologies mobile (for example, running an MRI at room temperature) are formulated options to extend the applicability and to enable screening on location. The technologies are envisioned by one participant as follows:

> I think, in 2040 you just go to the general practitioner, located in large practices, and there you go under the hairdryer [mobile scanner visualized as kind of hairdryer].

14.3.5.3 Health Care Organizational Barriers

The current clinical focus is on treatment. Implementation of neuroimaging applications in the field of prevention and diagnosis requires a shift in the organization of the health system from secondary to primary care; from treatment to prevention. Barriers to these shifts comprise changed competencies of health professionals to enable them to perform new and different tasks. The profession of radiologist is envisioned to disappear due to these developments; while the profession of intervention radiologist will increase. New professions will emerge that understand the technical, scientific and clinical side of disorders as well as neuroimaging technologies in order to apply neuroimaging technologies in the clinical setting. This is already visible with the increase of university study directions focusing on technical medicine and medical natural sciences. Furthermore, a shift in behavior is required from (potential) patients. As one of the participants put it:

> So, the individual (....) becomes a kind of manager of his own health and that requires a different attitude.

14.3.5.4 Societal Barriers

At the societal level, the participants articulated expectations of the general public about the possibilities and limitations of neuroimaging as a barrier. Many participants felt that the public has very high, unrealistic expectations. This barrier is considered to be caused by hypes relating to neuroimaging technologies in the mass media and non-experts claiming they are neuroimaging experts. An example of an unrealistic expectation is that the general public might think that mind-reading will become possible, resulting in fear and rejection of the use of neuroimaging applications. Interesting to note is that the participants differed in opinion whether mind-reading will become possible in the future. Part of the difference in opinion might be caused by the various definitions of mind-reading that were used, such as mind-reading as the 'reading' of conscious thoughts and mind-reading as the 'reading' of underlying, unconscious thoughts.

To come to a responsible embedding, more correct information about the possibilities and limitations of these technologies to society, including the government, is required according the scientists and technology developers. In this way, most of the participants would prevent hype and horror stories in the media and to manage the expectations of the public. They mentioned a significant role for the government to arrange this, but the scientific community should also have a part in correcting 'false' stories and proving information to the public and end-users. Moreover, many participants agreed that the role of the government, besides the financing of neuroimaging research, is to develop clear visions on desirable directions of neuroimaging developments and how concerns should be addressed.

14.3.6 Undesirable Directions

Many of the mentioned undesirable developments relate to broader issues that currently dominate debates in health care, including the violation of privacy by misusing information, the determination of a predisposition or diagnosis for which treatment does not yet exist and the fear of stigmatization and discrimination as a result of predisposition or diagnosis. This is mentioned by almost all participants in relation to psychiatric disorders. They consider that psychiatric disorders are less accepted by society than neurodegenerative disorders. With the availability of faster and better diagnostic methods for these disorders, psychiatric patients are supposed to treat themselves in order to function as expected within society. The participants feared that this could potentially lead to a reduction of the autonomy of psychiatric patients.

Furthermore, the participants articulated several undesirable developments of neuroimaging that are not related to the use of neuroimaging in the field of health care. New developments to manipulate the identity of people, a possible result of increasing knowledge and technological options to intervene efficiently and relatively easily in the brain, were seen as undesirable. Furthermore, the application of these technologies as a lie detection method in the domain of justice was seen as undesirable. Also the use of neuroimaging to assess brain capacities to create profiles as criteria for employment assessment purposes was considered undesirable. In this case, it is envisioned that besides current selection criteria, like appearance and work performance, the capacity of the brain will also play a role in work, dating, and social contacts. They imagined that societal sub-categories would develop based on brain information, or that admission to a university only becomes possible when people have the right brain profile.

14.4 Conclusions and Discussion

The guiding visions presented in this article provide empirical information, from scientists' and technology developers' perspectives, on desirable directions of future medical neuroimaging developments, how these applications are envisioned to be used in health care and their impacts on individuals, the health care setting and society. The visions illustrate what knowledge (for example, how does the 'normal' brain function) and technological developments (for example, specific biomarkers and mobile imaging technologies) are required in order to realize the desirable applications. They also indicate who will be the potential users of these applications and who will be potentially affected, both positively and negatively, by the implementation of these applications. The visions show that a shift in health care, from secondary to primary health care, is needed to implement the envisioned applications. This shift is not only required at an organizational level (for example, how to maintain the affordability and accessibility and educate and incorporate new professions to interpret neuroimaging data), but also requires a shift in attitude and behavior of both health professionals and (future) patients towards anticipation on their future health prevention.

14.4.1 Basic Assumptions

Although the participating scientists and technology developers did not explicitly articulate the basic assumptions underlying their guiding visions, these were implicitly mentioned. In striving towards technological developments that can control and prevent disease, a society is envisioned in which more health care options and prevention programmes become available; a society in which everyone is 'normal', in which we prevent individuals from developing disorders and, if they do develop a disorder, it can be detected, treated and cured as soon as possible. Subsequently, this striving towards control and manipulability, potentially results in a shift of the boundary between health and disease – discourse in which the dominant vision might eventually be that everyone can and should enhance their (brain) functions and capacities. However, the desirability of the use of neuroimaging for enhancement of healthy people was contested by most participants in our study.

14.4.2 Guiding Visions and Barriers

The identified guiding visions and barriers can be understood as formulations from two different, recurrent positions resulting from what Moreira and Palladino (2005) call the 'regime of hope' and the 'regime of truth'. These regimes can according to Pickersgill (2011: 460) be understood as *interlocking regimes (. . .) that shape contemporary biomedicine*: hope that new and better health care options and preventive programmes will result from neuroimaging research (the identified guiding visions), and looking at what is positively known, rather than what can be, because most health care options resulting from (for example) neuroimaging technologies are less effective than promised, clinically fail, and/or are ethical onerously (the identified barriers). This ambivalence is also visible in hype-horror and promise-disappointment cycles. In order to attract political and financial support for further research and development, scientists formulate promises from the regime of hope (the scientist as entrepreneur), whilst from the regime of truth they realize that it is possible that the innovation could (partly) fail, is less effective then intended or could be ethically problematic (the scientists as knowledge creator) (Brown and Michael 2003).

Identifying desirable directions of medical neuroimaging with the aim to steer these directions towards shared, responsible directions does not imply that these will dominate the innovation processes. The formulated innovative opportunities in this research resulted in a space in which the participants could articulate their ideas in a way that they were not bound to rules, regulations and procedures of the regime they adhere to in their daily working life: the guiding visions are articulated from a regime of hope (Moreira and Palladino 2005). Roelofsen (2011:150) showed in a similar CTA process on ecological genomics that the formulated opportunities in this research *were not placed high on the agenda of the participants, since they did not fit in established ways of working and thinking* – the options did not fit the current dominant regime (resulting from a transition perspective, Geels and Kemp 2000). In other words, a CTA process should be combined with a system perspective.

Kloet et al. (2013) used the Multi-Level Perspective (MLP) of Geels (2010), which provides an overview 'of the multidimensional complexity of changes in socio-technical systems', to capture real-time dynamics surrounding large-scale interdisciplinary research programmes. Viewed from this (transition) perspective, factors such as money, rules, regulations, procedures and selection mechanisms present in the current dominant discourse determine which innovations succeed and which fail. Analyzing the dynamics and the socio-technical system surrounding medical neuroimaging is beyond the scope of this article but, by framing the identified guiding visions within the MLP model, we are able to estimate how realistic the identified guiding visions are.

The guiding visions regarding diagnosis and treatment options are most likely to be more realistic compared to the guiding visions concerning preventive options, since options for diagnosis and treatment fit within the current structure of the health system (regime level) and require ('only') knowledge and technological improvements to be realized. In other words, diagnosis and treatment options are relative quick-wins and can be realized on the short term. The articulated preventive options, on the other hand, do not fit within the current health system and require transitions both at the level of the current organization of the health system (regime level) as well as at the societal level (landscape level). These transitions imply major changes, including macro-economic (for example, the redistribution of money from cure to care and new type of professions) and deep cultural changes (for example, low threshold to access the health system from general practitioner to mobile scan possibilities; and a shift in behavior from (potential) patients, see also Sect. 14.3.5.3). Changes at landscape level occur slowly and are beyond the direct influence of regime and niche actors (Kloet et al. 2013; Geels and Schot 2007). Moreover, regimes are large and stable communities which will not easily change unless they are destabilized and a 'window of opportunity' opens up (Kingdon 1995; Kloet et al. 2013; Geels and Schot 2007). The identified guiding visions concerning preventive options are therefore long-term options and, viewed from the current regime, unrealistic and ideal solutions. This is in line with how the participants articulated the preventive options: as a desirable situation/solution. However, through the current economic situation in The Netherlands and rising trends in chronically ill patients (both landscape level developments), we notice a growing awareness of policy makers that major changes in the health system are envisioned to be necessary, and a 'window of opportunity' might occur through which the change from treatment towards prevention and, moreover, from patient towards manager of own responsibility, becomes economically necessary (Arentshorst et al. forthcoming a).

14.4.3 Undesirable Directions

The identification of undesirable directions shows the contrast with the desirable neuroimaging applications. Remarkable is that none of these undesirable directions are envisioned as a direct effect of neuroimaging applications in the field of health care, but relate to general health discussions and particularly applications outside the domain of health care. This pinpoints structures of the current dominant regime in which many scientists and technology developers do not feel accountable for the indirect implications of the innovations they establish. As Brown and Michael (2003) describe, scientists perform two roles (or identities) in innovation cycles: those of researchers and those of entrepreneur. In the role of researcher, scientists create knowledge, and in the role of entrepreneur, scientists use this knowledge in order to attract investments (via promises). They describe the dynamics of the future market in which promises (expectations) attract short-term shared values, *but without any necessary requirement for [scientists in the role of] entrepreneurs to fulfill their longer-term promises* (Brown and Michael 2003: 13). This refers to a long known resistant theme, the non-accountability of researchers towards indirect implications of the innovations they establish which is an urgent problem in transition management. Addressing this denial/cognitive dissonance is crucial for developing responsible medical neuroimaging applications. Therefore, in our research, we aim to seduce scientists and technology developers to include potential indirect impacts directly into their research in order to establish innovations with 'better' results, because they can benefit for example more actors, and to establish an easier and more broad embedding of their innovations. This implies action, externally driven, since scientists and technology developers are not the ones who include potential implications in their research. In other words, knowledge management is needed.

14.4.4 Management of Innovations

Regarding the management of innovations towards responsible innovations, this study does not imply that in order to manage innovations, phenomena as 'promise', 'expectation', or 'hype' should be avoided. Contrarily, we observe that the driving forces behind expectations and promises, namely the guiding visions, should be made explicit and assessed on their realistic value, before these change into hypes, horror-stories, monstrous images and/or disappointment. Therefore, these visions should be acknowledged and steered in an early phase towards desirable directions according to all relevant stakeholders in which positive impacts are maximized and potential negative impacts are minimized. Our research shows that the consulted scientists and technology developers recognize and articulate the mechanisms of hype-horror and promise-disappointment cycles. They expressed fear of creating hypes and the resulting unrealistic expectations of the general public (see Sect. 14.3.5.4). Because of this fear, they found it difficult to formulate their guiding visions. The results show that interviews combined with focus groups offer the possibility to identify and articulate the implicit steering mechanisms behind expectations and promises (the guiding visions), resulting in the establishment and understanding of concrete and contextualized visions. In other words, a CTA-process combined with vision assessment offers the opportunity to investigate critically the mechanism of hype-horror and promise-disappointment cycles, the social-technical system surrounding innovations, and might result in a rational design that includes these mechanisms and dynamics.

In sum, the findings of this research illustrate the relevance of identifying guiding visions and barriers that need to be overcome: besides technical information, potential implications of neuroimaging in areas of application and related societal and policy challenges are made explicit. Moreover, the findings show which aspects need further exploration and follow-up activities in order to develop neuroimaging in more desirable, responsible directions.

In the next phase of this research, a wider circle of relevant stakeholders of medical neuroimaging will be invited to reflect upon the guiding visions and develop own guiding visions, and ideally results in insights whether the desirable directions as formulated by the scientists and technology developers are also desirable from the perspective of other relevant stakeholders. This process should result in the identification of alternative visions of desirable future neuroimaging applications and possible barriers that were not envisioned by the scientists and technology developers (Grunwald 2004).

The following steps of our research show that citizens do not primarily frame health prevention from a macro public health problem but rather from an individually centered micro perspective. They immediately relate the outcome of preventive neuroimaging to their private life, such as: what is the impact of this (uncertain) knowledge for me as an individual, a parent, and as an employee (Arentshorst et al. forthcoming b)? This contrast seems to be a challenge to bridge in order to come towards responsible innovations of medical neuroimaging.

The subsequent integration of different stakeholder perspectives through mutual learning in dialogue meetings, aims to result in mutual understanding and, where appropriate, adjustment of the guiding visions towards responsible future directions of medical neuroimaging (Broerse and Bunders 2000).

References

Akrich, M. 1992. The description of technical objects. In *Shaping technology/building society: Studies in sociotechnical change*, ed. W.L. Bijker. Cambridge: MIT Press.

Arentshorst, M. E., T. de Cock Buning and J.E.W. Broerse. Forthcoming a. Exploring responsible neuroimaging innovation: vision from a societal actor perspective.

Arentshorst, M. E., T. de Cock Buning, and J.E.W. Broerse. Forthcoming b. Exploring responsible innovation: Dutch public perceptions of the future of medical neuroimaging technology.

Broerse, J.E.W., and J.F. Bunders. 2000. Requirements for biotechnology development: The necessity for an interactive an participatory innovation process. *International Journal for Biotechnology* 2(4): 275–296.

Broerse, J.E.W., T. de Cock Buning, A. Roelofsen, and J.F. Bunders. 2009. Evaluating interactive policy-making on biotechnology: The case of the Dutch Ministry of Health, Welfare and Sport. *Bulletin of Science, Technology & Society* 29(6): 447–463.

Brown, N., and M. Michael. 2003. A sociology of expectations: Retrospecting prospects and prospecting retrospects. *Technology Analysis & Strategic Management* 15(1): 3–18.

Brown, N., A. Rip, and H. Van Lente. 2003. *Expectations in & about science and technology. A background paper for the 'expectations' workshop of 13–14 June 2003*.http://www.york.ac.uk/satsu/expectations/Utrecht%202003/Background%20paper %20version%2014May03.pdf. Accessed September 2012.

Chilvers, J., and P. Macnaghten. 2011. *The future of science governance: A review of public concerns, governance and institutional response.* BIS/Sciencewise-ERC.

Collingridge, D. 1981. *The social control of technology*. Milton Keynes: Open University Press.

Dickstein, S.G., K. Bannon, F.X. Castekkanos, and M.P. Milham. 2006. The neural correlates of attention deficit hyperactivity disorder: An ALE meta-analysis. *Journal of Child Psychology and Psychiatry* 47(10): 1051–1062.

Fisher, E., C. Mitcham, and R. Mahajan. 2006. Midstream modulation of technology: Governance from within. *Bulletin of Science, Technology & Society* 26: 485–496.

Fuchs, T. 2006. Ethical issues in neuroscience. *Current Opinion in Psychiatry* 19: 600–607.

Geels, F.W. 2010. Ontologies, socio-technical transitions (to sustainability), and the multi-level perspective. *Research Policy* 39(4): 495–510.

Geels, F., and R. Kemp. 2000. *Transities vanuit socio-technisch perspectief.* Enschede/Maastricht: CSTM/MERIT.

Geels, F.W., and J. Schot. 2007. Typology of sociotechnical transition pathways. *Research Policy* 36(3): 339–417.

Glahn, D. 2008. Psychiatric neuroimaging: Joining forces with epidemiology. *European Psychiatry* 23(4): 315–319.

Glannon, W. 2006. Neuroethics. *Bioethics* 20(1): 37–52.

Grin, J., and A. Grunwald. 2000. *Vision assessment: Shaping technology in 21st century society; towards a repertoire for technology assessment*. Berlin: Springer.

Grunwald, A. 2004. Vision assessment as a new element of the FTA toolbox. In EU-US seminar: 885 New technology foresight, forecasting & assessment methods, 53–67. Seville, 13–14 May 2004. http://foresight.jrc.ec.europa.eu/fta/papers/Session%204%20What%27s%20the %20Use/Vision%20Assessment%20as%20a%20new%20element%20of%20the%20FTA %20toolbox.pdf

Hagendijk, R., and A. Irwin. 2006. Public deliberation and governance: Engaging with science and technology in contemporary Europe. *Minerva* 44: 167–184.

Illes, J., and E. Racine. 2005. Imaging or imagining? A neuroethics challenge informed by genetics. *American Journal of Bioethics* 5(2): 5–18.

Kingdon, J. 1995. *Agendas, alternatives and public policies* (1st ed. 1984). New York: Harper Collins.

Kleiner, K. 2008. The backlash against biofuels. *Nature Report Climate Change* 2: 9–11.

Kloet, R.R., L. Hessels, M.B.M. Zweekhorst, J.E.W. Broerse, and T. de Cock Buning. 2013. Understanding constraints in the dynamics of a research program intended as niche innovation. *Science and Public Policy* 40(2): 206–218.

Laney, K. 2006. *Biofuels: Promises and constraints*. International Food & Agricultural Trade Policy Council. IPC Discussion Paper.

Malhi, G.S., and J. Lagopoulos. 2007. Making sense of neuroimaging in psychiatry. *Acta Psychiatrica Scandinavica* 117(2): 100–117.

Mambrey, P., and A. Tepper. 2000. Technology assessment as metaphor assessment. Visions guiding the development of information and communication technologies. In *Vision assessment: Shaping technology in 21st century society; towards a reportoire for technology assessment*, ed. J. Grin and A. Grunwald, 33–51. Berlin: Springer.

McGuire, P., O.D. Howes, J. Stone, and P. Fusar-Poli. 2008. Functional neuroimaging in schizophrenia: Diagnosis and drug discovery. *Trends in Pharmacological Sciences* 29(2): 91–98.

Mokyr, J. 1990. *The lever of riches: Technological creativity and economic progress*. New York: Oxford University Press.

Moreira, T., and P. Palladino. 2005. Between truth and hope: On Parkinson's disease, neurotransplantation and the production of the 'self'. *History of the Human Sciences* 18(3): 55–82.

Petrella, J.R., R.E. Coleman, and P.M. Doraiswamy. 2003. Neuroimaging and early diagnosis of Alzheimer disease: A look to the future. *Radiology* 226(2): 315–336.

Pickersgill, M. 2011. 'Promising' therapies: Neuroscience, clinical practice, and the treatment of psychopathy. *Sociology of Health & Illness* 33(3): 448–464.

Rip, A., T. Misa, and J. Schot. 1995. *Managing technology in society: The approach of constructive technology assessment*. London: Pinter.

Roelofsen, A. 2011. *Exploring the future of ecogenomics: Constructive technology assessment and emerging technologies*. Ridderkerk: Ridderprint.

Roelofsen, A., J.E.W. Broerse, T. de Cock Buning, and J.F.G. Bunders. 2008. Exploring the future of ecological genomics: Integrating CTA with vision assessment. *Technological Forecasting and Social Change* 75: 334–355.

Roelofsen, A., R.R. Kloet, J.E.W. Broerse, and T. de Cock Buning. 2010. Guiding visions in ecological genomics: A first step to exploring the future. *New Genetics and Society* 29(1): 19–36.

Rosas, H.D., A.S. Feigin, and S.M. Hersch. 2004. Using advances in neuroimaging to detect, understand, and monitor progression in Huntington's disease. *The Journal of the American Society for Experimental NeuroTherapeutics* 1(2): 263–272.

Swierstra, T., and A. Rip. 2007. Nano-ethics as NEST-ethics: Patterns of moral argumentation about new and emerging science and technology. *Nanoethics* 1: 3–20.

Willmann, J.K., N. van Bruggen, L.M. Dinkelborg, and S.S. Gambhir. 2008. Molecular imaging in drug development. *Nature Reviews – Drug Discovery* 7: 591–607.

Wilsdon, J., and R. Willis. 2004. *See-through science, why public engagement needs to move upstream*. London: Demos.

Chapter 15
Optimization of Complex Palliative Care at Home via Teleconsultation

Jeroen Hasselaar, Jelle Van Gurp, Martine Van Selm, Henk J. Schers, Evert van Leeuwen, and Kris Vissers

Abstract Palliative care involves the care for patients with a life threatening disease, often advanced cancer, aiming at an optimal quality of life for the patient and his/her family. Although many patients with advanced cancer live at home in the last phase of disease, hospital transfers are often performed increasing burdening of patients and families and health care costs. Teleconsultation may be able to bring hospital expertise to the patient's home, thereby supporting home care and fostering continuity of care. This research will combine qualitative and quantitative research to investigate whether teleconsultation will contribute to aspects of symptom management and quality of life in palliative patients residing at home. Research objectives will not only cover domains of quality of care but will also include ethical and communication aspects of teleconsultation for palliative patients.

J. Hasselaar, Ph.D. (✉) • J. Van Gurp • K. Vissers, MD, Ph.D.
Department of Anesthesiology, Pain and Palliative Medicine, Radboud University Nijmegen Medical Centre, PO Box 9101, 6500 HB Nijmegen, The Netherlands
e-mail: J.Hasselaar@anes.umcn.nl

M. Van Selm
Amsterdam School of Communication Research (ASCOR), University of Amsterdam Kloveniersburgwal 48, 1012 CX Amsterdam, The Netherlands

H.J. Schers, MD, Ph.D.
Department of Primary and Community Care, Radboud University Nijmegen Medical Centre, PO Box 9101, 6500 HB Nijmegen, The Netherlands

E. van Leeuwen, Ph.D.
Department of IQhealthcare; Ethics, Radboud University Nijmegen Medical Centre, PO Box 9101, 6500 HB Nijmegen, The Netherlands

J. van den Hoven et al. (eds.), *Responsible Innovation 1: Innovative Solutions for Global Issues*, DOI 10.1007/978-94-017-8956-1_15,

15.1 Introduction

Palliative care is an important public health issue since the past decade (World Health Organization 2004). The ageing of the population contributes to this development. Additionally, most diseases people suffer from have changed from acute illnesses towards chronic illnesses (World Health Organization 2004; Murray et al. 2005; Shugarman et al. 2009). Finally, advances in medical knowledge and technology increase treatment possibilities at the end of life. Therefore, these epidemiological transitions have led to a growing need for palliative care in the last phase of life (Seale 2000).

15.1.1 Palliative Care

The primary goal of palliative care is to ensure the best possible quality of life of patients and their families facing a life threatening illness (World Health Organization 2004; Cohen et al. 2001). Most people in their end-stage of life, regardless of their initial disease, want to be cared for and want to die at home (Beccaro et al. 2006; Higgingson and Sen-Gupta 2000). Therefore, place of death is considered an indicator of quality of end-of-life care (Teno et al. 2004). However, research in Belgium and in the Netherlands has shown that 30–40 % of palliative patients are transferred from home to a hospital or health care institution in the last week of their lives (Van den Block et al. 2007; Klinkenberg et al. 2005). This trend is also seen internationally (Burge et al. 2005). Transitions in the location of care are often extremely stressful for patient and caregivers (Burge et al. 2005) and can pose a challenge for the continuity of care (Burge et al. 2005; Meier and Beresford 2008).

Place of death has also become a topic of wider interest for public health policy, due to the focus in health care on cutting costs in acute care settings (McCorkle and Pasacreta 2001). Many European countries have implemented policy measures to reduce the number of acute care hospital beds as a means to restrict hospital expenditure (Cohen et al. 2001). With this shift in location of care for the seriously ill from hospital to home, the reliance on family caregivers to support patients with terminal illness at home is growing (McCorkle and Pasacreta 2001). These family caregivers are of vital importance for those wanting to die at home. Without them, staying at home in the last phase of life would be impossible for many patients (Bainbridge et al. 2009; Ramirez et al. 1998). However, caregiving for terminally ill patients can be burdensome for informal caregivers, possibly leading to burn-out (Aoun et al. 2005; Van Ryn et al. 2011).

15.1.2 Teleconsultation

Due to a growing number of palliative patients and the desire for less institutionalized care, community-based palliative care will become a big challenge

(Ingleton et al. 2011). The development of innovative approaches to deliver good quality of care at home is therefore necessary, e.g. mobile care teams, case management, and advanced care planning. Another type of approach is the use of telemedicine. Telemedicine involves the use of telecommunications and information technologies to share and maintain patient health information and to provide clinical care and health education to patients and professionals at a distance (Bashsur 1995). Teleconsultation is a specialized form of telemedicine that uses technology to provide real-time visual and audio patient assessment (Kitamura et al. 2010). Teleconsultation is an instrument to transfer expertise from the hospital into primary healthcare. Teledermatologic consultation has been one of its first known applications. Literature shows that teleconsultation reduces the number of traditional face-to-face consultations with a dermatologist (Eminovic et al. 2009; Knol et al. 2006; Wootton et al. 2000; Whited et al. 2002). In the field of palliative homecare however, few studies have been performed (Bowles and Baugh 2007; Jordhoy et al. 2000; Laila et al. 2008; Hebert et al. 2006; Melin-Johansson et al. 2010). Recognized problems in telemedicine research in palliative care concern small sample sizes, comparability of intervention and control groups and the handling of drop-outs (Bowles and Baugh 2007; Smeenk et al. 1998).

15.1.3 Medical, Societal and Ethical Concerns

The development of telemedicine gives rise to several medical, societal and ethical considerations. Van Wijnsberghe and Gastmans (2008), following Tronto (1993), have pointed to the values of attentiveness, responsibility, competence, and responsiveness. It should be considered whether telemedicine is able to *attend* to the patient as a whole (all dimensions of caregiving) and not be restricted to the physical needs of the patient. The evaluation of the medical effectiveness of palliative teleconsultation should take this holistic consideration into account. Furthermore, it should be evaluated how responsibilities evolve in a process of care at a distance. Via telecommunication, the expert at a distance becomes part of the local care system with consequences for medical, ethical, professional, and legal responsibility, accountability, and collaboration. Moreover, the communication at a distance may influence the communication within the patient-physician relationship. Also, telemedicine is often said to empower the patient by increasing patient possibilities for self-management and autonomy. However, it should be questioned whether the vulnerable patient in the last phase of life and his relatives are able to function in such a paradigm. Moreover, it should be considered whether telemedicine contributes to a ‘good’ death in contrast with an undesired medicalisation of dying. The ethical principle of proportional care and the balance between right indication and possible benefits and harms for the patient will be important to evaluate by caregivers. Finally, the ethical principle of equal access to care should be taken into account, which includes issues of equal availability, implementation, and financial reimbursement. In this perspective, the cost-effectiveness of telemedicine is an important topic to consider (Smith 2007; Verhoeven et al. 2007).

Telemedicine comprises a communication tool for participants in the practice of home based palliative care, and it is reasonable to think that the application will have an effect on existing communication patterns in this practice or will even facilitate new ways of communication. One of these changes may involve a shift from a (strongly hierarchical) transmission pattern to a consultation- or even a conversation pattern (Bordewijk and Van Kaam 1982). Furthermore, the Telemedicine (TM)-application may transform communication genres, which are 'typified communicative actions invoked in response to recurrent situations' (Yates and Orlikowski 1992; Yates et al. 2008). One classic example of such a genre in health care is the so-called asymmetry of information, implying that the doctor has the lead in the communication with the patient. Genres are guided by a set of –often implicit- historical and cultural rules on the substance and the form of communication. Genres can be modified in three ways due to the use of innovations in existing communicative praxes. Firstly, existing genres may remain unaffected by the innovation. Secondly, the communication patterns may become slightly adapted due to new conditions, for example a new communication medium, but 'without substantially departing from those genre rules'. Here, the genres will become elaborated and enriched without affecting 'shared' considerations of caring practice. Thirdly, the existing genres may become challenged, for example because the new medium causes symmetry of information between doctor and patient. In this case, communication genres will become modified and with this modification (normative) perspectives on caring may be transformed, too. The information resulting out of these analyses will enable ethical reflection on the role of telehealth in palliative care in relation to normative notions on living and dying with advanced cancer.

15.1.4 Taking Responsibility Seriously

Innovation involves a process of intellectual creativity combined with physical, creative action and the dissemination of something new (Burge et al. 2005; Meier and Beresford 2008). In health care, these innovations are always centered around patient health. In line with Hellström (McCorkle and Beresford 2001), health care innovations are considered to involve both a process approach and a product approach. Telemedicine is therefore not only about the supply and functionality of a technical device, but also -and more importantly- about the design and implementation of an optimal process of palliative care in which ICT devices support care-giving and care-receiving. Responsibility is often discussed in the context of *someone taking responsibility*. A central element here is an actor "who is responsible for a certain situation or a state of affairs towards a certain person or institution" (Ramirez et al. 1998). Responsibility in health care innovation is not only the concern for the designer of products and processes, but also a major concern for health care providers who use the innovation for caring and for researchers who interact in this process. Additionally, also the patient and his relative have an important role because they receive care and finally judge whether this type of care is suitable for their situation. Being responsible for health care innovation involves that designers, researchers, caregivers and patients need to show (a) awareness of their interpretations of reality on which their acts are based, (b) acknowledgment of normative rules currently present in the care practice, and (c) commitment to improve future practice (Burge et al. 2005; Ramirez et al. 1998; Aoun et al. 2005).

15.2 Description of Research

15.2.1 Project Organization

This study is divided into two research projects that are strongly connected. The first project will mainly address the medical effectiveness of palliative teleconsultation in terms of symptom control and psycho-social wellbeing. Also, health care utilization due to the use of telemedicine will be investigated. This study is based on a quantitative research design (randomized trial). The second research project will concentrate on social-ethical issues concerning the use of telemedicine in palliative patients. This project relies on a qualitative research design (in-depth interviews and observational research). Both research projects are formally accepted by the ethical committee of Arnhem/Nijmegen, and the trial is internationally registered. Apart from the research component, a part of the project is dedicated to process redesign and preparation of implementation. The multidisciplinary project team consists of researchers in the field of medicine, ethics, and communication science, with a background in both quantitative and qualitative research methodology. The project is coordinated by the Department of Anesthesiology, Pain and Palliative Medicine of the Radboud University Nijmegen Medical Centre. This department works in close collaboration with the Department of Primary and Community Care. In addition, ZZG Zorggroep, a large regional homecare is involved. Finally, an ICT-installation company (FocusCura) supports the technical development and installation of the telemedicine application at the patient's home.

15.2.2 Description of Teleconsultation Intervention and Device

After completing of the baseline measurement, a telemedicine computer (e.g. iPad) will be installed at the patient's home. Soon after the installation, the nurse practitioner of the consultation team contacts the patient to make an appointment for the first teleconsultation. During this first digital screen-to-screen contact between the patient and the nurse practitioner, the nurse checks for problems in palliative care following a consultation inventory instrument (e.g. multidimensional analysis of problems: physical, social, psychological and spiritual needs, coordination of care). After the first teleconsultation, the nurse practitioner, in cooperation with the palliative care specialist of the palliative consultation team, advises the GP on the treatment policy for the patient. During this trajectory, the GP continues to be the coordinator of medical care. The teleconsultations will return every week, but more or less frequent contact is possible when the patient and the team agree on this. There are no installation or internet costs for the patient and also the use of the telemedicine device is for free.

The telemedicine application itself is a computer including a screen, a microphone/speaker and a camera. The interface involves large and easy to understand pictograms. Weekly consultations between the nurse of the hospital and the patient are planned and performed. The telemedicine application will not be used in emergency situations due to safety restrictions. Below, the specific research aims and methods of both research projects will be described more in detail.

15.2.3 Research Project A: Medical Effectiveness of Palliative Teleconsultation

Study A aims to evaluate the effectiveness of teleconsultation in palliative homecare. The primary goal is to evaluate the effectiveness of teleconsultation on the burden of physical and psychological suffering of palliative patients at home. Secondary objectives are (1) to investigate whether teleconsultation influences the number of hospital admissions by acting more pro-active on escalating problems of patients, (2) to consider if the burden of the family caregiver ameliorates by giving them a better opportunity to address their needs and problems, (3) to study the patient experienced continuity of medical care in the last phase of life, and (4) to assess patient and caregiver satisfaction with the teleconsultation contact.

The study consists of a two-armed cluster randomized controlled trial, aiming at patients with an incurable progressive stage of cancer. The symptom burden of the patient and also the secondary outcomes between the two study arms will be compared. On the moment of inclusion, the patient's GP acts as the coordinator of medical care, and patients reside at their homes. Patients unable to give informed consent and patients with an active psychotic disorder or a serious cognitive disorder are not eligible for inclusion. The protocol of the present study was approved by the Central Committee on Research Involving Human Subjects (CCMO) Arnhem/Nijmegen.

15.2.4 Research Project B: A Socio-ethical Study of Palliative Teleconsultation

One of the main purposes of this research project B is to consider to what extent telemedicine alters existing communication patterns in palliative care and to what extent this affects underlying moral conceptions on dying and caregiving. In study B, existing – often implicit- normative conceptions surrounding palliative care and telemedicine will be explicated by analyzing (a) changes in verbal and non-verbal communication patterns between participants while using telemedicine at the homes of very ill patients, and (b) the participants' reflections on the experienced teleconsultations.

The primary objective is to investigate how patients, proxies, general practitioners, and medical specialists experience a multifunctional telemedicine-application at the patient's home and whether (and why) they find this telemedicine-application acceptable. Secondary objectives involve: (1) To investigate how the use of the TM-application mediates communication between patients, proxies, and caregivers; (2) To investigate whether and how the people involved in the process of care create new daily routines; (3) To investigate how the use of a telemedicine-application for the purpose of palliative home care relates to normative ideas about 'dying well' and 'good palliative care'; (4) To investigate whether and how the telemedicine-application empowers the patient to stay in control of his/her own care.

This study comprises a media ethnography (La Pastina 2005), and builds on two methodological approaches: **interviewing** and doing **observations**. Interviews will be conducted with all participants in the practice of home based palliative care. Each participant is asked to participate in a sequence of short interviews dispersed over several weeks. There will be (a) one-on-one interviews between the interviewer and a participant and (b) multi person interviews, while patients and their families live closely together in these last months. This dispersion guarantees that change and process are captured during the research. Moreover, a constant reflection on collected data can take place, which will generate new conversation topics for further interviews. All interviews will have a semi-structured character, which means that the researcher will use an interview guide with a few general topics to structure the interviews to a certain extent (Mason 2002). In addition, the researcher will conduct observations in the patients' homes, the doctors' offices, and in the hospital, which are always planned around weekly teleconversations. After the conceptualization of themes, normative reflection will start upon these themes (see also Charmaz 2009; Corbin and Strauss 2008).

15.2.5 Preliminary Findings

In the preparation phase of the clinical study, interviews and an expert meeting have been performed with the members of the palliative care hospital team about their expectations of palliative telecare (Van Gurp et al. 2013a, b). One of the worries was that teleconsultation 'at a distance' will limit aspects of care related to physical proximity making accurate anamnesis and 'getting a feel of a patient' more difficult. An initial face to face contact with the patient to establish a relationship was considered important. At the same time however, they expected that telemedicine will make it easier to monitor a patient at a low threshold. Also, teleconsultation will give patients more space in the treatment process to bring in their own expectations and wishes. Team members expected therefore that weekly teleconsultation will be beneficial in itself, because patients receive more attention and time. In this regard telemedicine will give the opportunity to share more aspects of (visual and oral) information than regular phone calls. Interestingly, team members also noticed that with teleconsultation, the hospital makes a step out of their own building into the houses of patients. This will probably open new dimensions of caregiving, also with regard of the role of the general practitioner in transmural care. All together, teleconsultation is expected to influence the patient professional relationship due to the new communication modes. Also, telecare will challenge the rather static location based distinction between home care and hospital care into a more dynamic relationship between hospital caregivers and general practitioners based on shared care and mutual expertise.

In later stages of the study, the transmural care dimension of telecare will be further explored, together with the patient perspective.

15.3 Discussion

The telemedicine-application in its mediating form is about expanding and integrating home care and clinical care. This fits current health policy aims, at least in the Netherlands, to organize care near the patients home and to give the patient a more central role in the delivery of health care. It is important to gather data to reveal the reliability of technology in relation to quality of palliative care. Although video-communication involves a complete different way of health care technology compared to MRI or EEG scans, it is also about shaping images of a patient. In fact, video communication will influence caregiving by amplifying certain aspects of a patient and reducing others. Ihde observed that although information and communication technologies do not have a consciousness like humans, they do have intentionality to some extent (Ihde 1979, 1990; Verbeek 2008a, b). Orlikowski, who studied different practices within professional organizations, considered this 'recursive intertwining of humans and technology' to be generally overlooked by social scientists (Orlikowski 2007). For example, rituals may evolve around the telescreens that influence social life (Morley 2007). In addition, patients can feel comforted by the presence of a screen that may provide a sense of safety, but may also be discomforted because the screens continuously remember them of their ill status. Here, ethical convictions on good care and 'dying well' will presumably be influential.

However, there are also several challenges in this study. For study A, a first challenge will be to enroll a sufficiently large sample of patients to make sure that differences in symptom alleviation between the intervention group and the control group can be detected. This can be challenging in a group of vulnerable patients. For the qualitative study, including the perspectives of patients will be extremely valuable, in particular as older patient may not be familiar with a virtual reality.

This research project stimulates collaboration between primary care and hospital care in order to optimize the continuity of care in advanced illness. Collaboration at the 'digital' workfloor may raise questions for both hospital specialists and general practitioners and may sometimes involve a 'clash' of different cultures of caring. However, it may also break possible dividing walls between first and second line care and provide meaningful and innovative forms of transmural care. Transmural collaboration may also strengthen patient participation. To what extent patients and professionals are ready for such a 'virtual' paradigm shift will be revealed further in our study.

Acknowledgements We would like to thank Ms. Duursma for her involvement in an earlier draft version of this chapter.

Appendix

Questionnaires used in the trial study (A)
Patient questionnaires
Administered at baseline
Basic demographic information (7 questions)
Administered at baseline and every week
ESAS (Edmonton Symptom Assessment System)
10 items on symptom assessment
Administered at baseline and every 4 weeks
PNPC-sv (Problems and Needs in Palliative Care – short version)
33 questions on experienced problems and needs for care
HADS (Hospital Anxiety and Depression Scale)
14 items on anxiety (7 items) and depression (7 items)
NCQ (Nijmegen Continuity Questionnaire)
28 items within 3 subscales on continuity of care
Family caregiver questionnaire
Administered at baseline and every 2 weeks
EDIZ (one dimensional assessment of care burden)
9 items on the experienced burden from informal care
Patient, GP and a member of the palliative consultation team
Administered after the first two teleconsultations
PSQ (Patient Satisfaction Questionnaire)
5 questions on satisfaction with the teleconsultation

References

Aoun, S.M., L.J. Kristjanson, D.C. Currow, and P.L. Hudson. 2005. Caregiving for the terminally ill: At what cost? *Palliative Medicine* 19: 551–555.

Bainbridge, D., P. Krueger, L. Lohfeld, and K. Brazil. 2009. Stress processes in caring for an end-of-life family member: Application of a theoretical model. *Aging & Mental Health* 13: 537–545.

Bashsur, R.L. 1995. On the definition and evaluation of telemedicine. *Telemedicine Journal* 1: 19–30.
Beccaro, M., M. Costantini, P.G. Rossi, G. Miccinesi, M. Grimaldi, and P. Bruzzi. 2006. Actual and preferred place of death of cancer patients. Results from the Italian survey of the dying of cancer (ISDOC). *Journal of Epidemiology and Community Health* 60: 412–416.
Bordewijk, J.L., and B. Van Kaam. 1982. *Allocutie. Enkele gedachten over communicatievrijheid in een bekabeld land*. Baarn: Bosch & Keuning nv.
Bowles, K.H., and A.C. Baugh. 2007. Applying research evidence to optimize telehomecare. *The Journal of Cardiovascular Nursing* 22: 5–15.
Burge, F.I., B. Lawson, P. Critchley, and D. Maxwell. 2005. Transitions in care during the end of life: Changes experienced following enrolment in a comprehensive palliative care program. *BMC Palliative Care* 4: 3.
Charmaz, K. 2009. *Constructing grounded theory. A practical guide through qualitative analysis*. London: Sage.
Cohen, S.R., P. Boston, B.M. Mount, and P. Porterfield. 2001. Changes in quality of life following admission to palliative care units. *Palliative Medicine* 15: 363–371.
Corbin, J.M., and A.L. Strauss. 2008. *Basics of qualitative research techniques and procedures for developing grounded theory*. Los Angeles: Sage.
Eminovic, N., N.F. de Keizer, J.C. Wyatt, G. ter Riet, N. Peek, H.C. van Weert, C.A. Bruijnzeel-Koomen, and P.J.E. Bindels. 2009. Teledermatologic consultation and reduction in referrals to dermatologists: A cluster randomized controlled trial. *Archives of Dermatology* 145: 558–564.
Hebert, M.A., R. Brant, D. Hailey, and M. van der Pol. 2006. Potential and readiness for video-visits in rural palliative homecare: Results of a multi-method study in Canada. *Journal of Telemedicine and Telecare* 12(Suppl 3): 43–45.
Higgingson, I.J., and G.J.A. Sen-Gupta. 2000. Place of care in advanced cancer. A qualitative systematic literature review of patient preferences. *Palliative Medicine* 3: 287–300.
Ihde, D. 1979. *Technics and praxis. A philosophy of technology*. Dordrecht: D. Reidel Publishing Company.
Ihde, D. 1990. *Technology and the lifeworld from garden to earth*. Bloomington: Indiana University Press.
Ingleton, C., J. Chatwin, J. Seymour, and S. Payne. 2011. The role of health care assistants in supporting district nurses and family carers to deliver palliative care at home: Findings from an evaluation project. *Journal of Clinical Nursing* 20: 2043–2052.
Jordhoy, M.S., P. Fayers, T. Saltnes, M. Ahlner-Elmqvist, M. Jannert, and S. Kaasa. 2000. A palliative care intervention and death at home: A cluster randomised trial. *Lancet* 356: 888–893.
Kitamura, C., L. Zurawel-Balaura, and R.K.S. Wong. 2010. How effective is video consultation in clinical oncology? *A Systematic Review. Current Oncology* 17: 17–27.
Klinkenberg, M., G. Visser, M.I. van Groenou, G. van der Wal, D.J. Deeg, and D.L. Willems. 2005. The last 3 months of life: care, transitions and the place of death of older people. *Health & Social Care in the Community* 13: 420–430.
Knol, A., Th.W. van den Akker, R.J. Damstra, and J. de Haan. 2006. Teledermatology reduces the number of patient referrals to a dermatologist. *Journal of Telemedicine and Telecare* 12: 75–78.
La Pastina, A.C. 2005. Audience ethnographies: A media engagement approach. In *Media anthropology*, ed. E.W. Rothenbuhler and M. Coman, 139–148. Thousand Oaks: Sage.
Laila, M., V. Rialle, L. Nicolas, C. Duguay, and A. Franco. 2008. Videophones for the delivery of home healthcare in oncology. *Studies in Health Technology and Informatics* 136: 39–44.
Mason, J. 2002. *Qualitative researching*. London: Sage.
McCorkle, R., and J.V. Pasacreta. 2001. Enhancing caregiver outcomes in palliative care. *Cancer Control* 8: 36–45.
Meier, D.E., and L. Beresford. 2008. Palliative care's challenge: Facilitating transitions of care. *Palliative Medicine* 11: 416–421.
Melin-Johansson, C., B. Axelsson, F. Gaston-Johansson, and E. Danielson. 2010. Significant improvement in quality of life of patients with incurable cancer after designation to a palliative homecare team. *European Journal of Cancer Care* 19: 243–250.

Morley, D. 2007. *Media, modernity and technology the geography of the new*. London: Routledge.

Murray, S.A., M. Kendall, K. Boyd, and A. Sheikh. 2005. Illness trajectories and palliative care. *BMJ* 330: 1007–1011.

Orlikowski, W.J. 2007. Sociomaterial practices: Exploring technology at work. *Organization Studies* 28(9): 1435–1448.

Ramirez, A., J. Addington-Hall, and M. Richards. 1998. ABC of palliative care: The carers. *BMJ* 316: 208–211.

Seale, C. 2000. Changing patterns of death and dying. *Social Science & Medicine* 51: 917–930.

Shugarman, L.R., S.L. Decker, and A. Bercovitz. 2009. Demographic and social characteristics and spending at the end of life. *Journal of Pain and Symptom Management* 38: 15–26.

Smeenk, F.W.J.M., J.C.M. van Haastregt, L.P. de Witte, and H.F.J.M. Crebolder. 1998. Effectiveness of home care programmes for patients with incurable cancer on their quality of life and time spent in hospital: Systematic review. *BMJ* 316: 1939–1944.

Smith, A.C. 2007. Telepaediatrics. *Journal of Telemedicine and Telecare* 13: 163–166.

Teno, J.M., B.R. Clarridge, V. Casey, L.C. Welch, T. Wetle, R. Shield, and V. Mor. 2004. Family perspectives on end-of-life care at the last place of care. *JAMA* 291: 88–93.

Tronto, J.C. 1993. *Moral boundaries: A political argument for an ethic of care*. New York: Routledge.

Van den Block, L., R. Deschepper, J. Bilsen, V. Van Casteren, and L. Deliens. 2007. Transitions between care settings at the end of life in Belgium. *JAMA* 298: 1638–1639.

van Gurp, J., M. van Selm, E. van Leeuwen, and J. Hasselaar. 2013a. Transmural palliative care by means of teleconsultation: A window of opportunities and new restrictions. *BMC Medical Ethics* 14(March 7): 12.

van Gurp, J., J. Hasselaar, E. van Leeuwen, P. Hoek, K. Vissers, and M. van Selm. 2013b. Connecting with patients and instilling realism in an era of emerging communication possibilities: A review on palliative care communication heading to telecare practice. *Patient Education and Counseling* 93(3): 504–514.

Van Ryn, M., S. Sanders, K. Kahn, C. van Houtven, J.M. Griffin, M. Martin, A.A. Atienza, S. Phelan, D. Finstad, and J. Rowland. 2011. Objective burden, resources, and other stressors among informal cancer caregivers: A hidden quality issue? *Psycho-Oncology* 20: 44–52.

van Wynsberghe, A., and C. Gastmans. 2008. Telesurgery: An ethical appraisal. *Journal of Medical Ethics* 34(10): e22.

Verbeek, P.P. 2008a. Cultivating humanity: Towards a non-humanist ethics of technology. In *New waves in philosophy of technology*, ed. J.-K. Berg Olsen, E. Selinger, and S. Riis, 241–266. Hampshire: Palgrave MacMillan.

Verbeek, P.P. 2008b. Morality in design: Design ethics and the morality of technological artifacts. In *Philosophy and design: From engineering to architecture*, ed. P.E. Vermaas, A. Light, and S.A. Moore, 91–103. Dordrecht: Springer.

Verhoeven, F., L. van Gemert-Pijnen, K. Dijkstra, N. Nijland, E. Seydel, and M. Steehouder. 2007. The contribution of teleconsultation and videoconferencing to diabetes care: A systematic literature review. *Journal of Medical Internet Research* 9: e37.

Whited, J.D., R.P. Hall, M.E. Foy, L.E. Marbrey, S.C. Grambow, T.K. Dudley, S. Datta, D.L. Simel, and E.Z. Oddone. 2002. Teledermatology's impact on time to intervention among referrals to a dermatology consult service. *Telemedicine Journal and E-Health* 8: 313–321.

Wootton, R., S.E. Bloomer, R. Corbett, D.J. Eedy, N. Hicks, H.E. Lotery, C. Mathews, J. Paisley, K. Steele, and M.A. Loane. 2000. Multicentre randomised control trial comparing real time teledermatology with conventional outpatient dermatological care: Societal cost-benefit analysis. *BMJ* 320: 1252–1256.

World Health Organization. 2004. *Palliative care: The solid facts*, ed. E. Davies and I.J. Higginson. Copenhagen: World Health Organization.

Yates, J., and W.J. Orlikowski. 1992. Genres of organizational communication – A structurational approach to studying communication and media. *Academy of Management Review* 17(2): 299–326.

Yates, J., W.J. Orlikowski, and A. Jackson. 2008. The six key dimensions of understanding media. *Mit Sloan Management Review* 49(2): 63–69.

Chapter 16
Privacy Aspects of Video Recording in the Operating Room

Claire B. Blaauw, John J. van den Dobbelsteen, Frank Willem Jansen, and Joep H. Hubben

Abstract A variety of applications of video recording in health care is growing including the development of a Digital Operation Room Assistant (DORA). This is a monitoring system, based on the analysis of video templates, that, similar to the black box in aviation, automatically records events in the operating room (OR). After the completion of the surgical procedure the video images may also be useful for other purposes such as the evaluation of the surgical technique or as an aid in training and education of medical staff and students. Yet, under Dutch privacy law, video images, once they are stored, are considered as personal data and legal demands have to be met to keep, use or delete these images.

While the value of video recording in the OR, – the enhancing of the quality of the surgical procedure and the patient's safety – is widely acknowledged, concerns have risen. Medical professionals claim to be unfamiliar with the legal demands that have to be met and fear to unlawfully violate the patient's privacy. Another concern is that the video images of the surgical procedure will be used in court proceedings and as such might lead to the physician's liability or a disciplinary measure, such as an official warning.

In the Netherlands research was done on the legal implications of a Digital Operation Room Assistant. The aim of the study is to provide a legal framework comprising the prerequisites for the storage, application and deletion of video

C.B. Blaauw (✉) • J.H. Hubben
Department of Health Law, University Medical Center Groningen, Groningen, The Netherlands
e-mail: Claireblaauw@gmail.com

J.J. van den Dobbelsteen
Department of Biomechanical Engineering, Delft University of Technology, Delft, The Netherlands

F.W. Jansen
Department of Biomechanical Engineering, Delft University of Technology, Delft, The Netherlands

Department of Gynecology, Leiden University Medical Center, Leiden, The Netherlands

J. van den Hoven et al. (eds.), *Responsible Innovation 1: Innovative Solutions for Global Issues*, DOI 10.1007/978-94-017-8956-1_16,

images of the surgical procedure. In addition the legal position of the medical professional should be clarified.

Legal demands related to the processing of personal data, such as the limitation, the quality, the specification of the purpose, the limited use and the openness of the video images, have to be taken into account. In addition measures should be taken to guarantee the safety of and the accountability for the video images. Video images of surgical procedures are considered as personal data concerning a person's health and should be treated with even more care. The patient's consent for video recording is obviously the most legitimate basis for video recording in this context. To make legal implications more explicit three situations are discerned in which video registration in the OR will take place. The application of the video images in the discerned situations has consequences for the patient's consent for the video recording. As a rule the video images shall be deleted as soon as they have achieved the purpose they have been set up for. This is not the case when the video images shall be added to the medical file. It is advisable to set up a protocol about the correct procedure in this matter.

The data subject is not only the patient but also the medical professional who is filmed. The video recording should take place with his consent. By all means it must be clear that the video data shall not be used for other purposes (such as the assessment of the medical professional) than the original purpose for video registration.

16.1 Introduction

Video recording is becoming increasingly important in health care (Xiao et al. 2007). For example, video cameras are introduced in emergency rooms for surveillance purposes, in delivery rooms to evaluate the delivery (van Balen et al. 2010) or during psychiatric consultations (Brandsma et al. 2007). Video registration during surgical procedures can serve various purposes as well. Numerous studies on improving quality of care by video registration are found in literature (Mackenzie et al. 2007; Blom et al. 2007; Aggarwal et al. 2007; Weinger et al. 2004; Verdaasdonk et al. 2007). Video images can be an important tool to train and teach students and professionals about the technique used during the surgical procedure, for instance by a remote real-time video system (Hahm et al. 2007). They can serve as a guide or illustration of best practice to prepare professionals for a difficult procedure. In addition they can be used for in-depth review after the procedure is completed if the video recordings of an entire surgical or medical procedure is stored. Another application in the operating room is to provide information (i.e. about the patient's vital signs, and the status of equipment and staff) for coordination of the workflow, thus improving the organization of the whole surgical procedure (Hu et al. 2006).

A new application of video images in health care is the Digital Operation Room Assistant (DORA); an instrument to improve the quality and safety of care during surgical procedures. The Delft University of Technology, the Netherlands, is doing research on DORA, which can be described as a monitoring system which, similar to a black box device in aviation, automatically records all processes in the operating room (OR). DORA uses video images to detect adverse events related to technical problems. Additionally direct information about technical problems is send to the OR personnel so that necessary adjustments can be made in time. Furthermore, once the operation is finished, the video images can be used for other purposes such as the evaluation of the procedure or for teaching and training of the professionals.

Although a Digital Operating Room Assistant can be a powerful tool to establish a safe and efficient operating room (OR) environment, legal concerns have prevented video recording from being part of the OR routine. One concern relates to unfamiliarity of the medical professionals with the legal framework regarding privacy issues (see for instance Verdaasdonk et al. 2007). Video images that have been stored and are accessible for further consultation, are as such legally considered as personal data (Blaauw and Hubben 2011). According to Dutch privacy laws certain conditions concerning the lawful processing of the video images, have to be taken into account. Video images of surgical procedures, once they are stored, are considered as personal data concerning a person's health and should be treated with even more care. A clear description of the legal framework in this context is becoming more urgent, since systematic video recording is increasing.

Another concern is that video images of the surgical procedure may be used for punitive or controlling purposes, leading to the physician's liability or an official warning. The legal position of the physician should be clarified and the consequences of the incorporation of video images in the medical file should be described. The section Health Law of the University Medical Centre in Groningen, the Netherlands has performed research on the legal implications of video registration in the OR. This study provides an extensive description of the legal framework in this context and gives insight in the legal demands and consequences regarding video recording in the OR (Blaauw and Hubben 2011). Some results of this study are touched on in this paper. The research questions addressed in this paper, are restricted to the clarification and description of the legal requirement of the patient's consent for video recording; the legal requirements for the documentation and deletion of the video images and the legal position of the medical professional.

16.2 Methods

The relevant sections of law such as international law, private law, administrative law, criminal law and self regulation, were investigated and screened for further study. This resulted in elaborate study of law, literature and jurisprudence on the international as well as the national level. Sources such as the European Data Protection Treaty, the 'Guidelines on the protection of Privacy and Transborder Flows of Personal Data' and the European Guideline 95/46/EG as well as international jurisprudence on the processing of video images and the protection of privacy were investigated and served in building a legal framework. On the national level sources, such as the Data Protection Act, the Individual Health Care Professions Act and the Medical Treatment Contracts Act and the guidelines on the transfer of medical data of the Royal Dutch Medical Association (KNMG 2010) were incorporated in the study. National jurisprudence on the use of video images in health care and privacy protection was studied and discussed in the context of DORA. In addition an expert meeting was organized with professionals of several medical specialties. The outcome of the expert meeting is included in the results (Blaauw and Hubben 2011).

16.3 Results

Video registration of patients touches on their privacy. The Dutch *Personal Data Protection Act* deals with the safe and adequate processing of personal data thus protecting the privacy of the person concerned. Video images that have been stored - contrary to images that have faded away -, are accessible for further consultation and for this reason are considered as personal data under this law. The legal demands entail the processing of personal data in accordance with the law and in a proper and careful manner. The conditions for the processing of personal data relate to the limitation (no excessive data), the quality, the specification of the purpose, the limited use and the openness of personal data, in this context the video images (Personal Data Protection Act art. 6–14). The video images must be collected for specific, explicitly defined and legitimate purposes. In addition measures should be taken to guarantee the safety of and the accountability for the video images. The processing of personal data shall be based on a legitimate ground which, in the context of DORA, is the unambiguously given consent of the patient.

16.3.1 The Patient's Consent

Due to the fact that video images of surgical procedures relate to a person's health and as such are considered as special personal data, they must be treated with even more care.[1]

Ways of dealing with issues such as the patient's consent, the incorporation in the medical file and matters of storage and deletion, often depend on the circumstances in which the video images are used. To make the discussion of the legal implications of video recording in the OR more explicit, three situations are discerned in which video registration takes place, (a) as an essential part of treatment (endoscopic

[1]Special personal data are defined as data regarding a person's religion, race, political conviction, health, sexual life and data regarding the membership of a union. For an extended list and regulations see Personal Data Protection Act art. 16.

surgery, for example), (b) to improve the quality of the procedure, and (c) for the purposes of peer assessment or education. In the first situation (endoscopic surgery), the video images are closely related to the performance of the surgical procedure in such a way that the video images are an indispensible part of the surgical procedure. The procedure cannot be carried out without the use of video images. It is therefore accepted that once the patient has given consent for the operation, there is no supplementary consent from the patient necessary for the (temporary) storage of video images as a technique which is being used during the procedure, provided that the patient has been well informed about the procedure in advance. In the other two situations the video recording is not an indispensible part of the procedure. In these situations the procedure can be carried out without the use of video images. This leads to the conclusion that supplementary consent from the patient for the video registration is necessary. In case the video images are used for educational purposes, explicit consent from the patient to use the video images for this purpose is required.

In addition to the patient the medical professional who is filmed is a data subject as well. In that case he shall give his consent for video recording and the further use of the video images.

16.3.2 Documentation and Deletion of the Video Images

From the Data Protection Act arises the general rule that personal data shall not be kept longer than necessary to achieve the purposes for which they have been processed or collected (the limitation principle) (Personal Data Protection Act art. 10). Consequently, the video images of medical procedures shall be deleted as soon as the purpose has been achieved. An example of this is the deletion of video images of a delivery in Obstetric ward directly after the physician's morning meeting (van Balen et al. 2010). According to the Dutch Medical Treatment Agreement Act the physician shall document all information that relates to the patient's treatment and the condition of his health and add it to the patient's file, in accordance with his duty as a good care provider. Consequently, the video images that give information about the treatment and the patient's health shall be incorporated in the file (e.g. critical view of safety in Minimally Invasive gallbladder surgery). As a consequence of the limitation principle excessive storage of video images should be avoided (Personal Data Protection Act art. 11). The three situations as mentioned above (the patient's consent) are relevant for the documentation of the video images as well. In the first situation (endoscopy) the close relation between the video images and the performing of the surgical procedure justifies that the images are incorporated in the medical file. According to the limitation principle (which entails that the processing of personal data should be adequate, relevant and not excessive; Personal Data Protection Act art. 11), a selection of images is appropriate. It is a topic of discussion which images the physician selects for documentation (see below). In the other two situations (to improve the quality of the surgical procedure, peer assessment and education) the relation between the use of video images and the performance of the surgical procedure is more distant.

The medical file (with the video images included) shall be kept for at least 15 years (Medical Treatment Contracts Act art. 454). It should be noted that the patient has the right of access to and a copy of his medical file. When a copy of the medical file (included the video images) is in the patient's possession it can be used in court proceedings.

16.3.3 The Legal Position of the Medical Professional

Video registration in the OR may contain images of the medical professional and other professionals assisting the procedure. As mentioned above they should give their consent for video registration as well. Covert video recording of an employee is a breach of privacy and essentially punishable by law. In this context it should be considered that in the near future consent for video registration is given in advance, for instance at start of employment.

It should be noted that in exceptional cases, others next to patient and physician may have access to the video images as well. This may be the Public Prosecuter who, under certain circumstances, as a criminal investigator, has access to the medical file including the video images. In addition, the Health Care Inspectorate can claim the video images for investigation, in view of a demand in the Dutch Quality of Health Care Act that any calamity (an unintended adverse event resulting in death or serious harm of a patient) which occurs in a medical institution must be reported to the Health Care Inspectorate. In case an offence is identified by the Inspectorate, the Public Prosecuter must be informed.

According to the Dutch Personal Data Protection Act the purpose of the video recording shall be explicitly defined. By all means it must be clear for all persons involved that video recordings may not be used to assess the professional performance in order to dismiss the physician involved.

16.4 Discussion

The results show that in case of an endoscopic procedure, the video images shall be added to the medical file. This flows from the duty of the care provider to document all information that relates to the patient's treatment and the condition of his health and add it to the patient's file. Yet, due to the limitation principle, the physician shall refrain from unnecessary documenting of video images. This leads to the question which video images should be selected and incorporated in the file and which images should be (automatically) deleted. It can be concluded that at least the crucial operation steps shall be incorporated. These should not be limited to situations in which complications occur. A crucial operation step can also be a decisive point during the operation. An example of this is the Critical View of Safety (CVS) during a cholecystectomy when the triangle of Calot is revealed before the cystic duct is transected, in order to avoid bile duct injuries. The Dutch Society for Endoscopic Surgery states in the 'best practice for laparoscopic cholecystectomies' that the CVS has to be documented (Lange and Stassen 2006). It is up to the professional societies to define other CVS's for other operations or specific procedures. Practical guidelines should define which are the crucial operation steps to be included for documentation.

In case the video images are used to improve the quality of care and for peer assessment, the video recording is not necessary for the performing of the procedure. Nonetheless it may happen that video images in this context reveal incidental findings and complications. In that case the video images should be incorporated in the file as well because they are relevant for the further treatment of the patient. This is a consequence of the physician's duty to treat the patient according to the standard of a good care provider. The same we advice when video images are used for education purposes and a medical relevant outcome is discovered by chance. Before video registration can be used as a tool for quality evaluation, optimal performance must be clearly defined by the medical professionals.

Another point of discussion is that medical professionals may be reluctant to introduce video registration in clinical practice because the images might reveal a wrong decision or an incorrect procedure during surgery. Under these circumstances the video images can be used by the patient as proof of evidence in court. Unfortunately this cannot be avoided as it is a consequence of the principle that the medical professional shall document the procedure (even if the procedure has an adverse outcome). On the other hand, the images might show that the right surgical procedure is followed. An illustration of this is the documentation of the CVS in laparoscopic procedures.

In case the images are used for peer review, attention should be paid to the fact that the purpose of the video recording is to improve the quality of care. It must be clear for the medical staff involved that the video recordings may not be used to assess the professional performance in order to dismiss the person involved. When this is clear to employer and employee, reluctance to video registration in the OR may disappear.

For the latter situation, from the technical point of view it might be a safe option to automatically make the video data anonymous. The advantage of this measure may be attractive but is contrary to the purpose of the video registration; the treatment of the patient which cannot be done anonymously.

It is advisory for health care institutions to take measures to meet legal demands prior to the implementation of a video registration technique in the OR. Measures should be taken to guarantee the safety of and the accountability for the video images.

It is advisable to set up a protocol about the correct procedure in this matter. The results of this paper as well as the full study can be incorporated in the protocol (Blaauw and Hubben 2011).

This study is relevant for other European countries as well, due to the fact that all European Member States are bound by the European guideline 95/46/EG. This guideline deals with the processing and transfer of personal data including medical data. The national laws of the European Member States are based on this guideline.

Acknowledgements This research was supported by the Netherlands Organization for Scientific Research (NWO Grant 313-99-003).

References

Aggarwal, R., T. Grantcharov, K. Moorthy, T. Milland, P. Papasavas, A. Dosis, F. Bello, and A. Darzi. 2007. An evaluation of the feasibility, validity, and reliability of laparoscopic skills assessment in the operating room. *Annals of Surgery* 245: 992–999.

Blaauw, C.B., and J.H. Hubben. 2011. *Video in de operatiekamer vanuit gezondheidsrechtelijk perspectief*. Den Haag: SDU.

Blom, E., E. Verdaasdonk, L. Stassen, H. Stassen, P. Wieringa, and J. Dankelman. 2007. Analysis of verbal communication during teaching in the operating room and the potentials for surgical training. *Surgical Endoscopy* 21: 1560–1566.

Brandsma, G.M., A.J. Hondius, and J.H. Hubben. 2007. Video- en geluidopnames van de patiënt. Hoe kan de hulpverlener de privacy van de patiënt waarborgen? *Journaal GGZ en recht* 8: 158–163.

Hahm, J., H. Lee, S. Kim, S. Shimizh, H. Choi, Y. Ko, K. Lee, T. Kim, J. Yun, and Y. Park. 2007. A remote educational system in medicine using digital video. *Hepato-Gastroenterology* 54: 373–376.

Hu, P., Y. Xiao, D. Ho, C. Mackenzie, H. Hu, R. Voigt, and D. Martz. 2006. Advanced visualization platform for surgical operating room coordination: Distributed video board system. *Surgical Innovation* 13: 129–135.

KNMG (Royal Dutch Medical Association). 2010. Guidelines for the use of medical data. http://www.knmg.nl/publicatie/medischegegevens.

Lange, J.F., and L.P. Stassen. 2006. Best practice: De techniek van de laparoscopische cholecystectomie (Critical view of safety [CVS] in 7 stappen). *Minimaal Invasieve Chirurgie. Plan van aanpak en beleid*. NVEC 2009: 56–59.

Mackenzie, C., Y. Xiao, F. Hu, F. Seagull, and M. Fitzgerald. 2007. Video as a tool for improving tracheal intubation tasks for emergency medical and trauma care. *Annals of Emergency Medicine* 50: 436–442.

van Balen, M., J.H. Hubben, M. Groenewout, G.G. Zeeman, L.R. van Lonkhuijzen, and P.P. van den Berg. 2010. Video op de verloskamer. *Medisch Contact* 65(13): 590–591.

Verdaasdonk, E.G., L.P. Stassen, M. Elst, T.M. van der Karsten, and J. Dankelman. 2007. Problems with technical equipment during laparoscopic surgery. An observational study. *Surgical Endoscopy* 21(2): 275–279.

Weinger, M., D. Gonzales, J. Slagle, and M. Syeed. 2004. Video capture of clinical care to enhance patient safety. *Quality & Safety in Health Care* 13: 136–144.

Xiao, Y., S. Schimpff, C. Mackenzie, R. Merrell, E. Entin, R. Voigt, and B. Jarrell. 2007. Video technology to advance safety in the operating room and perioperative environment. *Surgical Innovation* 14: 52–61.

Chapter 17
Assessing the Future Impact of Medical Devices: Between Technology and Application

Neelke Doorn

Abstract The aim of this contribution is to see how interdisciplinary collaboration in the development of medical technologies can enhance ethical reflection on the social and moral impact of new medical devices. On the basis of a so-called "ethical parallel research," it was investigated how the social impact of a new medical device, based on Ambient Intelligence, could be assessed during the process of technology development. The study indicates that technical researchers tend to make a sharp distinction between technologies and applications; the former supposedly being "neutral." They framed their own work in terms of neutral technology. This one-sided focus on technology may hamper the assessment of social and moral impact and the prevention of negative side-effects. The case suggests that the assessment of the social and moral impact of new medical devices requires expertise that researchers themselves indicate to be lacking. The involvement of ethicists or social scientists in the development of these devices may encourage technical researchers to bridge the gap between applications and technologies, such as to effectively encourage socially responsible technology development.

17.1 Introduction

Although most often aimed at the advancement of human well-being, the introduction of new technologies in our society is not without risks. Especially so-called emerging or converging technologies (that is, technologies that combine

The Brocher Foundation is acknowledged for providing the author the opportunity to work on this paper during her stay at the Centre as visiting researcher in fall 2011.

N. Doorn (✉)
Delft University of Technology, PO Box 5015, 2600 GA Delft, The Netherlands
e-mail: n.doorn@tudelft.nl

J. van den Hoven et al. (eds.), *Responsible Innovation 1: Innovative Solutions for Global Issues*, DOI 10.1007/978-94-017-8956-1_17,

previously distinct fields) create opportunities that were until then unknown.[1] The accompanying changes to society are so fundamental that we may be faced with new ethical challenges (Kulinowski 2004; McCray 2005). One of the problems is that the development of these technologies is often running ahead of legislation. By their very nature, the implications of these technologies are not yet fully known (McGregor and Wetmore 2009). Due to the newness of these implications, the ethical challenges are often phrased in rather abstract terms (e.g., privacy, human enhancement, safety) with sometimes little bearing on the physical applications themselves (Van den Hoven and Vermaas 2007). We therefore need new approaches that are able to discern "specific contextualized ethical issues raised by specific technological and scientific developments" (1) while "the technology is still in its very early phases of development" (2) and "do so in such in way that these reflections can inform the process of technological development itself" (3) (Van de Poel 2008: 26; in the remainder of the text I will call these the methodological demands).

The aim of this contribution to see whether interdisciplinary collaboration can help meeting the methodological demands; more particular, to see how interdisciplinary collaboration in technological project can enhance "real-time" ethical reflection on the social on moral impact of newly developed devices. I present the results of a so-called ethical parallel research that was carried out in the context of the ALwEN (Ambient Living with Embedded Networks) project, aimed at the development of an in-house monitoring application for people with COPD based on ambient intelligence technology.[2]

The outline of this contribution is as follows. Following this introduction, I first briefly describe ambient intelligence, the technology concerned. I then sketch an overview of this recent trend of interdisciplinary collaboration, followed by a brief description of the ALwEN project. I then discuss the results of the ethical parallel research. In a concluding section, I summarize the findings and I come back to the question whether interdisciplinary collaboration is able to meet the "methodological demands" mentioned above.

17.2 Ambient Intelligence

The European term Ambient Intelligence (AmI) – or equivalently ubiquitous and pervasive computing (USA) or ubiquitous networking (Japan) – reflects a vision of the future of ICT in which 'intelligence' is embedded in virtually everything

---

[1]The following terms are also common: NBIC technologies, where NBIC is an acronym for the convergence of Nanotechnology, Biotechnology, Information technology, and Cognitive science (originally coined in a report commissioned by the National Science Foundation (Mihail and Bainbridge 2002)).

[2]A more elaborate discussion of the methodology of ethical parallel research can be found in Van de Poel and Doorn (2013).

around us. This intelligence is built into tiny processors and sensors which are integrated into everyday objects (such as clothes and furniture) and which are able to communicate directly with each other without the need of traditional PC input and output media (Mattern 2004).

In the health care sector, for example, AmI technology is considered a promising way to address the problem of population ageing and the associated increasing costs of providing care (Steg et al. 2006). Not only are industrialized countries facing a demographic challenge with a population that is becoming increasingly older, due to improvements in the treatment of (chronic) diseases like Alzheimer, cardiovascular diseases, COPD and diabetes, the demand by this elderly population on the health care system shifts from providing cure to care. At the same time, people want to live independently as long as possible, within their health restrictions (Steg et al. 2006).

Taken together, these socio-economic and medical trends ask for innovative solutions to provide basic care in and around the home, within available personal and monetary budgets. Ideally, these solutions should address the improvement of the quality of life and reduction of prolonged care at the health care institutions, while ensuring a high quality of life, autonomy and security. AmI based technologies carry the promise to provide such a solution exactly because of their ubiquitous and unobtrusive analytical, diagnostic and monitoring functionality (Korhonen and Bardram 2004; Bardram and Mihailidis 2007; Lina et al. 2008). Applied in the context of health care and wellness, the broad range of (electronic) devices as well as services that are providing unobtrusive support for daily life based on and adapted to the assisted person in his or her own context are often referred to as Ambient Assisted Living (AAL). In general, AAL technologies are aimed at providing assistance to carry out daily activities, health and activity monitoring, enhancing safety and security, getting access to social, medical and emergency systems, and facilitating social contacts (Steg et al. 2006).

The introduction of AmI or AAL in our society is not without risks. The obvious and most widely discussed risks are those related to the collection, storage and processing of (personal) data (Lyytinen and Yoo 2002; Duan and Canny 2005; Gadzheva 2008; Joshi et al. 2008; Lahlou 2008; Neitzke et al. 2008; Park et al. 2009; Spiekermann and Langheinrich 2009). These risks are related to the very nature of AmI. In order to deliver personalized services to a user, the user's personalized profile must be gathered and stored, which raises the risk of abuse, either accidentally or intentionally (Wright 2008). Comparatively few papers discuss social issues that go beyond these privacy and security threats, such as the digital divide and the delegation of control (Bohn et al. 2004; Wright et al. 2008), or environmental concerns (Köhler and Erdmann 2004; Köhler and Som 2005; Kräuchi et al. 2005).

The importance of addressing these risks is recognized by both policy-makers and researchers, witnessing the attention for the study of Ethical, Legal, and Socio-economic Aspects (ELSA) of AmI within the *European Commission Seventh Framework Programme* and the development of the so-called ISTAG scenarios, a set of AmI-based scenarios created in commission by the Information Society Technologies Advisory Group (ISTAG) of the European Union to explore the social and technical implications of long-term developments in ICT.[3] However, so far, the implementation of ELSA and the ISTAG scenarios in actual research practice remains limited (Fisher 2005). This implementation failure is not limited to AmI but is witnessed in the broader field of Health Technology Assessment (HTA) (Drummond and Weatherly 2000).

One possible remedy is to shift the management focus to the phase of technology development itself and encourage the researchers to reflect on the impact of their work "on the work floor."

17.3 Interdisciplinary Collaboration

In the last decade, interdisciplinary collaboration between researchers from the social sciences or humanities on the one hand and researchers from the natural and applied sciences on the other has received increasing attention. This collaboration provides an opportunity to modulate research decisions according to social concerns, so it is argued (Fisher 2007). The field is booming due to the fact that multi-billion initiatives like the National Nanotechnology Initiative in the US and the 7th framework program in Europe require ELSA research alongside and in the early phases of the development of new science and technology. In the last decade, science policies in the US, Europe and elsewhere have called for "responsible innovation" in science and technology, implying that social and ethical considerations be integrated with R&D processes (Schuurbiers et al. 2013).

One of the institutional contexts to perform such third generation or real-time Technology Assessment (RTTA) is on the work floor (or, more prosaically, in the laboratory). Contrary to earlier TA approaches, the focus of third generation TA approaches is on "opening up the innovation process, rather than managing it after-the-fact" (Sarewitz 2005). Similar to second generation TA approaches, RTTA seeks to build learning into the implementation process, but by staying close to the technological development process itself these newest approaches have more impact on technological development. Instead of merely addressing the impacts of technology, these RTTA approaches aim at "shaping the trajectory of technological development" (Wilsdon 2005) in order to improve both the societal consequences and the decision making about science and technology (Sarewitz 2005). Although different approaches exist, in most of them an "embedded" researcher (who can be someone with a humanities or a social sciences background) visits the workplace and interacts with the natural and technical scientists. The embedded researcher may ask questions or just report her observations. The results of these reflections and studies feed back into the ongoing research. The aim may vary from making the researchers simply aware of social impact of the work to raising reflexive

[3]In the creation of the ISTAG scenarios both industrial stakeholders and science policy officials were involved (see also http://cordis.europa.eu/fp7/ict/istag/home_en.html).

awareness and sometimes even deliberately modulating research into a particular direction. Ultimately, this may lead not only to more societally responsible research but possibly to more efficient and effective research as well.

In the Netherlands, the more common term for referring to this interdisciplinary collaboration is "ethical parallel research." In the beginning of the twenty-first century, the most important public financer of technology research STW and the Netherlands Organization for Scientific Research (NWO) started a pilot of four ethical parallel research projects that were carried out parallel to technical research projects (Van der Burg 2009). In the ethical project described in this contribution, the insights gathered with those pilot projects are used to further develop the methodology for ethical parallel research. Since the case presented in this contribution concerns a Dutch research project, I use the term "ethical parallel research" in the remainder of this contribution to refer to this kind of interdisciplinary collaboration.[4]

17.4 Case and Methodology

The ethical parallel research described in this contribution concerns the ALwEN project,[5] which is aimed at developing a prototype in-house monitoring application based on Wireless Sensor Networks (WSN) technology, the combination of body sensors, ambient sensors and wireless networks. The project started in 2007 and continued until the end of 2011. Already with the composition of the research team, the ALwEN consortium tried to differentiate itself from other projects by capturing the whole trajectory of fundamental research to the development of a prototype application and ultimately commercial exploitation. In order to do so, four universities, two independent industrial research institutes, one clinical partner and a consortium of 12 SMEs cooperated. At the start, the ambitions of the ALwEN team were high. In the project proposal, the ALwEN consortium had set itself the goal of bringing the engineering science for such a technology to the level of commercial product viability. The aim was to develop a prototype Ambient Assisted Living (AAL) type application to monitor and assist the activities of the elderly in the context of an elderly home. In this pilot application, so it was mentioned in the project proposal, concepts and techniques required to safeguard security and privacy of the information collected through use of WSNs could be tested and further developed. Rather than focusing on isolated aspects of the technology, the ALwEN consortium aimed at a more systematic and integral approach to scientifically understand all interactions, interferences, and cross-relations of WSN technology, such as to find the right balance and trade-offs on the system level.

---

[4]Doorn et al. (2013) provides a state-of-the-art volume on interdisciplinary approaches.

[5]For a more elaborate description of the project, see Doorn (2012).

The ethical parallel research consisted of studying documents, attending and observing meetings, semi-structured interviews with the team members and the organization of a workshop.[6] The people were interviewed approximately 16 months after the start of the project. The workshop was organized halfway through the project. The interviews and personal workshop data were approved by the participants. Regarding the interpretation of the data, the relevant participants were asked whether or not they agreed with the specific interpretation of their views (and in the one case of disapproval, the text was adapted according to the interpretation given by the researcher himself). The results of the ethical investigations were fed back to the researchers in the project meetings and a formal presentation.

In the workshop, insights from political philosophy were used to structure the discussions, especially on the normative topics. In a team like the ALwEN project team, people may have different views on what to include in the project and by whom it should be addressed. By discussing these issues in terms of a philosophical framework (rather than in terms of opposing interests), it may be possible to reach a consensus which is considered fair by all people involved.

17.4.1 Social Acceptance

At the start of the project "social acceptance" was identified as one of the crucial points for the successful implementation of the technology. In addition to technical and economic goals, the project consortium had therefore set itself the following two goals related to the social acceptance of the application:

- Quality of life: the project will develop a pilot application to monitor and assist the activities of the elderly in the context of an elderly home. The main societal criterion for the success of this application is that it contributes to the quality of life of the elderly, in the sense that it helps them to maintain their independent living.
- Security and privacy: even though personal information may be pervasively collected and distributed over wireless communication channels, the security of the information and the privacy of the patient must be guaranteed.

These two goals related to "social acceptance" were used as discussion topic in the interviews and the workshop.

---

[6]In order to gain the trust of the project members, some informal meetings and site-visits were attended as well. Trust is indeed an important issue in this kind of research. A combination of personal skills and institutional safeguards is probably required to deal with the challenge of being able to raise critical issues, while at the same time being recognized in the team. In this particular project, it helped that the involved ethicist had an engineering background as well. For a more elaborate discussion of these issues, see Van de Poel and Doorn (2013) and Doorn and Nihlén Fahlquist (2010).

17.5 Results

17.5.1 General Observations

The ALwEN project fits within a longer cooperation between the consortium of SMEs and the different universities. The cooperation between these partners went therefore relatively smoothly.

The remaining partners, the clinical partner (a rehabilitation research institute) and the industrial research institutes, were less at the core of the project. They became involved in the project at a later stage. Especially the role of the rehabilitation research institute remained somewhat inarticulate at the start of the project. The primary role of the rehabilitation institute was to contribute to the development of the prototype application. Their task was to write a realistic and feasible use case on the basis of which the prototype application could be developed and to perform some pilot studies in real-life situations (including a requirements analysis and an evaluation scheme). This use case was intended to serve as an example of what can be done with WSNs and to focus the work of the demonstration activities of the project. The eventual use case described a situation of in-house monitoring of the daily activities of a patient with COPD.

Halfway through the project, the overall work of the project was still mainly focused on development of the technology and not so much on development of the prototype application, which made the position of the rehabilitation institute in the project team somewhat disconnected. In the interview series, the technical researchers mentioned the limitation of the use case to monitoring patients with a non-life threatening disease as the most tangible result of the involvement of the clinical partner. Monitoring cardiovascular patients, for example, would have been too demanding in terms of Quality-of-Service requirements. In practice, the use case characteristics had little bearing on the actual technical work.

17.5.2 Interview Series

After 16 months of research time, a selection of the project team members was interviewed about the moral and social implications of this particular project. The interviews revealed that the researchers make a strict distinction between technology and application. This applied to both the fundamental researchers at university and the more applied researchers at the SMEs. They all considered themselves to be working on the development of a technology rather than an application. This application may have social and moral implications, the technology itself is considered neutral, the interviewees indicated.

Since “social acceptance” was, in the original research proposal, identified as a crucial element of the success of the project, this notion of “social acceptance” was chosen as the starting point of the author’s ethical investigations, including the necessary conditions for getting the technology socially accepted. In the interviews, the representatives of the different institutional partners involved in the project were invited to brainstorm on the relevant "moral issues" pertaining to the project. The interviewees were asked to think of "moral issue" in as broad a way as possible: anything related to risks and moral values (e.g., social acceptance, human well-being, privacy, society, and sustainability) was considered relevant. According to the technological researchers, these issues should be addressed in order to gain social acceptance.[7]

Although the "social acceptance" of the technology was defined as the explicit goal of the project, its interpretation was still rather vague at the start of the project. In the interviews, The participants were therefore asked how they conceived of this notion of social acceptance.

It is interesting to note the differences between the technical researchers and engineers on the one hand, and the clinical researchers on the other. For the former, social acceptance (as a goal of the project) was primarily conceived as social *acceptability*; that is, a prospective quality that was to be determined by experts, by whom they meant the clinical researchers and possibly the ethicist as well. For these researchers, the involvement of clinical and ethical experts was therefore crucial for achieving the goal of social acceptance.

The clinical partners, on the other hand, defended a participatory approach, the result of which would lead to social acceptance. As one of the clinical researchers argued:

> In a way you could say "the proof of the pudding is in the eating": if the service is being used by the intended users (including all technical imperfections, user unfriendliness and the fact that they have to pay for it), it is socially acceptable. That means that the use of the service for some other than the intended purpose is still OK. The acceptability lies in the final use.

In this view, the technology's acceptability is defined in terms – or maybe one could say, is constructed by – its acceptance. For these researchers, the inclusion of end users was therefore of primary importance.

In the remainder of the text, I use the term "social acceptance" to refer to both the social acceptance and the social acceptability.

From the interviews it followed that most researchers considered the prototype application merely a demonstration tool and not a service that is to be commercially exploited. Consequently, they deemed the ethical issues and social acceptance of the prototype application not so urgent. Most researchers thought that a health technology assessment should be done only if some application would be commercially exploited.

---

[7]It is realized that this description of moral issue is not as well-defined as some philosophers would like it to be. However, since the interviews and the workshop were explicitly aimed at tracing the opinions of the engineers themselves, they were not given any constraints on what counts as a moral issue nor were there any issues introduced that were not mentioned by the engineers themselves. For a more well-wrought description of when a value can be considered a *moral* value, see Nagel (1979).

Most researchers agreed that privacy and security are important issues to address when AmI based technologies are applied to human beings. However, since most of the SMEs involved are not primarily interested in the health care or wellness domain, they rendered this prototype application rather ambitious compared to their own companies' (future) commercial applications (which they for reasons of strategy did not disclose).

The technical researchers recognized that the social acceptance of a future application may be related to particular technological choices but they did not consider it urgent to address these issues at this stage of the development.

17.5.3 Workshop

Six months after the end of the interview series, a workshop was organized for which all the interviewees were invited. (Of the 13 interviewees, 5 people were unable to participate in the workshop. With the remaining 8 participants, all institutions participating in the project were represented.) On the basis of the interview results, a list of eight salient "tasks" was established and put on the agenda for discussion. This list included tasks like "making sure that the application does not interfere with everyday life," and "identifying how technological choices affect social acceptance." The workshop participants were asked to indicate during which project activity the different tasks were to be addressed.

In the workshop, the participants showed a broader conception of the scope of the project than in the interviews. Most participants recognized the multi-faced (i.e., technical and social) character of social acceptance. The participants were allowed to say that certain tasks were beyond the scope of the project (or not to be addressed at all) and only few participants actually labeled some of the tasks as "beyond the scope of this project." Those participants who did tended to change their opinion in the course of the workshop. This means that ultimately most participants considered it the project team's responsibility (either on an individual basis or collectively) to address the moral tasks, also the broader societal ones (e.g., addressing legal questions related to data storage and data access).

It was generally agreed that the focus of the project should shift towards laboratory and clinical experimentation with a (prototype) application in order to better investigate and address the moral and social issues pertaining to the project. Some tasks prompted particular discussion because the participants disagreed about the question where and by whom this task should primarily be carried out. In their evaluation of the workshop, most participants indicated that they had become more aware of certain moral issues. The technical researchers, for example, realized that including the clinical partner in the project is in itself not enough to have the end users represented, but that they needed to involve end users themselves. There was a general agreement that most moral issues span several activities within the project and that it is therefore difficult to single out one project activity where it should primarily be addressed. The primary responsibility was in those cases ascribed to the project management for coordinating this joint effort, to the experimentation phase where all activities were supposed to come together, or to the clinical partner. Some participants explicitly mentioned that this workshop made them realize that some moral issues were currently not addressed adequately. The idea that the work should shift from research towards experimentation prompted a refocus of the work and soon after the workshop, a brainstorm meeting was scheduled in which the requirements for clinical experimentation were discussed in more detail. In this meeting, both the technical and non-technical requirements pertaining to the use case were discussed. In 2010, these real-life experiments were prepared and in the first half of 2011, clinical experiments were carried out in a home in Enschede. In these experiments, people volunteered to live in a house equipped with a sensor network to monitor their behavior. The people stayed in the house for 5 days (24 h a day). This refocus to experimentation enabled the researchers to take the user experiences into account and adapt the design accordingly.

Ten months after the workshop, another set of interviews was held with two team members with a formal role in the management of the project. They were asked after their experiences with the ethical parallel investigations. They both expressed their appreciation of the involvement of an ethicist in the project and they argued that it had helped them giving "ethics" a more profound role in the project. One interviewee argued that the involvement of an ethicist can help making technical people more aware of things they otherwise tend to overlook. Regarding future projects, they both thought it should be common practice to give an ethicist a formal role in technical projects during the whole course of the project. Both interviewees indicated that they see ethics as a relevant, but for themselves unknown, field of expertise. Since they considered themselves lacking the ethical expertise, they thought future projects would gain in quality by composing multidisciplinary teams. They considered ethics not as instrumental to successful technology implementation but rather as an end in itself. Ideally, ethics should be seen as a "non-functional requirement that you cannot ignore," one of the interviewees remarked.

17.6 Discussion

From the interviews it followed that there were two main obstacles for addressing the social and moral implications of the application. The first obstacle concerns the cooperation between the clinical and the technical partners. Whereas the former did not have a clear insight from what to expect from the technology, the latter did not know how technical choices affected the eventual application and its social implications. From the start, it was unclear who should take the initiative to bring the technical and clinical work together. Although the ALwEN consortium had set itself the goal of covering the whole trajectory from fundamental to clinical research, including clinical experimentation, the cooperation between the different partners proved difficult in practice. Especially regarding the cooperation between the clinical partner and the technical partners, the team members adopted an attitude of waiting. The technical partners seemed to be waiting for instructions "how to establish social acceptance," whereas the clinical partners seemed to be unaware of the possibilities of WSN technology. The introduction of the COPD use case improved the communication somewhat, but the cooperation remained difficult. Ultimately, during the workshop, initiatives were taken to make the cooperation run more smoothly.

The second obstacle concerns the way the team members framed their work: this was about technology rather than application. This distinction between technology and application fits in the Mertonian view on science and technology, which states that research should be driven by the norms of universalism, communism, disinterestedness, and organized skepticism (Merton 1942 [1996]). This view leaves only room for epistemic values like predictive accuracy, coherence, consistency (Ruse 1999). It is the researcher's "moral duty" to avoid the involvement of ethical and social considerations in research practice. In the last decades, this view has been criticized by philosophers, who argue that this "neutrality view" is untenable (cf. Winner 1980; Ihde 1990; Verbeek 2005). If we take the design of a technological artifact, several choices are made which carry with them moral implications (Van de Poel and Van Gorp 2006). Technical choices as to the encryption techniques, communication protocols, and energy use may influence the visibility and ease-of-use of future applications and thus its social acceptance. Although the technical researchers recognized that the social acceptance of a future application may be related to particular technical choices, they did not consider it urgent to address these issues at this stage of the development. They tried to leave as many technical options open as possible, such as to give future technology producers the freedom to develop an application with ample opportunity to optimize, for example, the "privacy settings." However, in practice not everything could be left open and choices *were* made, including morally-relevant (or, more loosely, value-laden) choices. To frame the work in terms of "neutral" technology rather than application seems therefore a false escape from the moral domain. In the second interview round, the technical researchers indicated that it was partly due to the involvement of the ethicist in the project that they came to realize that they should shift their focus from research to clinical experimentation in order to assess the impact of their technical choices.

17.7 Conclusions

The experience with the ALwEN project shows that the involvement of an ethicist during technology development can be an effective means to address ethical issues; that is, an involvement that can actually steer the direction of technology development and provide contextualized feedback. Whereas the more traditional TA approaches are sometimes prone to the Collingridge or control dilemma (Collingridge 1980) – they come either too early, when there is little known, or they come too late, when all decisions are already made – collaboration between TA consultants, ethicists, or social scientists on the one hand and technical or applied researchers on the other may be a fruitful way to make researchers think in terms of applications rather than technology. This helps making the moral issues more "concrete" (cf. the three methodological demands mentioned in the introduction).

For the assessment of the ethical implications of technology, it is of paramount importance that the researchers do not stick to the Mertonian view of neutral technology but that they think in terms of applications as well. It is desirable that the social and moral impact of technological work is tested in as early a stage as possible. However, the question remains to what extent these issue could and should be addressed by engineers themselves, or whether external researchers should be invited, like social scientists or – as in the current project – ethicists. And if external researchers are invited, how to make sure that the reflective awareness for the social and moral impact of technological work is sustained. For one, funding organizations increasingly recognize the importance of addressing these issues. And was it 10 years ago maybe enough to write a paragraph on ELSA in funding proposals (with the risk of reducing it to a mere "checkbox ethics"), new funding programs like the MVI program[8] require genuine cooperation between different disciplines. In addition to this requirement from funding organizations, it is of paramount importance that (prospective) engineers are trained in recognizing moral issues during their professional work. Courses like engineering ethics or value sensitive design should therefore be part of every engineering curriculum. Whether this will make the role of TA consultants, ethicists, or social scientists completely replaceable is doubtful, but it will probably make engineers more prone to inviting these people with a non-technical background in their project if they need their advice.

Regarding the cooperation between the technical and clinical researchers, it is desirable that the development should be seen as a cyclic process rather than a linear trajectory from fundamental to applied research to design. Only then can clinical experts clarify their wishes and have them implemented in future medical devices.

References

Bardram, J.E., and A. Mihailidis (eds.). 2007. *Pervasive computing in healthcare*. Boca Raton: CRC Press.

Bohn, J., V. Coroama, et al. 2004. Living in a world of smart everyday objects—Social, economic, and ethical implications. *Human and Ecological Risk Assessment* 10(5): 763–785.

Collingridge, D. 1980. *The social control of technology*. New York: St. Martin's Press.

Doorn, N. 2012. Exploring responsibility rationales in R&D. *Science, Technology & Human Values* 37(2): 180–209.

[8]MVI is the acronym of "Maatschappelijk Verantwoord Innoveren" (in English; responsible innovation). This program, funded by the Netherlands Organization for Scientific Research NWO, is aimed at multidisciplinary research concentrating on current issues with both scientific and societal relevance. The current chapter is published in the first edited volume published in the framework of this program.

Doorn, N., and J.A. Nihlén Fahlquist. 2010. Responsibility in engineering. Towards a new role for engineering ethicists. *Bulletin of Science, Technology & Society* 30(3): 222–230.

Doorn, N., D. Schuurbiers, I.R. Van de Poel, and M.E. Gorman (eds.). 2013. *Early engagement and new technologies: Opening up the laboratory*. Dordrecht: Springer.

Drummond, M., and H. Weatherly. 2000. Implementing the findings of health technology assessments – If the CAT got out of the bag, can the TAIL wag the dog? *International Journal of Technology Assessment in Health Care* 16(1): 1–12.

Duan, Y., and J. Canny. 2005. Protecting user data in ubiquitous computing. Towards trustworthy environments. *Lecture Notes in Computer Science* 3424: 167–185.

Fisher, E. 2005. Lessons learned from the ELSI program: Planning societal implications research for the National Nanotechnology Program. *Technology in Society* 27: 321–328.

Fisher, E. 2007. Ethnographic invention: Probing the capacity of laboratory decisions. *NanoEthics* 1(2): 155–165.

Gadzheva, M. 2008. Privacy in the age of transparency – The new vulnerability of the individual. *Social Science Computer Review* 26(1): 60–74.

Ihde, D. 1990. *Technology and the lifeworld: From garden to Earth*. Bloomington/Indianapolis: Indiana University Press.

Joshi, A., T. Finin, et al. 2008. Security policies and trust in ubiquitous computing. *Philosophical Transactions of the Royal Society A-Mathematical Physical and Engineering Sciences* 366(1881): 3769–3780.

Köhler, A., and L. Erdmann. 2004. Expected environmental impacts of pervasive computing. *Human and Ecological Risk Assessment* 10(5): 831–852.

Köhler, A., and C. Som. 2005. Effects of pervasive computing on sustainable development. *IEEE Technology and Society Magazine* 24(1): 15–23.

Korhonen, I., and J.E. Bardram. 2004. Guest editorial introduction to the special section on pervasive healthcare. *IEEE Transactions on Information Technology and Biomedicine* 8(3): 229–234.

Kräuchi, P., P.A. Wager, et al. 2005. End-of-life impacts of pervasive computing. *IEEE Technology and Society Magazine* 24(1): 45–53.

Kulinowski, K.M. 2004. Nanotechnology: From "wow" to "yuck"? *Bulletin of Science, Technology & Society* 24(1): 13–20.

Lahlou, S. 2008. Identity, social status, privacy and face-keeping in digital society. *Social Science Information Sur Les Sciences Sociales* 47(3): 299–330.

Lina, C.C., R.G. Leeb, et al. 2008. A pervasive health monitoring service system based on ubiquitous network technology. *International Journal of Medical Informatics* 7(7): 461–469.

Lyytinen, K., and Y.J. Yoo. 2002. Introduction to the special issue: Issues and challenges in ubiquitous computing. *Communications of the ACM* 45(12): 62–65.

Mattern, F. 2004. Ubiquitous computing: Scenarios for an informatized world. In *E-merging media: Communication and the media economy of the future*, ed. A. Zerdick, A. Picot, K. Schrape et al., 155–174. Berlin: Springer.

McCray, P.W. 2005. Will small be beautiful? Making policies for our nanotech future. *Journal of History and Technology* 21(2): 177–203.

McGregor, J., and J.M. Wetmore. 2009. Researching and teaching the ethics and social implications of emerging technologies in the laboratory. *NanoEthics* 3(1): 17–30.

Merton, R.K. 1942 [1996]. The ethos of science. In *On social structure and science*, ed. P. Sztompka, 267–276. Chicago: University of Chicago Press.

Mihail, C.R., and W.S. Bainbridge. 2002. *Converging technologies for improving human performance: Nanotechnology, biotechnology, information technology and cognitive science*. Arlington: U.S. National Science Foundation.

Nagel, T. 1979. *Mortal questions*. Cambridge: Cambridge University Press.

Neitzke, H.P., M. Calmbach, et al. 2008. Risks of ubiquitous information and communication technologies. *GAIA – Ecological Perspectives for Science and Society* 17(4): 362–369.

Park, J.H., S. Gritzalis, et al. 2009. Intelligent ubiquitous computing: Applications and security issues. *Internet Research* 19(2): 133–135.

Ruse, M. 1999. *Mystery of mysteries: Is evolution a social construction?* Cambridge, MA: Harvard University Press.

Sarewitz, D. 2005. This won't hurt a bit: Assessing and governing rapidly advancing technologies in a democracy. In *The future of technology assessment*, ed. M. Rodemeyer, D. Sarewitz, and J. Wilsdon, 14–21. Washington, DC: Woodrow Wilson International Center for Scholars.

Schuurbiers, D., N. Doorn, I.R. Van de Poel, and M.E. Gorman. 2013. Mandates and methods for early engagement. In *Early engagement and new technologies: Opening up the laboratory*, ed. N. Doorn, D. Schuurbiers, I.R. Van de Poel, and M.E. Gorman, 3–14. Dordrecht: Springer.

Spiekermann, S., and M. Langheinrich. 2009. An update on privacy in ubiquitous computing. *Personal and Ubiquitous Computing* 13(6): 389–390.

Steg, H., H. Strese, et al. 2006. *Ambient assisted living – European overview report. Europe is facing a demographic challenge ambient assisted living offers solutions*. Berlin: VDI-VDE-IT.

Van de Poel, I.R. 2008. How should we do nanoethics? A network approach for discerning ethical issues in nanotechnology. *NanoEthics* 2(1): 25–38.

Van de Poel, I.R., and N. Doorn. 2013. Ethical parallel research: A network approach. In *Early engagement and new technologies: Opening up the laboratory*, ed. N. Doorn, D. Schuurbiers, I.R. Van de Poel, and M.E. Gorman, 111–136. Dordrecht: Springer.

Van de Poel, I.R., and A.C. Van Gorp. 2006. The need for ethical reflection in engineering design: The relevance of type of design and design hierarchy. *Science, Technology & Human Values* 31(3): 333–360.

Van den Hoven, M.J., and P.E. Vermaas. 2007. Nano-technology and privacy: On continuous surveillance outside the panopticon. *Journal of Medicine and Philosophy* 32(3): 283–297.

Van der Burg, S. 2009. Imagining the future of photoacoustic mammography. *Science and Engineering Ethics* 15(1): 97–110.

Verbeek, P.P. 2005. *What things do: Philosophical reflections on technology, agency, and design*. University Park: Pennsylvania State University Press.

Wilsdon, J. 2005. Paddling upstream: New currents in European technology assessment. In *The future of technology assessment*, ed. M. Rodemeyer, D. Sarewitz, and J. Wilsdon, 22–29. Washington, DC: Woodrow Wilson International Center for Scholars.

Winner, L. 1980. Do artifacts have politics? *Daedalus* 109(1): 121–136.

Wright, D. 2008. Alternative futures: AmI scenarios and minority report. *Futures* 40(5): 473–488.

Wright, D., S. Gutwirth, et al. (eds.). 2008. *Safeguards in a world of ambient intelligence*. Dordrecht: Springer.

Chapter 18
Video-Surveillance and the Production of Space in Urban Nightlife Districts

Irina van Aalst, Tim Schwanen, and Ilse van Liempt

Abstract This chapter is based on a research project that examines if and how technologically mediated forms of surveillance and policing improve the safety and wellbeing of nightlife consumers whilst at the same time also contributing to processes of socio-spatial exclusion of particular groups. By interrogating the triad of surveillance and policing, wellbeing and exclusion in nightlife districts in Dutch city centers we found that the effects of video-surveillance on the production of space are complex and ambiguous. Storylines used by local policy-makers with regard to CCTV differ considerably between cities and tend to overestimate the benefits of CCTV surveillance. Moreover, consumers' awareness and knowledge of CCTV tends to be limited and only a few experiences a real sense of enhanced safety and wellbeing because of the presence of technology alone. At the same time, the effects of surveillance and policing on the exclusion of certain groups from nightlife districts are not equivocally supported by our initial findings either.

Contribution on the basis of the research project: '*Surveillance in Urban Nightscapes: The Socio-Spatial Effects of Video-Surveillance in Urban Nightlife Districts*'Grant 313-99-140; Responsible Innovation Programme; Netherlands Organization for Scientific Research (NWO)

I. van Aalst (✉) • I. van Liempt
Department of Human Geography and Planning, Faculty of Geosciences, Utrecht University, Utrecht, The Netherlands
e-mail: i.vanaalst@uu.nl; i.c.vanliempt@uu.nl

T. Schwanen
Transport Studies Unit, School of Geography and the Environment, University of Oxford, Oxford, UK
e-mail: tim.schwanen@ouce.ox.ac.uk

J. van den Hoven et al. (eds.), *Responsible Innovation 1: Innovative Solutions for Global Issues*, DOI 10.1007/978-94-017-8956-1_18,

18.1 Introduction: The Rise of Night-Time Economies

Across Western Europe districts of nightlife entertainment are attracting increased attention in urban policy and governance, because these spaces are unique configurations of economic opportunity, pleasure and excess. In response to globalization, neo-liberalism and the decentralization of governmental power from the national to the local level, European cities have become more proactive in enhancing competitiveness and stimulating economic growth (Harvey 1989; Hall and Hubbard 1998). By trying to make the city centre a site of spectacle, consumption and pleasure, policymakers, corporate actors and other urban stakeholders hope to attract tourists, business travelers, students and others; to keep young, middle-class families from moving to the suburbs; and to become a magnet for businesses (Judd and Fainstein 1999; Miles and Paddison 2005; Schmid et al. 2011). Thus, the organization of festivals and the development of spatial clusters of bars, clubs, restaurants and cinemas are familiar governmental strategies for improving a city's attractiveness and livability. The term night-time economy, which is commonly used in the UK-based scholarly literature, is telling with regard the obvious links between nightlife, profitability and inter-urban competitiveness (Shaw 2010).

Nonetheless, compared to other forms of consumption, the governance of urban nightlife is imbued with profound ambiguity. Whilst stimulated for economic reasons, nightlife is also kept under (increasingly tight) control in an attempt to mitigate real and imagined excesses. The urban night is after all a distinctive space-time (Hubbard 2005; Williams 2008) that offers a wide range of intense emotional experiences – from pleasure, excitement and adventure to fear and distress – and myriad opportunities for the transgression of otherwise taken-for-granted social norms. Regarding such transgression, the emphasis is usually on binge-drinking, vandalism and violence (Winlow and Hall 2006; Roberts and Eldridge 2009). However, more positive forms of transgression, such as overcoming the restraint to approach strangers or impediments to free self-expression, are also significant. They allow forms of sociality and conviviality to emerge that are not normally encountered during daylight (see also Jayne et al. 2011).

The most common governmental response to the complex entanglements of economic opportunity, pleasure and excess has been the intensification of surveillance and policing in nightlife districts (Helms 2008; Roberts and Eldridge 2009): police agents, private security firms and Closed Circuit Television (CCTV) systems are among the many techniques employed to enhance the safety and wellbeing of the various stakeholders involved in urban nightlife, including (benevolent) consumers, bar and club owners and staff, police officers and ambulance personnel. Wellbeing is a widely used but elusive term that is often taken to refer to the level of happiness, pleasure and satisfaction individuals experience (Diener 2009). The meaning of wellbeing is, however, broader than personal enjoyment. Building on recent work in geography (Conradson 2005; Fleuret and Atkinson 2007; Atkinson et al. 2012), we understand wellbeing as an individually experienced but socially produced and intrinsically spatial phenomenon, emerging from – in our case – the interactions between consumers, police officers, bouncers, club and bar owners, CCTV systems and the built environment as well as collective norms, values and customs. Space is thus taken to be actively involved in the production of wellbeing (or the lack thereof); it is not simply a passive background to the actions and perceptions of individual agents. This means that spaces of wellbeing are spaces that offer joy, self-fulfillment, self-esteem, protection from harm, and/or restoration from stress and forms of ill-health. Such spaces can also contribute to emancipation, mutual valuing and inclusivity, for instance through the reworking of prejudices about certain social groups (Fleuret and Atkinson 2007).

In the Netherlands the surveillance and policing of urban nightlife districts is increasingly undertaken in the context of what since the mid-1990 has become known as Safe Nightlife Policies [*Veilig Uitgaan Beleid*]. These policies are framed around the idea that (local) government cannot monitor and police nightlife districts on its own; club and bar owners, residents, consumers and other actors also have to take responsibility and contribute to this form of nodal governance (Van Aalst and Van Liempt 2011). Another key trend has been increased technological mediation of the surveillance and policing of nightlife districts and city-centers more generally. It is not simply that CCTV systems have become more widespread; new technological hardware, software and procedures have been introduced and piloted. Mobile cameras, computer code to manage recorded data streams, the continuous tracking of specific individuals moving through an area and real-time feedback from CCTV operators to police and bouncers 'on the ground' are obvious examples.

These forms of technological mediation are widely claimed to be successful in reducing crime and disorder by politicians, policymakers and the popular press alike (Webster 2009). Systematic reviews of CCTV evaluations suggest, however, that the effectiveness of CCTV has consistently been overrated (Armitage 2002; Welsh and Farrington 2003). Concerns have also been raised in the academic literature about the extent to which technologically mediated forms of surveillance and policing may marginalize and disadvantage particular social groups: CCTV has been considered a masculine technology unable to register and respond to the forms of (verbal) harassment that tend to intimate women in particular (Koskela 2002); research among CCTV operators has suggested that their decisions about who to monitor are often informed by racist and ageist prejudices (Norris and Armstrong 1999); and computer code used to automatically detect behavior considered deviant or for facial recognition may also embody social stereotypes about race/ethnicity and particular youth cultures (Graham 2005).

The main objective of this chapter is to examine if and how technologically mediated forms of surveillance and policing really improve the safety and wellbeing of nightlife consumers whilst at the same time also contributing to the socio-spatial exclusion of particular groups from nightlife districts. Our research project interrogates the triad of surveillance and policing, wellbeing and exclusion by focusing on the different actors involved in the production of the spaces of nightlife districts in the Dutch cities of Rotterdam, Utrecht and Groningen. These three cities have been selected on the basis of differences in population composition, spatial structure of the nightlife district, and surveillance and policing practices (for more information, see Schwanen et al. 2012). We use a mixed-method approach that considers as many relevant agents as possible, including but not limited to (potential) consumers, nightlife entrepreneurs, policymakers, CCTV operators, bouncers and police officers. Their perspectives and views are gauged and articulated via a range of research methods: repeated on-site observations during nighttime in selected nightlife districts, in-depth interviews with consumers and other stakeholders, analyses of policy documents, questionnaires among (non-) visitors of urban nightlife districts and series of participatory workshops with consumers and other stakeholders.

After outlining some of the theoretical notions and commitments guiding our analysis, we chart three complexities regarding video-based surveillance in nightlife districts. We will firstly consider how different discourse coalitions (Hajer 2005) emerged around CCTV and contributed to different surveillance practices in Rotterdam and Utrecht. Secondly, we examine the nuanced experiences and understandings consumers have regarding CCTV and how these differ from policy discourses. Finally, we discuss how the increased use of mobile devices equipped with cameras among consumers has the potential to disrupt and rewrite the relations between the watcher and the watched and introduce fundamental novelty in surveillance routines. We then briefly discuss some initial findings regarding the relations between surveillance and policing more generally and the dynamics in the character of nightlife districts as spaces of pleasure and excess, before drawing some conclusions.

18.2 Theoretical Background: An Assemblage Approach

The project draws on and brings together a range of theoretical registers from human geography, science and technology studies, sociology, urban studies and cultural studies. For the purpose of this chapter it suffices to highlight three starting points that are central to the study:

- The surveillance and policing of nightlife districts need to be understood as the outcome of distributed assemblages.
- Discourses about and the practice of such surveillance may not coincide with each other.
- It is not immediately apparent that surveillance and policing practices make nightlife districts safer and/or more enjoyable for all actual and potential nightlife consumers.

Following Deleuze and Guattari (1987) and DeLanda (2006), we understand an assemblage as a collective whose properties emerge from the relations between its heterogeneous parts. Heterogeneity is crucial: it is from the interactions of different components – human bodies, technological artifacts, codes, built structures, symbols, ideas, energies, emotions, and so on – that assemblages come into being and effects are generated. Adopting this assemblage approach has many advantages, one of which is that there are no restrictions on the character of the elements that can become part of an assemblage. As such the notion of the assemblage does not privilege the discursive, the material or indeed any other ontological realm. Another attractive feature is that any assemblage, on Deleuze's view, is characterized by both stability and instability. This implies, among others, that the properties of assemblage are open to change: there is always an immanent possibility of ambiguity, novelty or something unexpected happening. And, as shown below, the surveillance and policing of nightlife districts is indeed an arena where continuity and change coalesce and where interactions between people and camera technologies are a source of novelty and ambiguity. We use the adjective 'distributed' in distributed assemblage in a dual sense. Not only are competencies, capacities, actions, events, meanings, and so on, usually distributed across sets of multiple elements; these elements also tend to be distributed geographically. Thus, the capacity to monitor a nightlife district weaves together many different elements, from the cameras hanging on buildings and in public spaces to IT networks through which information is transported to the control room (which is sometimes located in another city) where software developed by engineers in locations that can be as far away as Bangalore and embodied skills acquired over CCTV operators' life-course are crucial to the decoding and interpretation of the footage by those operators.

Surveillance assemblages in urban spaces in the Netherlands and elsewhere have undergone two key changes: spatial extension and technological advancement. 77 % of Dutch cities with more than 100,000 inhabitants now have (standard and static) surveillance cameras in public spaces (Schreijenberg et al. 2009) and the tendency towards 'blanket surveillance' is set to continue. Interest in and use of 'smart' cameras and 'smart' algorithms to handle and interpret data flows are also increasing. Cities are experimenting with mobile cameras (e.g. Rotterdam) and cameras equipped with sensors for recording sounds (e.g. Groningen), although success has so far been mixed (Gemeente Groningen 2011). There have also been experiments with the use of algorithms in CCTV control rooms that reduce data stream in such a way that multiple cameras can be monitored simultaneously on a single screen. For the future much is expected from facial recognition software and algorithms for the automatic detection of deviant behavior.

The increased role of technologies in surveillance and policing means that the capacities, competencies and actions of surveillance and policing assemblages are likely to change with potentially significant effects for public spaces. Whatever the nature of such effects, it is important to be attentive to differences between discourses about surveillance and policing and practices 'on the ground'. Now, any discourse – i.e. the ideas, meanings and practices through which surveillance and policing are made understandable – is multiple and differentiated (Foucault 2002; Hajer 2005), and this is also true of contemporary surveillance in general and of CCTV in particular. Utopian understandings foregrounding the crime-reducing and safety-enhancing capacities of CCTV exist side by side with dystopian variants that emphasize the risks of increased social sorting (Lyon 2003), enhanced social stratification, privacy issues and the production of sterile urban spaces. It is sometimes also suggested that CCTV has relatively little effect on events in urban spaces. At the same time, understandings of video-surveillance as a techno-fix for all kinds of urban problems remain widespread, at least in the popular media and political rhetoric. Relatively little is known about the discourses around CCTV and surveillance in nightlife districts and at the city level more generally. Given that in the Netherlands policies regarding the surveillance and policing of nightlife are formulated and implemented at the city level, one of project's goal is to develop a better grasp of local differences in the social shaping of video-surveillance.

Notwithstanding their differentiation and multiplicity, the discourses about CCTV that circulate through the popular press, political institutions and evaluative reports prepared by consultants are unlikely to emulate the complexity and ambiguity of the actual practices of (video-)surveillance and policing in Dutch nightlife districts. One objective of the research program, and indeed this chapter, is to map out part of that complexity and ambiguity. The point here is not to celebrate complexity for its own sake. We rather seek to identify and contribute to the development of new or hitherto underappreciated possibilities to make nightlife districts spaces of wellbeing as defined above for consumers, police officers, ambulance personnel, club and bar owners and staff and other relevant stakeholders.

There is an extensive literature in the social sciences in support of the notion that contemporary nightlife districts may not be places of well-being, at least not for everybody. Here we are thinking of work not only on nightlife's excesses, such as binge-drinking, alcohol-fuelled violence and vandalism (Winlow and Hall 2006; Roberts and Eldridge 2009; Jayne et al. 2011) but also on processes of social exclusion. Research has shown, for instance, that since 1990 nightlife in the centers of London and Manchester has become homogenized along lines of class and race through a variety of processes, including bouncer practices, price setting, online reservation and screening systems, dress codes, prejudices about non-western youth cultures, and the licensing practices of local authorities (Talbot and Böse 2007; Measham and Hadfield 2009). Fears have been expressed that such processes will intensify with a further shift within surveillance and policing assemblages towards mediation by advanced digital technologies (*cf.* Graham 2005), especially when in the future CCTV footage of individuals can be coupled in real time to their 'data-doubles' – the digital information on them that is stored in the databases of public authorities, corporations and possibly other actors (Haggerty and Ericsson 2000). Many of these claims, however, demand detailed scrutiny and this is another area where our project intends to make a contribution. Further analysis of exclusionary processes in urban nightlife is also warranted because the existing literature is dominated by evidence from the UK. In that country the commercialization and corporatization of nightlife premises, which is often cited as a cause for social exclusion in urban nightlife (Chatterton and Hollands 2003; Talbot 2007), is more profound than elsewhere in Europe.

In short, much is unclear about the extent to which technologically mediated surveillance and policing contribute to the production of safe and enjoyable nightlife spaces, who – (potential) consumers across different ethnic, class and other social categories; bar and club officers; staff of nightlife establishments; police officers and ambulance personnel; local public authorities; and so on – benefits and in what ways, and who/what is excluded. Below we present some initial results and thoughts regarding these issues.

18.3 Locally Differentiated Discourses: On the Geographies of CCTV's Role in Surveillance Assemblages

The spatial extension of CCTV in city centers, including nightlife areas, is well documented in the academic literature (McCahill 2002; McCahill and Norris 2002; Hempel and Töpfer 2004; Welsh and Farrington 2009). Less is known, however, about the rationalizations and legitimizations of installing and using video-cameras in public spaces in nightlife district. We analyzed the discourses embedded in policy documents prepared by city-level and national authorities and mobilized during in-depth interviews with experts involved in Rotterdam's and Utrecht's Safe Nightlife Policies (Van Liempt and Van Aalst 2012). The focus on discourses follows from the recognition that the ideas and concepts of Safe Nightlife Policies cannot be imposed in a top-down manner and are contested in struggles about their meaning, interpretation and implementation. The fact that multiple actors debate safe nightlife in shared terms does not mean that they all have the same ideas and understandings about it. The assumption of mutual understanding that is at the base of these policies is often misplaced and tends to conceal much discursive complexity. Regarding video-surveillance, we suggest that locally differentiated discourse coalitions – ensembles of storylines (narratives in which metaphors play an important role), actors articulating these storylines and practices that are consistent with them (Hajer 2005) – came into existence around CCTV in Rotterdam and Utrecht, which has led to differences between these cities in the role video-surveillance plays in wider local policy. Utrecht and Rotterdam provide strongly contrasting examples: In the latter the camera came to be understood as an 'extra' eye on the street that is constantly watching, but in Utrecht the camera was also discussed in terms of the 'spy' putting non-criminals under surveillance. Because of this contrast, and the unequal development of CCTV in both cities, we limit the discussion to these two cities in this part of the chapter.

18.3.1 Rotterdam: Watching CCTV Footage 24/7

Rotterdam is the second largest city in the Netherlands with a specific local political landscape that has shifted drastically since 2000. Pim Fortuyn, who was murdered in 2002, started his political career in the city of Rotterdam and had a major influence on the shift in the city's political landscape from a strong socio-democratic tradition to a landscape dominated by a populist party (*Leefbaar Rotterdam*). Pim Fortuyn, together with the former mayor and minister of Safety and Justice Ivo

Opstelten – nicknamed 'the Dutch Giuliani' – promoted a policy of 'zero tolerance' to make Rotterdam safer. Zero tolerance is not unique to Rotterdam but the city is one of the few in the Netherlands that is openly communicating and embracing this approach. Some typical Rotterdam examples of zero tolerance policy are the introduction of so-called City Marines (*Stadsmariniers*[1]) who have the power and financial means to solve concrete problems and/or to manage unsafe areas (Tops 2007), and Rotterdam's slogan 'Rotterdam Presses On' – a point of reference for many other Dutch cities intent on implementing restrictive safety policies during day or night time.

In the summer of 2000 the mayor, the chief of police, the chief public prosecutor and a representative of Promotion Stadhuisplein signed the first Covenant Safe Nightlife for Stadhuisplein – the most important spatial concentration of nightlife premises in Rotterdam's city centre. The covenant contained agreements to increase safety on the square. In the same year the first public cameras were installed in Rotterdam. The Euro 2000 and preceding football riots sped up this decision and convinced critics of its necessity. Today Rotterdam is the city in the Netherlands with the largest number of publicly installed CCTV cameras (350) (Van Schijndel et al. 2010). Camera images are watched 24/7 seven days a week and there is immediate contact between the control room and police officers on the ground. Local government has opted for standard cameras without many bells and whistles; the emphasis is on human resources rather than technological advancement. Cameras are not seen as a replacement for the police but more as an 'extra' eye on the street. This metaphor and the emphasis on the importance of follow-up to the viewing of CCTV footage are crucial to the discourse coalition that has emerged around CCTV in Rotterdam. Visitors to the CCTV control room are shown a film of a criminal arrested (in a rather aggressive style) thanks to live watching of CCTV footage and quick and efficient follow-up by policemen on the ground. Successes are being emphasized.

Another important element of the discursive way in which Rotterdam's CCTV policy is described is the focus on quantitative information and 'results'. In the city of Rotterdam as a whole around 60 incidents are observed every day using 281 CCTV cameras. For the main nightlife district, Stadhuisplein, the number is around 4 incidents per day with 14 CCTV cameras. The majority (2/3) of these observations are followed by actions on the ground by immediate assistance teams (Van Schijndel et al. 2010). In some ways the focus in Rotterdam on no-nonsense, pragmatism and efficacy in the sense of follow-up and 'hard figures' appears factual and scientific. However, research has shown CCTV to not be very effective in curtailing street crime and violence that occurs impulsively, such as when alcohol and/or drugs

[1]The Dutch word 'stadsmarinier' has been invented by a Dutch psychologist, Diekstra, who argued that when policing unsafe areas the City Council should deploy the best people who should be given authority, power and financial support. He made the comparison with the military which also sends its best people to the front.

are involved (Welsh and Farrington 2009). In response to similar findings on the effectiveness of CCTV in the Netherlands by the Netherlands Institute for Social Research (Van Noije and Wittebrood 2009), the Municipal Officer responsible for CCTV in Rotterdam argued: "when cameras images are not watched, yes they are ineffective, but you cannot argue that cameras are not effective, that only means that not watching results in ineffectiveness". In an interview with a Dutch expert on CCTV systems and policies in Rotterdam the emphasis on no-nonsense and efficacy was emphasized: "In Rotterdam we do not want to create an illusion of safety. We do not have a policy of empty boxes as in other cities. One very important pillar of our safety policy is that we watch the video images 24/7. If we think a camera is needed we put one in and once it is there we use it".

The metaphor of the empty box is used to refer to cities that have cameras in public space but where more meaning is ascribed to the symbolic meaning of the camera than to the actual practice. In terms of technology the understanding that technology can prevent and/or reduce crime is present and produces specific effects. In Rotterdam's control room, for instance, people are increasingly trained to recognize deviant behavior and to use the 'extra eye' on the street in the most efficient way possible. At present smart software is being developed to help operators select and interpret the data, although it is also recognized that using such software constitutes a real challenge in nightlife districts with many people passing by and impulsive behavior.

18.3.2 Utrecht: The Camera as a Spy

Utrecht is the fourth largest city in the Netherlands. Utrecht's municipal council consisted at the time of writing (2011) of a coalition between the social democrats (*PvdA*), the social liberal democrats (*D66*) and the Green Party (*Groen Links*), and is more reluctant to implement restrictive safety measures than Rotterdam has been. CCTV practices in Utrecht's nightlife district are not very different from Rotterdam in the sense that there is immediate contact between police officers on the ground and the operators in the control room. The local political discussion about CCTV is nonetheless very different from the one in Rotterdam. Privacy arguments continue to be emphasized in Utrecht and surveillance technology is often understood as dangerous and risky. The metaphor of the camera as a 'spy' was clearly embedded in political discussions at the start of Utrecht's camera project. When the first public camera was installed in the city-centre in 2001, it was decided that the images would only be watched live on clubbing nights (Thursday, Friday and Saturday) in order to prevent the targeting of the 'wrong' people. This policy was supported using the following argument: "In Utrecht we do not want to spy on innocent citizens, we only watch camera images if there is a considerable risk that something might happen" (Municipal officer, Utrecht).

The argument of the 'considerable risk' made it difficult for the city council to sell this policy. The first evaluation of camera surveillance in Utrecht showed that the target that was set at the beginning of the video surveillance project – a 10 % drop of crime rates – was not met (Gemeente Utrecht 2002). With this finding the legitimacy of the CCTV policy was immediately contested and challenged. The relative low frequency of violence related to going out in Utrecht made it difficult to continue this policy. On the other hand, when during student induction week a student was partially paralyzed as a result of a serious fight in 2008, the 'solution' was immediately framed around CCTV: the boy's parents, for instance, claimed in the media that their son could have been saved had a camera been in place. In fact, there had been a camera covering the location of the accident, but on a Wednesday evening the images were not watched live. After this incident the mayor increased the surveillance hours for CCTV so that images are now watched every night of the week (Mo–Wed 6:00 PM–2:00 AM, Thu–Sa 2:00–6:00 PM, Su 2:00 PM–2.00 AM). This provides one example of how the human impact frame (Barnard-Wills 2011) is often used effectively to legitimize surveillance measures. Emphasis is in that case placed on people who would have been saved by surveillance.

Although camera surveillance, especially in public spaces, has been an important focal point in the public debate on privacy in the Netherlands in general, the Dutch have more or less accepted the phenomenon (Nouwt et al. 2005). Even if camera surveillance contributed little to a reduction in crime rates, the sense of security among citizens did appear to increase and in this way camera-surveillance may have enhanced wellbeing. The argument of greater security has also been used by politicians in Utrecht to continue CCTV surveillance of public space. In Utrecht there were at the time of writing 87 public cameras. Nonetheless, the decision in 2009 by Utrecht's city council to freeze this number and to discuss more intensively their necessity, effectiveness and the safeguarding of legal rights shows that the storyline around the metaphor of the camera as a spy has persisted and continues to generate effects. The general impression that cameras were never removed after installation was an important trigger for this ruling. The number of incidents observed by cameras in Utrecht is not published, which in itself is already an interesting difference between the two cities. Unpublished data from Utrecht police show that the rate (20 %) of follow-up activity by assistance teams on the ground is rather low (20 %) and that the majority of observed incidents were disorder related, including among others public urinating and public drunkenness.

In short, we have identified different discourses in the cities of Rotterdam and Utrecht. Using Barnard-Wills' (2011) terminology, we can describe these as discourses of 'appropriate surveillance' and 'inappropriate surveillance', respectively. The first draws on discourses of crime prevention and safety and security, the latter on privacy and personal liberty. In Rotterdam CCTV has become a municipal safety policy tool supported by the police, policy officials as well as the mayor. CCTV is considered an additional tool in daily policing that generates few constraints. In Utrecht opposition to CCTV is much more embedded in local policies. The main political actors, including the mayor, are openly communicating their criticism on CCTV.

18.4 Nuanced Understandings of CCTV: On How Consumers Experience and Understand Video-Based Surveillance

As shown above, it is often assumed that CCTV has a direct impact on behavior, safety and wellbeing in public spaces. There is, however, very little in-depth understanding of how CCTV is actually experienced and perceived in the midst of action by users of public spaces. We therefore designed short on-site interviews in which 84 participants in Rotterdam's and Utrecht's nightlife were directly confronted with various forms of video-surveillance, including CCTV, between 10:00 PM and 2:00 AM on several Thursday, Friday and Saturday nights in 2010. Participants were confronted with the availability of CCTV surveillance in situ. We first examined their awareness of CCTV and then asked whether the fact that cameras had been pointed out to them altered their feelings of safety. From the responses to this question we were able to derive valuable insights about how participants thought CCTV worked and affected their safety (more details available in Brands et al. 2014; Timan and Oudshoorn 2012). The results from this part of our research indicate a number of contradictions regarding CCTV.

First, participants' awareness of CCTV turned out to be more layered than initially thought. CCTV awareness cannot be understood in a crisp and dichotomous manner and is better conceptualized as having multiple gradations (Brands et al. 2014). There were consumers who: (1) had no knowledge of CCTV presence; (2) assumed CCTV would be present but had no clue as to where or when; (3) knew there was CCTV on the square where the experiments took place but could not pinpoint any; and (4) could pinpoint individual cameras. Secondly, participants' knowledge of how CCTV worked was often limited, although a few had a deep understanding of CCTV practices. The limitations on participants' knowledge are evident from the observation that few of them knew if and/or when footage was watched live. This was even true of Rotterdam in spite of this city's 24/7 watching policy.

Thirdly, and most significantly, only a small subset of the participants experienced a sense of clearly enhanced safety because of CCTV presence; indifference to this form of video-surveillance was the most common response. Most participants understood CCTV as a passive 'recording' device that is instrumental to catching a perpetrator after a crime has taken place but that can do little 'in the heat of the moment' of an unpleasant encounter or as a safeguard against crime. This finding concurs with previous studies (Koskela 2003; Klauser 2007). On the other hand, the majority of participants believed that CCTV is most beneficial in terms of enhancing safety when the images are watched live and immediate action is taken. One participant from Utrecht who highlighted the importance of live watching explained that "such a camera, if it is watched continually, then you know that it's safer here". Another participant said: "I think that it does make a difference for people if they see and experience that filming actually has an effect (. . .) Not so much the immediate film but more the feeling of safety it gives that immediate [human] action will follow" (Brands et al. 2014).

These quotes illustrate that from a consumer perspective technology alone cannot reduce nightlife's excesses. The eye of the camera needs to be complemented by human vision. It is the real-time presence and activity of a human-machine hybrid that is required, for it is the operator who can mobilize police offers and others who can provide true assistance and enhance perceptions of safety. It appears that only if competencies are distributed in a dual sense that video-surveillance can help to enhance safety: the non-human camera needs to work in tandem with the human operators and police officers, and the CCTV operator room 'far' away from the consumer needs to collaborate with police officers in close proximity of that consumer. From a policy and governance perspective, thinking about the relations between CCTV and experienced safety through the lens and logic of assemblage theory may assist in maximizing the safety benefits of video-surveillance.

18.5 Redefining Vision: Consumers' Own Surveillance Practices in Nightlife Districts

So far the discussion has focused on static CCTV cameras but this is only part of the story: personal media devices (PMDs), including mobile phones equipped with cameras and pocket-size photo and film cameras, are used increasingly by police officers and private security guards as well as nightlife consumers and have the potential to act as surveillance technologies. In Rotterdam, for instance, police vans and cars and the helmets worn by police officers on bikes are increasingly equipped with mobile cameras. Consumers' use of PMDs to record images has, of course, been discussed by academics before. Mann et al. (2003), for instance, coined the terms sousveillance and inverse surveillance to denote the watching by citizens rather than institutionalized organizations. The use of PMDs devised by citizens is often, and perhaps usually, for leisure rather than surveillance purposes. Consumers can, however, use PMDs to monitor the practices of specific people in a nightlife district, such as fellow consumers, police officers and private security guards (including bouncers). The multiplication of recording devices in nightlife districts has potentially profound consequences for the surveillance enacted by distributed assemblages. On the basis of initial research within our project (as described in Timan and Oudshoorn 2012), it can be argued that this multiplication inserts a dual openness in existing surveillance assemblages.

A first sense of openness pertains to the destination and use of camera footage. One interesting result reported by Timan and Oudshoorn (2012), who compare the experience of various forms of video-recording in public space by nightlife consumers in Rotterdam, is that the destination of CCTV footage was clear to participants. However, the recording of images with a mobile camera triggered uneasy responses, primarily because the destination of footage was uncertain. This raised privacy concerns (which did not exist for CCTV) among participants and made them feel more 'surveilled'. Indeed, Timan and Oudshoorn contend that PMD usage by consumers needs to be thought of as Open-Circuit Television (OCTV) because recordings may travel much further than CCTV recordings: OCTV footage may remain stored within the mobile device, sent to others, downloaded to a personal computer or uploaded to the Internet. Whilst CCTV recordings usually move from the public sphere (the nightlife district) to the private (the control room), OCTV footage tends to travel in the opposite direction: from more intimate and private situations in specific public or private spaces to the public domain. However, the distinction is not always so sharp: after riots at Rotterdam's Stadhuisplein in the summer of 2012, CCTV footage was for example broadcasted on TV and uploaded to the internet, which led to a number of youngsters turning themselves in at the police station voluntarily. In general, however, the juxtaposition of CCTV by public authorities and OCTV by consumers offers a useful heuristic, among others because of the much stricter legal requirements and protocols with regard to video-surveillance by public authorities.

A second sense of openness that mobile cameras introduce into surveillance assemblages pertains to agency and subjectivity. A critical difference between CCTV and OCTV is that the latter grants greater agency to nightlife consumers and citizens more generally. In CCTV technologies consumers and citizens are configured as passive subjects, whereas OCTV cameras configure them as active participants. The shift in capacities due to the invasion of PMDs into nightlife districts means that the traditional relation between the watcher and the watched is rewritten with potentially profound consequences. OCTV can be used to complement and extend the 'official' surveillance assemblage. This is at least what the Netherlands Ministry of Interior seeks to achieve though a publicity campaign to convince citizens who witness violence against relief workers, such as ambulance personnel, to submit film footage of these wrongdoings to the authorities. OCTV footage has also been instrumental in reconstructing what exactly had happened during riots at a beach party in Hoek van Holland (Flight and Hulshof 2010). However, OCTV can also be used to criticize and question the legitimacy and justice of the actions of police officers, bouncers and other formal surveillance agents against consumers and other citizens. In June 2012, for instance, a video clip of the actions of a female police officer in Rotterdam was published on YouTube. The footage showed clearly how the officer repeatedly kicked a drunken man who did not defend himself. Her male colleague stood on the side, watched and did not interfere. The clip caused considerable public outrage and led to an internal inquiry by Rotterdam's police force. With the further growth of OCTV new forms of accountability for institutionalized organizations may come into existence.

However, the democratic potential of OCTV should not be overrated (Timan and Oudshoorn 2012), at least not in the short term. During the previously mentioned on-site interviews with nightlife consumers in Rotterdam and Utrecht OCTV and CCTV were associated with different subject positions for the participants. With OCTV, more so than with CCTV, participants became passive victims of the unclear intentions of the person making the recording. The uncertainty regarding intentions and use of the footage resulted in a form of ambiguity and even subordination that may temper OCTV's democratic potential. While the use of PMDs in nightlife districts is likely to increase in the coming years, the wider range of intentions and possible uses of OCTV vis-à-vis CCTV may continue to complicate the extent to which mobile devices can contribute to the empowerment and wellbeing of nightlife consumers. Further research will have to demonstrate how PMDs help to shape consumers' experiences of nightlife districts and wellbeing in the present and near future.

18.6 Surveillance, Policing and the Production of Spaces in Nightlife Districts

Having considered differentiations and ambiguities in video-mediated surveillance assemblages in the previous sections, we now turn to how surveillance and policing more generally are implicated in the production of space of nightlife districts. As already indicated in the introduction, we understand space not as a passive and static container in which actions unfold and meanings are created. Space is rather the outcome of ongoing encounters and interactions of people, artefacts, buildings, other forms of materiality, ideas, symbols, emotions, and so on. It is an assemblage of assemblages (of which the surveillance assemblage is but one) and intrinsically dynamic: interacting changes occur at a wide range of time scales. Therefore, ethnographic observation of what happens and changes in a nightlife district over the period of a night and a week offers a useful and insight research method, and two researchers in our team carried out systematic observations at strategically selected sites with the nightlife districts of Groningen, Utrecht and Rotterdam between 10:00 PM and 5:00 AM during nine Thursday, Friday and Saturday nights in March-April 2010. Those nights were chosen because they attract the largest and different crowds: Thursday is the typical student night out, whilst Fridays and particularly Saturdays attract more school-going adolescents and (full-time employed) younger adults. Details of procedures and methods are available in Schwanen et al. (2012). Suffice it to say that the researchers systematically registered visitor characteristics (gender, ethnicity, age, etc.), features of the surveillance and policing in place, events that occurred, weather conditions, sounds, smells and expressions of disorderliness at four sites in various intervals during the night. The collected information provides rich and nuanced accounts of how the atmosphere and character of nightlife changes in the course of a night and is highly differentiated spatially within each nightlife district.

On one level it is tempting to conclude that Rotterdam's style of surveillance and policing is successful in enhancing safety and wellbeing. With the exception of public drunkenness, such disorderliness as vandalism, public urination, substance abuse and littering was observed less frequently than some discourses about urban nightlife's excesses would make us believe in either Groningen, Utrecht or

Table 18.1 Ethnicity, nightlife district visitors and surveillance agents, by city

	Share of resident population from non-Western descent[a]	Share of non-white visitors in total number, per 10-min interval			Average number of police officers, per 10-min interval		
		All non-white	Arabic	Afro-American	Police officers	Street wardens	Bouncers/ private security
Groningen	10 %	15.4 %	7.5 %	3.4 %	1.35	0.00	1.95
Utrecht	22 %	11.2 %	5.3 %	2.4 %	1.45	0.02	1.55
Rotterdam	37 %	42.2 %	18.7 %	11.1 %	1.35	0.27	5.33

[a]Obtained from municipal websites

Rotterdam. They were nonetheless observed least frequently in Rotterdam. This difference coincided with much greater numbers of street wardens, private security guards and bouncers were much more common in Rotterdam than in the other two cities (the number of police officers did not differ much across the three cities). Interventions into the behavior of consumers by police and bouncers also occurred rather infrequently and these were distributed evenly across the three cities, but the character of interventions was different in Rotterdam. A pro-active, zero-tolerance approach was visible on the street and the surveillance assemblage clearly orchestrated via modern communication technologies and CCTV.

At another level our research also indicates that causal relations between surveillance and policing and the character of nightlife may be more complex. Rotterdam's nightlife district attracts considerably fewer consumers than Groningen's although more than Utrecht's the average number of observed consumers per 10-min interval was 124 in Rotterdam against 268 in Groningen and 92 in Utrecht. Rotterdam's lower level of disorderliness vis-à-vis Groningen appears to be at least in part a consequence of smaller crowds. Also relevant in this regard is that consumers in Rotterdam dwelled and traversed the nightlife district in cars rather than on foot or by bike than in Utrecht and Groningen.

Our analysis does not support the notion that more surveillance and policing in general leads to exclusion in nightlife districts along lines of ethnicity, given that Rotterdam is also the city with the most ethnically diverse consumer population (Table 18.1). On the other hand, further analysis of the collected information reported in Schwanen et al. (2012) indicates that surveillance and policing are related in complex ways with the ethnic diversity of nightlife district's consumer populations. Controlling for differences in location (city and site within each city), we found higher levels of ethnic diversity among consumers to be associated with more police officers but also with a lower presence of bouncers. Inferring causality from these findings is not straightforward; however, on the basis of qualitative research in the Dutch cities of Apeldoorn and Arnhem (Van Aalst and Schwanen 2009), they may well indicate that to avoid trouble, youth from Arabic and Surinamese/Antillean descent keep away from the surroundings of nightlife premises where access is controlled by bouncers. Whilst more in-depth and ethnographic research into these matters is needed, our work so far suggests that it is important to consider police and security guards (including bouncers) separately when addressing questions of exclusion in nightlife districts.

We emphasize nonetheless that the effects of surveillance and policing on the exclusion of certain ethnic groups from nightlife districts appears to be relatively modest in the cities we have examined. Turning to Utrecht once more is instructive here. Table 18.1 shows that its nightlife district is disproportionally 'white', in particular compared to the city's resident population. The type of nightlife premises on offer is a key factor here. More so than in Rotterdam and Groningen bars, clubs and restaurants are oriented towards students and younger urban professionals who live in and around the city and are largely white. The area around the Neude and Janskerkhof squares offers very few premises specifically targeting consumers from Arabic or Antillean/Surinamese descent, and the few pubs with a vernacular style in the vicinity of the Neude do not attract as many people as the student-oriented bars and clubs. The orientation of Utrecht's nightlife on students is no coincidence. One club owner in Utrecht we interviewed was very clear about the type of customer he preferred: '*I like students to come to my bar. They know how to handle alcohol, they know their limits, they are quite mature and they know how to make a good party*'. Interviews show that the city council supports this orientation, given that keeping 'troublemakers' out of Utrecht's nightlife district is one of its top priorities. In sum, our research so far suggests that the exclusion of non-white youth from Utrecht's nightlife district operates more through the supply of nightlife premises than through surveillance and policing.

18.7 Conclusions

It is evident that technologically mediated surveillance and policing, and video-surveillance more specifically, are no techno-fix that helps cities to successfully juggle the economic opportunity, pleasure and excess dimensions of nightlife districts. Our research complements and extends previous research that has argued that the effectiveness of video-surveillance in reducing disorderliness and enhancing safety and wellbeing in urban spaces is often overrated. It does so by highlighting three sets of complexities and ambiguities. The first of these pertains to the policy arena: the storylines (Hajer 2005) used by policy-makers with regard to video-surveillance differ considerably between cities. As a result of this, the belief in and overrating of CCTV's effectiveness among policymakers and other stakeholders in urban governance vary across geographical space. Our analysis of Utrecht and Rotterdam suggest that the histories of local political constellations (e.g. which party is leading the debate and city government at critical moments in time) and local issues and concerns are the drivers of this spatial variation. At any rate, our results indicate that the claim that policymakers in general overestimate the benefits of CCTV surveillance is best avoided.

With regard to nightlife consumers, our research shows a remarkable heterogeneity in terms of understandings of the relations between video-surveillance, safety and wellbeing. Indifference and (mild) skepticism nonetheless prevail when it comes to the extent to which CCTV surveillance can enhance perceived safety amongst our research participants. Whilst this aligns with some earlier work (Koskela 2002; Klauser 2007), our research also shows that there is a discord between the skeptic discourses among consumers and the storylines dominating the discourse of policy-makers and politicians in Utrecht: for consumers the issue is not so much privacy related but CCTV's (perceived) incapacity to intervene or reduce harm when they (are likely to) become a victim of crime. This is, we believe, a finding with clear policy relevance, for it suggests that there are limits to the degree to which surveillance and policing by humans 'on the ground' can be substituted by digital surveillance. From a consumer perspective, insofar as surveillance and policing are capable of making nightlife districts safer and spaces of wellbeing, it needs to consist of visibly human agents. Based on our results to date, we are tempted to argue that the role of CCTV cameras should be no more than a complement and source of support to the actions of the police and private security guards already present in (the vicinity of) the nightlife district.

The third set of complexities and ambiguities concern video technology itself. Whether a camera is static or mobile and who uses it is likely to have an influence on how it is perceived, understood and experienced by nightlife district consumers. With the mobilization and multiplication of cameras in nightlife districts and the increased use of mobile recording devices among consumers as well as police officers and private security guards, a focus on how CCTV helps to produce spaces in nightlife districts is limited at best; mobile cameras should be given equally sustained attention. The use of mobile cameras not only enhances the complexity of the relations between surveillance and policing, wellbeing and exclusion (especially when they are used by nightlife consumers); it is also potentially unsettling. It allows new configurations of watching and being watched to emerge and it raises concerns about the purpose and destination of the recorded footage. The latter not only redirects debates about video-surveillance and privacy; it may also complicate the relations between camera use in nightlife and wellbeing on a range of time-scales. Footage recorded by consumers on nights out can end up on the internet and shape an individual's opportunities for self-fulfillment and self-esteem at later points in time. Potential employees searching the internet for footage of applicants constitute only one example of how OCTV can shape the relationships between consuming nightlife and wellbeing across timescales that exceed the night out.

In short, the effects of video-surveillance on the production of space are complex and ambiguous, and thinking about these effects using the concept and logic of assemblage helps us to make sense of that complexity. The idea that (more) video-surveillance will enhance safety and wellbeing in nightlife districts for the (vast) majority of nightlife consumers is not consistent with our findings. At the same time, all too dystopian understandings of video-surveillance and policing as excluding certain groups from nightlife districts *tout court* are not equivocally supported by our initial findings either. Certain surveillance practices do seem to contribute to the social exclusion of non-white youth from nightlife districts but such effects are geographically differentiated: they appear to vary from city to city and between sites and premises within a single city. In a way the relations between exclusion and surveillance are similar to those between safety and wellbeing on the one hand and surveillance on the other: the effects of surveillance are local, context-dependent, set in wider place-specific processes and difficult to generalize across space and time.

It is nonetheless clear that further research into the relations between video-surveillance and policing, wellbeing and safety, and socio-spatial exclusion in nightlife districts is required. Agents other than consumers, policy-makers and cameras should be considered, including police officers, bouncers, CCTV operators and technology developers. Further use of ethnographic methods is also needed, as are surveys among larger numbers of nightlife consumers than can be considered with in-depth interviews. Finally, the experiences of people who might potentially visit those districts but for some reason do not do so should also be explored. Our current research addresses these and other issues and will allow us to shed further light on the relationships between surveillance and policing, wellbeing and exclusion in urban nightlife districts.

References

Armitage, R. 2002. *To CCTV or Not to CCTV? – A review of current research into the effectiveness of CCTV systems in reducing crime*. London: Nacro.

Atkinson, S., S. Fuller, and J. Painter. 2012. *Wellbeing and place*. Farnham: Ashgate.

Barnard-Wills, D. 2011. UK news media discourses of surveillance. *The Sociological Quarterly* 52: 548–567.

Brands, J., T. Schwanen, and I. van Aalst. 2014. *What are you looking at? A visitors' perspective on CCTV in the night-time economy*. European Urban and Regional Research, doi:10.1177/0969776413481369.

Chatterton, P., and R. Hollands. 2003. *Urban nightscapes: Youth cultures, pleasure spaces and corporate power*. London: Routledge.

Conradson, D. 2005. Landscape, care and the relational self: Therapeutic encounters in rural England. *Health & Place* 11: 337–348.

Delanda, M. 2006. *A new philosophy of society: Assemblage theory and social complexity*. London: Continuum.

Deleuze, G., and F. Guattari. 1987. *A Thousand Plateaus. Capitalism and Schizophrenia*. Trans. Brain Massumi. Minneapolis: The University of Minnesota Press.

Diener, E. 2009. *The science of well-being: The collected works of Ed Diener*, vol. 1. Dordrecht: Kluwer.

Fleuret, S., and S. Atkinson. 2007. Wellbeing, health and geography: A critical review and research agenda. *New Zealand Geographer* 63: 106–118.

Flight, S., and P. Hulshof. 2010. *Roadmap Beeldtechnologie Veiligheidsdomein. Behoefte en Gewenste Innovaties voor 'Veilig door Innovatie'*. Amsterdam: DSP groep.

Foucault, M. 2002. *The Archeology of Knowledge*. Trans. A.M. Sheridan Smith. Abingdon: Routledge Classics.

Gemeente Groningen. 2011. *Evaluatie Agressiedetectie*. Groningen: Gemeente Groningen.

Gemeente Utrecht. 2002. *Convenant Veilig Uitgaan Binnenstad Utrecht 2002–2006*. Utrecht: Gemeente Utrecht.

Graham, S.D.N. 2005. Software-sorted geographies. *Progress in Human Geography* 29(5): 562–580.

Haggerty, K.D., and R.V. Ericsson. 2000. The surveillant assemblage. *British Journal of Sociology* 51(4): 605–662.

Hajer, M.A. 2005. Coalitions, practices, and meaning in environmental politics: From acid rain to BSE. In *Discourse theory in European politics: Identity, policy and governance*, ed. D. Howarth and J. Torfing, 297–315. Basingstoke: Palgrave Macmillan.

Hall, T., and P. Hubbard (eds.). 1998. *The entrepreneurial city*. Chichester: Wiley.

Harvey, D. 1989. From managerialism to entrepreneurialism: The transformation in urban governance in late capitalism. *Geografiska Annaler, Series B: Human Geography* 71: 3–17.

Helms, G. 2008. *Towards safe city centres? Remaking the spaces of an old-industrial city*. Aldershot: Ashgate.

Hempel, L., and E. Töpfer. 2004. CCTV in Europe; final report. *Urbaneye Working Paper* no. 15. Berlin: Centre for technology and Society, Technical University of Berlin.

Hubbard, P. 2005. The geographies of 'going out': Emotion and embodiment in the evening economy. In *Emotional geographies*, ed. L. Bondi, M. Smith, and J. Davidson, 117–137. Ashgate: Aldershot.

Jayne, M., G. Valentine, and S.L. Holloway. 2011. *Alcohol, drinking, drunkenness: (Dis)orderly spaces*. Farnham: Ashgate.

Judd, D., and S. Fainstein (eds.). 1999. *The tourist city*. New Haven: Yale University Press.

Klauser, F.R. 2007. Difficulties in revitalizing public space by CCTV: Street prostitution surveillance in the Swiss city of Olten. *European Urban and Regional Studies* 14: 337–348.

Koskela, H. 2002. Video surveillance, gender, and the safety of public urban space: "Peeping Tom" goes high tech? *Urban Geography* 23: 257–278.

Koskela, H. 2003. A two-edged sword – Public attitudes towards video surveillance in Helsinki. In *The European group conference paper*. Helsinki: University of Helsinki.

Lyon, D. (ed.). 2003. *Surveillance as social sorting: Privacy, risk and digital discrimination*. London/New York: Routledge.

Mann, S., J. Nolan, and B. Wellman. 2003. Sousveillance: Inventing and using wearable computing devices as data collection in surveillance environments. *Surveillance & Society* 1: 331–355.

McCahill, M. 2002. *The surveillance web: The rise of visual surveillance in an English city*. Devon: Willan.

McCahill, M., and C. Norris. 2002. CCTV in Britain. *Urbaneye Working Paper* no. 3. Berlin: Centre for technology and Society, Technical University of Berlin.

Measham, F., and P. Hadfield. 2009. Everything starts with an 'E': Exclusion, ethnicity and elite formation in contemporary English clubland. *Addiciones* 21: 362–386.

Miles, S., and R. Paddison. 2005. Introduction: The rise and rise of culture-led urban regeneration. *Urban Studies* 42: 833–839.

Norris, C., and G. Armstrong. 1999. *The maximum surveillance society: The rise of CCTV*. Oxford: Berg.

Nouwt, S., B.R. de Vries, and D. Van der Burgt. 2005. Camera surveillance and privacy in the Netherlands. In *Reasonable expectations of privacy? Eleven country reports on camera surveillance and workplace privacy*, ed. S. Nouwt, B.R. de Vries, and C. Prins, 115–165. The Hague: T.M.C. Asser Press.

Roberts, M., and A. Eldridge. 2009. *Planning the night-time city*. Abingdon: Routledge.

Schmid, H., W.D. Sahr, and J. Urry (eds.). 2011. *Cities and fascination. Beyond the surplus of meaning*. Aldershot: Ashgate.

Schreijenberg, A., J. Koffijberg, and S. Dekkers. 2009. *Evaluatie Cameratoezicht op Openbare Plaatsen, Driemeting, Eindrapport*. Amsterdam: Regioplan.

Schwanen, T., I. van Aalst, J. Brands, and T. Timan. 2012. Rhythms of the night: Spatiotemporal inequalities in the night-time economy. *Environment and Planning A* 44(9): 2064–2085.

Shaw, R. 2010. Neoliberal subjectivities and the development of the night-time economy in British cities. *Geography Compass* 4: 893–903.

Talbot, D. 2007. *Regulating the night: Race, culture ad exclusion in the making of the night-time economy*. Aldershot: Ashgate.
Talbot, D., and M. Böse. 2007. Racism, criminalization and the development of night-time economies: Two case studies in London and Manchester. *Ethnic and Racial Studies* 30: 95–118.
Timan, T., and N. Oudshoorn. 2012. New technologies of surveillance? How citizens experience the use of mobile cameras in public nightscapes. *Surveillance and Society* 10(2): 167–181.
Tops, P. 2007. *Regime Veranderingen in Rotterdam. Hoe een Stadsbestuur Zichzelf Opnieuw Uitvond*. Amsterdam: Uitgeverij Atlas.
Van Aalst, I., and I. van Liempt. 2011. Uitgaansstad onder spanning. *Justitiële Verkenningen* 37: 9–24.
Van Aalst, I., and T. Schwanen. 2009. Omstreden nachten; angstgevoelens van jongeren in de uitgaansgebieden van Arnhem en Apeldoorn. In *Omstreden Ruimte; over de Organisatie van Spontaniteit en Veiligheid*, ed. H. Boutellier, N. Boonstra, and M. Ham, 157–175. Amsterdam: Van Gennep.
Van Liempt, I., and I. van Aalst. 2012. Urban surveillance the struggle between safe and exciting nightlife districts. *Surveillance & Society* 9(3): 280–292.
Van Noije, L., and K. Wittebrood. 2009. *Overlast en verloedering ontsleuteld: veronderstelde en werkelijke effecten van veiligheidsbeleid*. Den Haag: SCP.
Van Schijndel, A., A. Schreijenberg, and G. Homburg. 2010. *Evaluatie Cameratoezicht Gemeente Rotterdam*. Amsterdam: Regioplan.
Webster, W. 2009. CCTV policy in the UK: Reconsidering the evidence base. *Surveillance & Society* 6: 10–22.
Welsh, B.C., and D.P. Farrington. 2003. The effects of closed-circuit television on crime. *The Annals of the American Academy of Political and Social Science* 587: 110–135.
Welsh, B.C., and D.P. Farrington. 2009. *Making public places safer, surveillance and crime prevention*. Oxford: Oxford University Press.
Williams, R. 2008. Night spaces: Darkness, deterrioralization and social control. *Space and Culture* 11: 514–532.
Winlow, S., and S. Hall. 2006. *Violent night: Urban leisure and contemporary culture*. Oxford: Berg.

Chapter 19
Responsibly Innovating Data Mining and Profiling Tools: A New Approach to Discrimination Sensitive and Privacy Sensitive Attributes

Bart H.M. Custers and Bart W. Schermer

Abstract Data mining is a technology that extracts useful information, such as patterns and trends, from large amounts of data. The privacy sensitive input data and the output data that is often used for selections deserve protection against abuse. In this paper we describe one of the main results of our research project on developing new privacy preserving and discrimination aware data mining tools, namely why the common measures for mitigating privacy and discrimination concerns, such as a priori limiting measures (particularly access controls, anonymity and purpose specification) are mechanisms that are increasingly failing solutions against privacy and discrimination issues in the novel context of advanced data mining and profiling. Contrary to previous attempts to protect privacy and prevent discrimination in data mining, we did not focus on new designs that better enable (a priori) access limiting measures regarding input data, but rather focused on (a posteriori) responsibility and transparency. Instead of limiting access to data, which is increasingly hard to enforce in a world of automated and interlinked databases and information networks, rather the question how data can and may be used was stressed.

B.H.M. Custers, (✉) Ph.D., M.Sc., LLM
eLaw@Leiden, The Centre for Law in the Information Society, Leiden University, Leiden, The Netherlands

WODC – Ministry of Security and Justice, The Netherlands
e-mail: b.h.m.custers@law.leidenuniv.nl

B.W. Schermer, Ph.D., LLM
eLaw@Leiden, The Centre for Law in the Information Society, Leiden University, Leiden, The Netherlands

J. van den Hoven et al. (eds.), *Responsible Innovation 1: Innovative Solutions for Global Issues*, DOI 10.1007/978-94-017-8956-1_19,

19.1 Introduction

The aim of data mining is to extract useful information, such as patterns and trends, from large amounts of data (Adriaans and Zantinge 1996; Fayyad et al. 1996; Mannila et al. 2001). In their fight against crime and terrorism, many governments are gathering large amounts of data to gain insight into methods and activities of suspects and potential suspects. This can be very useful, but usually at least part of the data on which data mining is applied is confidential and privacy sensitive. Examples are medical data, financial data, et cetera. This raises the question how privacy, particularly of those who are innocent, can be ensured when applying data mining. Furthermore, the results of data mining can lead to selection, stigmatization and confrontation (Custers 2004). False positives and false negatives are often unavoidable, resulting in the fact that people are frequently being judged on the basis of characteristics that may be correct for them as group members, but not as individuals as such (Vedder 1999; Solove 2004; Zarsky 2003). In the context of public security, false positives may result in investigating innocent people and false negatives may imply criminals remain out of scope.

A priori protection may be realized by protecting input data and access to input data. However, removing key attributes such as name, address and social security number of the data subject is insufficient to guarantee privacy; it is often still possible to uniquely identify particular persons or entities from the data, for instance by combining different attributes (Ohm 2010). Furthermore, a priori regulation may block the benefits of data mining from happening. Since the results of data mining are often used for selection, a posteriori protection is also desirable, in order to ensure that the output of data mining is only used within the imposed ethical and legal frameworks. This implies, for instance, that data mining results on terrorism, where data was collected within extensive jurisdiction of secret services, cannot be used just like that for shoplifting or car theft, where data is collected within limited jurisdiction of the police.

The aim of our project (a co-operation of Leiden University and Eindhoven University of Technology) was to investigate to what extent legal and ethical rules can be integrated in data mining algorithms. The focus was on the security domain. Key questions were: "How can legal and ethical rules and principles be translated in a format understandable for computers?" and "In which way can these rules be used in the data mining process itself?" A typical example of such an ethical and legal principle in this context concerns anti-discrimination. To reduce unjustified discrimination, it is prohibited to treat people differently on the basis of ethnic background or gender. Self-learning data mining algorithms can learn models to predict criminal behavior of existing and future criminals, based on historical data. However, these models may include discrimination of particular groups of people for discrimination in the past that can be found in historical datasets or for unbalanced datasets. To avoid such phenomena, self-learning algorithms must be 'made aware' of existing legal restrictions or policies.

In this paper, we describe one of the main results of our research project, namely why the common measures for mitigating privacy and discrimination concerns, such as a priori limiting measures (particularly access controls, anonymity and purpose specification) are mechanisms that are increasingly failing solutions against privacy and discrimination issues in the novel context of advanced data mining and profiling. For more results on the project itself, particularly for technological results such as new data techniques that we developed within the project, we refer to the project's website[1] and further literature (Calders and Verwer 2010; Kamiran and Calders 2010; Custers 2010), particularly the book resulting from this project (Custers et al. 2013).

First we start with pointing out the legal and ethical issues involved in data mining and profiling, particularly risks concerning discrimination and privacy (Sect. 19.2). Then we explain, by discussing some of our research results, that simply removing sensitive attributes in databases does not solve (most of) these problems (Sect. 19.3). We continue by explaining that another approach to discrimination sensitive and privacy sensitive attributes in databases is needed and provide suggestions for other approaches (Sect. 19.4). One of these new approaches focuses on new designs that better enable transparency and accountability, rather than on access controls to data, as we expected data access to be difficult to maintain in many situations (Sect. 19.5). To further explain this, we compare removing sensitive data from databases to avoid discrimination with removing identifiability from databases to avoid privacy infringements (Sect. 19.6). We conclude by discussing the limits of privacy (Sect. 19.7).

19.2 Ethical and Legal Issues Associated with Profiling and Data Mining

While profiling and data mining have proven to be very useful tools in dealing with the information overload, they are not without controversy. The reason for this is that there are a number of ethical and legal issues associated with profiling and data mining, some of which we shall describe in this section. We shall distinguish between the risks associated with profiling, and issues that may arise due to an incorrect application of data mining in the context of profiling (Schermer 2011).

Before discussing these risks and issues, it is important to note that the legal framework in the European Union is currently under revision. Currently, the collection and use of personal data is protected by Directive 95/46/EC, which has been implemented in national law in the member states of the European Union.[2]

[1] http://wwwis.win.tue.nl/~tcalders/dadm/doku.php

[2] Directive 95/46/EG of the European Parliament and the Council of 24th October 1995, [1995] OJ L281/31.

For more background on the current directive and its main privacy principles, we refer to Bygrave (2002). The proposed legal framework was published in January 2012 by the European Commission.[3] For more background on the proposed EU data protection regulation, we refer to Kuner (2012) and Hornung (2012). There are some significant differences between the existing and the proposed regulations and the proposed regulation does include a provision regarding profiling in Article 20. This provision may place restrictions on the way profiling is conducted. However, much of the terminology used is unclear and likely to be difficult to implement in practice (Kuner 2012, p. 7).

19.2.1 Risks Associated with Profiling

While data mining and profiling are for the most part conceptually framed as threats to informational privacy, it is our opinion that the current application of data protection laws obfuscates the actual risks that profiling and data mining may pose to groups and individuals. While Article 15 of the European Data protection directive (1995/46/EC) addresses the issue of automated profiling and the risks associated with it, it does not have a meaningful impact in addressing these issues in practice. It remains to be seen whether the proposed data protection regulation will change this. In our view the most significant risks associated with profiling and data mining are discrimination, de-individualization, information asymmetries and encroachment on moral autonomy.

19.2.1.1 Discrimination

Classification and division are at the heart of (predictive) data mining. As such, discrimination is part and parcel of profiling and data mining. However, there are situations where discrimination is considered unethical and even illegal. This can occur for instance when a data mining exercise is focused on characteristics such as ethnicity, gender, religion or sexual preference. But even without a prior desire to judge people on the basis of particular characteristics, there is the risk of inadvertently discriminating against particular groups or individuals. The reason for this is that predictive data mining algorithms may "learn" to discriminate on the basis of biased data used to train the algorithm.

A classifier (a data mining algorithm to establish whether a new object fits a previously established class) must be trained in order to classify data. When the training data is contaminated (for instance because it discriminates against a

---

[3]Proposal for a Regulation of the European Parliament and of the Council on the protection of individuals with regard to the processing of personal data and on the free movement of such data (General Data Protection Regulation), Brussels, 25.1.2012 COM(2012) 11 final 2012/0011 (COD).

particular group), the classifier will learn to classify in a biased way, strengthening discriminatory effects. Such cases occur naturally when the decision process leading to the labels was biased. An example in the area of law enforcement may help to explain this. When police officers have targeted an ethnic minority disproportionately in the past based on their own bias, it is likely that these minorities will feature more prominently in crime statistics. If these crime statistics are then used as training data for a classifier, chances are high that the classifier will learn that that there is a strong correlation between ethnicity and crime. This in turn will lead to discriminating results that can constitute the basis for future discrimination. This effect is further strengthened by the fact that a classifier will most likely not have access to all the important factors on which to base a prediction because, e.g., they are missing in the data. Therefore, the importance of those factors that are present in the data grows and will be even more important in prediction than they were in the input data.

19.2.1.2 De-individualization

In many cases data mining is in large parts concerned with classification and thus there is the risk that persons are judged on the basis of group characteristics rather than on their own individual characteristics and merits (Vedder 1999). Group profiles usually contain statistics and therefore the characteristics of group profiles may be valid for the group and for individuals as members of that group, though not for individuals as such. An example may illustrate this. For instance when people in Amsterdam are 3 % more criminal than people in the rest of the Netherlands, this characteristic goes for the group (i.e., people in Amsterdam), for the individuals as members of that group (i.e., randomly chosen people living in Amsterdam), but not for the individuals as such (i.e., for John, Mary and William who all live in Amsterdam). When individuals are judged by group characteristics they do not possess as individuals, this may strongly influence the advantages and disadvantages of using group profiles. Apart from the negative effects group profiling may have on individuals, group profiling can also lead to stigmatization of group members. Moreover, divisions into groups can damage societal cohesion. When group profiles, whether correct or not, become public knowledge, people may start treating each other accordingly. For instance, when people start believing that citizens of Amsterdam are more criminal, people may start to react and communicate with more suspicion towards citizens of Amsterdam, regardless of the correctness of such a profile.

19.2.1.3 Information Asymmetries

Data mining can lead to valuable insights for those parties employing it. When data mining is aimed at gaining more insight into individuals or groups, we encounter the problem of information asymmetry. Information asymmetries may influence the level playing field between government and citizens, and between businesses and consumers, upsetting the current balance of power between different parties.

In the context of the relation between government and citizens information asymmetries can affect individual autonomy. If data mining indeed yields actionable knowledge, the government will have more power. Moreover, the fear of strong data mining capabilities on the part of the government may "chill" the willingness of people to engage in, for instance, political activities, out of fear of being watched. For this panoptic fear to materialize, a data mining application does not even have to be effective (Schermer 2007).

Information asymmetries may obstruct the strive for level economic playing fields between consumers and businesses. Furthermore, there are instances where data mining can aid in making decisions about consumers that are considered unwanted, unethical or illegal. Examples of this would be excluding particular individuals or groups from goods and services based on their ethnicity, or singling them out for more intense security screenings (Zarsky 2006).

The issue of information asymmetry is exacerbated by the limited transparency of data mining. Since data mining (in particular predictive data mining) is used to make decisions about groups and individuals, people will be affected by data mining exercises. However, for the most part it will be unclear to persons why a particular decision has been made and on what grounds. This could lead to a sense of helplessness. A problem that is compounded by the fact that it is difficult to seek redress from automated decision processes. Solove (2004) has likened this situation with that of Josef K. in Kafka's Der Prozess (Solove 2004).

19.2.1.4 Encroachment on Moral Autonomy

An important aspect of privacy is that it enables us to preserve our moral autonomy. The right to privacy and the associated right to data protection provide invisible boundaries against the normative pressure of public opinion (van den Hoven 1997, p. 36) and gives us the seclusion necessary for self-evaluation and introspection (Westin 1967, p. 36). As such, privacy gives the individual room for the development of his or her personal identity.

Through profiling, a person may be confronted either directly or indirectly with the picture that an organization, or indeed society, has constructed of that person. Since the number of profiles used to categorize individuals is finite and tailored to the specific purposes of the entity doing the profiling, profiles are necessarily 'one-dimensional'. The abstraction of personhood to a one-dimensional profile and the subsequent confrontation with this profile may negatively affect the self-image of a person. For example, a negative credit score may reinforce feelings of helplessness or incompetence. In this way profiling reduces the possibilities for personal change and growth and leads to stigmatization and stereotyping, thereby encroaching on moral autonomy.

19.2.2 *Application Issues*

The risks described above may manifest themselves regardless of the fact that the data mining was applied correctly in a technical sense. But as we have seen in the example of discrimination, when data mining is applied incorrectly, the risks associated with profiling and data mining may be strengthened. Therefore we discuss several common application issues associated with data mining that may pose additional risks to groups and individuals.

19.2.2.1 Accuracy and Reliability

The success of a data mining exercise is dependent on the quality of the raw data being mined. If the data is inaccurate, the results will also be inaccurate. This is true for both descriptive and predictive data mining. In the area of predictive data mining issues with accuracy and reliability are particularly problematic, given the fact that the results of a predictive data mining exercise are often used to make (automated) decisions about individuals and/or groups. But even if the raw data is free of errors, accuracy and reliability remain an issue.

In particular there is the problem of "false positives" and "false negatives." This means that people that in fact do not fit the class are fitted in the class (a false positive), or people that fit the class are left out (false negative). False positives and false negatives occur for various reasons, one being that not all information is available. For example, the presence of attributes such as {yellow, beak, wings, tail} are strong indications that an animal could be classified as a canary. However, a duckling is also yellow and has a beak, wings and a tail, making it a false positive in our scenario.

19.2.2.2 Causation Versus Correlation

The goal of data mining is to find implicit and previously unknown relations between data. As such, data mining yields new knowledge about a given problem space. In descriptive data mining, this knowledge is based on the correlation between certain objects and attributes. However, while data mining can establish that there is a relationship between certain objects and attributes, it does not explain why this relationship exists. As such, it is important that we do not mistake correlation for causation (Pearl 2009). For instance, data mining may reveal that burglars use cannabis more often than other people. While it is tempting to point to cannabis as a cause for burglary, the data does not support such a conclusion.

Data mining experts warn for the fact that a correlation between particular attributes does not imply a causal relation, nor does it explain why there is a correlation between these attributes (Cocx 2009, p. 143). Such a warning needs to be heeded, given the fact that data mining efforts and statistics might provide input for policymaking. If the goal of policymaking is addressing the causes of an issue rather than fighting symptoms, it is important to know more about the background, emergence and causation of certain events. Particularly, unguided descriptive data mining is less suited, if not unfit, for the discovery of this information (Cocx 2009, p. 143).

19.2.2.3 Data Dredging

In unguided descriptive data mining we look for correlations in the data without using a pre-defined working hypothesis. Dependent on the size of the dataset and the "confidence interval" used to determine correlations, our data mining exercise will yield certain results. While these results might indeed be significant, there is also a chance that they are completely random. So, the results we find (and the hypothesis we formulate on the basis of these results) need to be validated to exclude the possibility that the correlation is in fact totally random. It is important to note that the dataset used in constructing the hypothesis cannot be used for the validation of that same hypothesis, since this data is by default "biased" towards supporting that particular hypothesis. The risk in data dredging is that we present the results of the initial data mining exercise as facts, rather than as a hypothesis that needs to be tested further. An example to illustrate this point is the following: suppose we want to test if some people have special mental abilities and are able to predict the future. We set up an experiment in which 1,000 volunteers need to predict which sequence of heads and tails will emerge from flipping an unbiased coin 10 times. Statistics teach us that if a participant makes random guesses, he or she still has a chance of 1 out of 1,024 to get the sequence fully correct. As such, we can expect that on average one participant will predict the outcome of all ten tosses correctly. Of course, this outcome does not support at all the claim that this participant has special mental abilities. First the participant needs to confirm his perfect scores in new controlled experiments. To link this example to data mining: one could consider the outcome of the prediction task as the input data. There are 1,000 hypotheses being tested: "participant 1 can predict everything correctly" till "participant 1,000 can predict everything correctly". As such the outcome of the experiment is used to describe the "pattern"; e.g., "participant 174 can predict everything correctly" that will be presented to the user. It is instructive to see that even a statistical test for significance will confirm the validity of this pattern.

19.2.3 A Positive Note

The issues mentioned above are possible ethical and legal issues that *may* occur due to the use of data mining and profiling techniques. However, it is important to note, first that it may well be that some or most of these issues will never materialize, and, second, that the use of data mining and profiling may also have a more positive effect on privacy and discrimination issues.

Starting with privacy issues, it may be argued that data mining and profiling techniques may help in selection target groups in ways that are less privacy invasive than other methods commonly used. For instance, in the area of criminal investigation, methods like body searches, car searches and house searches may be perceived as much more privacy invasive than data searches or internet searches. It is often argued, like we did above, that data mining and profiling may pose threats to privacy, but it may also be argued that these tools make other methods that are more privacy invasive redundant to some extent. If the numbers of false positives and false negatives are not very high, it may also be argued that profiling does not bother 'innocent people' (to stay within the criminal investigation context), but only bothers people who may have something to hide.

For discrimination issues, there is also an important positive aspect to note. It is often argued that discrimination is a characteristic shown by individual people that may be prejudiced (Withrow 2006; Del Carmen 2007; Meeks 2000). When discrimination is indeed caused by (subjective) prejudices that people may have, data mining and profiling techniques may be the objective approach to counter this. This may be, for instance, by showing that some of the prejudices are plainly wrong or not based on facts (such as people from particular ethnic backgrounds being more prone to criminal behavior than others), but also because data mining and profiling exclude the more subjective 'human factor' to some extent. The profiles discovered in the data mining processes will show which target groups to look at and may leave less discretion to individual officers in border control, policing, security checks, etc. In short, though data mining and profiling may cause concerns regarding discrimination, in some situations the alternative of continuing to rely on profiling by human beings may be more discriminatory.

19.3 Removing Sensitive Attributes

It may be suggested that removing sensitive attributes from databases is the best solution to prevent unethical or illegal discriminating data mining results. If sensitive attributes such as gender, ethnic background, religious background, sexual preferences, criminal and medical records are deleted from databases, or blocked by access controls, the resulting patterns and relations cannot be discriminating anymore, it may be argued. However, research in our project has shown that this assumption is not correct (Kamiran and Calders 2009). Classification models usually make predictions on the basis of training data. If the training data is biased towards certain groups or classes of objects, e.g., there is racial discrimination towards black people, the learned model will also show discriminatory behavior towards that particular community. This partial attitude of the learned model may lead to biased outcomes when labeling future unlabeled data objects.

For instance, throughout the years, in a particular organization black people might systematically have been denied jobs. As such, the historical employment information of this company concerning job applications will be biased towards giving jobs to white people while denying jobs from black people. Simply removing the sensitive attributes, or blocking them by access controls, from the training data in the learning of a classifier for the classification of future data objects, however, is not enough to solve this problem, because often other attributes will still allow for the identification of the discriminated community. For example, the ethnicity of a person might be strongly linked with the postal code of his residential area, leading to a classifier with indirect racial discriminatory behavior based on postal code. This effect and its exploitation is often referred to as redlining, stemming from the practice of denying or increasing services such as, e.g., mortgages or health care to residents in certain often racially determined areas. This claim is supported in other research: even after removing the sensitive attribute from the dataset discrimination persists (Pedreschi et al. 2008).

Removing data (or imposing access controls) does not prevent the discovery of discriminating patterns. As will be shown in Sect. 19.6, a similar argument can be made that anonymity does not prevent the discovery of privacy invasive patterns. Furthermore, the principle of purpose specification, from a technological perspective, does not prevent unauthorized use of particular data.

19.4 New Methods and Approaches

When removing sensitive attributes does not prevent unethical or illegal discrimination, other approaches are needed, since impartial classification results are desired or sometimes even required by law for future data objects in spite of having biased training data. We tackled this problem by introducing a new classification scheme for learning unbiased models on biased training data. Our method is based on massaging the dataset by making the least intrusive modifications which lead to an unbiased dataset. On this modified dataset we then learned a non-discriminating classifier.

In order to realize this, we translated laws and rules into quantitatively measurable properties against which the discovered models may be checked. Such a formalization enables verifying the correctness of existing laws and rules in the discovered models. We would like to stress here that it was not our ambition to build up a complete computerized and automated system for a code of laws. Rather we wanted to explore how some selected current legislation, e.g., anti-discrimination rules, translated into constraints that can be exploited computationally.

Next, we integrated these rules directly in the automated quest for suitable models. This was done by modifying the original dataset that was learned from, or by modifying the discovery strategies of the algorithms. This provided interesting challenges in the area of data mining research; current models hardly take into account ethical, moral and legal rules. As such, in the current situation the best that can be achieved is learning a model and verify it a posteriori (Pedreschi et al. 2008), and if the verification fails the model cannot be used. Another problem lies in the computational complexity of the verification. In the context of pattern mining it is, in general, computationally intractable to assess what can be derived from the output (Calders 2007, 2008). As such, it is not possible to guarantee privacy preservation afterwards. Several methods have been developed in the past to ensure privacy for data mining (Lindell and Pinkas 2002), but most of these methods are based on a statistical context and do not provide strict guarantees or are computationally very heavy. In our project we investigated whether anti-discrimination rules can be used as constraints in the model construction, aiming to remove or ignore a potential bias in the training data in favor of a deprived community.

We started with a dataset containing a sensitive attribute and a class label that is biased with relation to this sensitive attribute. For example, the dataset contains information about the frequency of certain crime types and the sensitive attribute indicated who inserted the data. This data was split into a training and a validation set to avoid that the efficiency of the approach is tested against the same data that was used for creating the model. For the training set, the bias is removed. Two methodologies were explored. In the first methodology we removed discrimination from the training dataset. For this purpose, ranking methods were used in order to predict the most likely victims of discrimination and to adjust their label. Based on the cleaned dataset, traditional classification methods can be employed. Notice that the combination of the bias removal step and the traditional model construction on the unbiased dataset can also be seen as a learning method applied directly on biased data. The advantage of this first approach is that it is independent of the model type being used, whereas the disadvantage clearly is the critical dependence of the overall result to the ability to accurately identify discrimination without the presence of training data for this task. A second methodology was to embed the constraints deeply inside model construction methods. When learning a Naïve Bayes classifier, e.g., we changed the probabilities to reflect "non-discrimination." The advantage here is that this method allows for a better control of the discrimination ration in the output models, whereas the disadvantage is that the method is less general than in the first approach.

19.5 Transparency and Accountability

By using the approach described above, we better enabled transparency and accountability of data mining and profiling because the *types of output* (not the output itself!) could be better predicted, i.e., enabled us to ensure that the patterns discovered caused less privacy and discrimination issues. This is an important shift in paradigm, since, in the past, much research has been done on a priori privacy protection (such as privacy enhancing technologies and privacy preserving data mining), our focus was on posteriori protection. The current starting point of a priori privacy protection, based on access restrictions on personal information (Chawla et al. 2002), is increasingly inadequate in a world of automated and interlinked databases and information networks, in which individuals are rapidly losing grip on who is using their information and for what purposes, particularly due to the ease of copying and disseminating information. In most cases, the output is much more relevant, since this is the information that is being used for decision-making. Therefore, it seemed useful to complement a priori access controls with a posteriori accountability. Such accountability requires transparency (Weitzner et al. 2006).

Transparency and accountability focus on the use of data rather than on the access to data. Transparency refers to insight in the history of collecting, processing and interpreting both the raw data and the results of data mining. Accountability refers to the possibility to check whether the rules of collecting, processing and interpreting both the raw data and the results of data mining were adhered to. New design principles for data mining may indicate how data can and may be used. In the context of discrimination awareness, new constraints can be imposed on the distribution over different population groups of the predictions by a model on future data. This target distribution can be significantly different from the distribution of the training data. Apart from this technological approach to transparency and accountability, we investigated (the limits of) a legal approach (see Sect. 19.7).

19.6 Anonymity and Privacy

Most of the classification models deal with all the attributes equally when classifying data objects and are oblivious towards the sensitivity of attributes. When the goal is to avoid or minimize discrimination, it may be useful to establish the sensitivity of attributes. However, when the goal is to avoid or minimize privacy infringements, it may be useful to establish the identifiability of attributes. From a technological perspective, identifiability is the possibility to single out a particular person from a group of potential candidates on the basis of a piece of information. Not every characteristic is equally useful for determining a person's identity. Similarly, not every characteristic is equally useful for selecting a person, nor is it equally allowed from an ethical or legal perspective.

Here an interesting corollary can be made between sensitivity and discrimination on the one hand and identifiability and privacy on the other hand. In most classification models, all attributes are treated equally. From a normative perspective, however, there may be reasons to treat certain attributes with reluctance. Some data are directly identifiable (e.g., name, address, and zip code), some data are indirectly identifiable (e.g., profession, residence), and some data are non-identifiable or anonymous (e.g., date of birth, gender). However, data from these categories may be combined and subsequently result in higher degrees of identifiability.

Similarly, some discrimination sensitive data are directly discriminating with regard to equal treatment when used for selection purposes (e.g., gender, ethnic background, sexual preferences, political views, union membership, criminal records, and medical records), some data are indirectly discriminating (e.g., income, social-economic status, zip code, first name, profession and level of education) and some data are not discriminating (e.g., shoe size, number of pets, and, in most cases, age). Again, data from these categories may be combined and subsequently result in higher degrees of sensitivity.

Anonymity may help to prevent privacy intrusions, but it may not prevent data mining and profiling. Next to techniques that provide (more) anonymity, techniques that limit linkability may be useful (Goldberg 2000). Linkability may be limited by restricting (the type of) access to the data items to be linked. This may be done with the help of multilevel security systems, requiring three types of information controls: access controls, flow controls and inference controls (Denning 1983). Access controls handle the direct access to data items. Information flow controls are concerned with the authorization to disseminate information. Inference controls are used to ensure that released statistics when combined do not disclose confidential data. As mentioned before, our previous research has shown that access controls are less successful with regard to sensitive data and anti-discrimination. Limiting (the influence of) sensitivity of indirectly discriminating attributes may provide more protection against discriminating data mining results.

19.7 Conclusion: The Limits of Privacy

In Sect. 19.2 we discussed several ethical and legal issues associated with data mining in the context of profiling. The current approach to mitigating these risks is by invoking the right to (informational) privacy. At the core of informational privacy is the notion that data subjects have the right and the ability to shield personal data from third parties or so-called informational self-determination (Westin 1967). Hence, (informational) privacy is often approached in the practical but limited meaning of personal data protection. Given the societal importance of the free flow of information, this right to informational privacy is balanced by the legitimate interests of third parties to process personal data. In Europe, the Data Protection Directive sets forth the rules under which data controllers are allowed to process personal data on individuals. A data controller is not allowed to process personal data unless he has a legitimate purpose (see article 7 of the Directive). Furthermore, the Directive sets forth some core data protection principles such as data quality, security safeguards and data minimization. While personal data protection is important and has provided individuals with protection from misuse and abuse of their personal data, there are a number of closely linked issues of a technological, legal and societal nature with this approach.

Although the current Data Protection Directive is under revision and some significant changes have been suggested, the basis of the proposed regulation remains the concept of informational self-determination and the focus remains on personal data and data protection principles. The proposed regulation does include a provision regarding profiling that may place restrictions on the way profiling is conducted, but since much of the terminology used is unclear and likely to be difficult to implement in practice (Kuner 2012), this provision may not offer protection against the major risks and issues of data mining and profiling. That does not imply that we should abandon the model of informational self-determination, which has many virtues, but it does imply that we should look for further models that address the risks and issues of data mining and profiling.

From a data mining perspective the primary issue with informational privacy is that by limiting the use of (particular) personal data, we run the risk of reducing the accuracy of the data mining exercise (Zarsky 2006). This does not only decrease the usefulness of data mining, it also increases the probabilities of false positives and false negatives. Moreover, data exclusion, anonymization and data minimization do not necessarily provide sufficient protection from the risks of profiling, because it is often still possible to uniquely identify particular persons from the data, even after key attributes such as name, address and social security number have been removed. This indirect identification is troublesome from a legal perspective, since personal data protection law is so dependent on the notion of personal data. Also, it is difficult to determine which pieces of data can lead to the identification of a person once combined. But even in those instances where identification is impossible, there is no guarantee that individuals are not adversely influenced. Take for instance the practice of behavioral targeting in online advertising. Actual identification is often not necessary for targeting, merely being able to individualize users on the basis of their IP-address or their browser cookies is enough for the targeting to be effective. Privacy and data protection law provide little protection from the problem of information asymmetry, since they are primarily focused on a priori privacy protection through the minimization of data.

It may even be so that invoking the right to informational privacy is counterproductive in some cases. Our research suggests that this is the case for anti-discrimination (Verwer and Calders 2010). In order to counter discrimination, the Data Protection Directive prohibits the processing of sensitive attributes such as ethnicity or religion. However, it is questionable whether this approach is effective, due to effects like redlining (see Sect. 19.3). Since discrimination may also be based on underlying "neutral" attributes such as area code and income, we need a mechanism to determine whether this is actually the case. Removing sensitive attributes from the dataset may deny us the means to check a posteriori whether an algorithm has indeed discriminated against a particular group. The reason for this is that by removing sensitive attributes from the data, we merely remove the key indicators of discriminatory results, but not other, more indirect indicators that may lead to similarly discriminating results.

A final issue concerns consent and the transparency of automated profiling. While the data subject may give his consent for the processing of personal data and the use of automated profiling, it will likely be unclear to the data subject what the actual extent and impact of profiling will be on his person. Moreover, consenting with automated profiling may yield direct benefits for consumers (such as free goods or services), whereas the risks of automated profiling may be less clear and tangible.

From the above considerations we may conclude that the concept of privacy and its application in data protection law does not provide adequate protection from the risks associated with automated profiling. While the data quality principle (article 6 of the Data Protection Directive) and the right not be subjected to a decision solely based on the automated processing of data (article 15 of the Data Protection Directive) may provide some measure of protection against problems associated with accuracy and reliability, the negative effects of data dredging and mistaking correlation for causation, in practice their use is limited. In part this can be attributed to the fact that the data mining community (adept at spotting application issues) and the legal community are not co-operating enough.

These technological and legal-technical issues highlight a key weakness in the current approach to data protection and informational privacy. Privacy and data protection law is based primarily on a priori protection, but has little in the way of a posteriori protection mechanisms. This has everything to do with the nature of privacy as a mechanism for hiding. We can say that privacy in many cases is a means rather than an end in itself (Schermer 2007). The actual underlying goals of privacy protection (autonomy, equal treatment, economic equality) are protected through the right to privacy. However, once this layer of a priori protection (i.e., the possibility to hide or shield information from observation) has been breached, there are few mechanisms for protecting the underlying interests. Because of the nature of information processing in today's hyperconnected network society, this layer of a priori protection is becoming weaker and weaker.

Moreover, a priori privacy protection also carries with it particular political issues. While privacy is a powerful meme in the political debate about fair information processing, it is also often seen as the antithesis of values that benefit from openness such as security, innovation and efficiency. Furthermore, given the fact that privacy is construed as an individual right it often loses out in the public debate, because it is positioned against the interests of society as a whole (Schermer 2007).

Acknowledgements The authors would like to thank the Netherlands Organization for Scientific Research (NWO) for enabling this research.

References

Adriaans, P., and D. Zantinge. 1996. *Data mining*. Harlow: Addison Wesley Longman.

Bygrave, L.A. 2002. *Data protection law; approaching its rationale, logic and limits*, Information law series, vol. 10. The Hague/London/New York: Kluwer Law International.

Calders, T. 2007. The complexity of satisfying constraints on transaction databases. *Acta Informatica* 44(7–8): 591–624.

Calders, T. 2008. Itemset frequency satisfiability: Complexity and axiomatization. *Theoretical Computer Science* 394(1–2): 84–111.

Calders, T., and S. Verwer. 2010. *Three Naive Bayes approaches for discrimination-free classification*. Data Mining and Knowledge Discovery, September 2010, Vol. 21, Issue 2, pp. 277–292.

Chawla, N.V., K.W. Bowyer, L.O. Hall, and W.P. Kegelmeyer. 2002. Smote: Synthetic minority over-sampling technique. *International Journal of Artificial Intelligence Research (JAIR)* 16: 321–357.

Cocx, T.K. 2009. *Algorithmic tools for data-oriented law enforcement*, PhD thesis, University of Leiden.

Custers, B.H.M. 2004. *The power of knowledge*. Tilburg: Wolf Legal Publishers.

Custers, B.H.M. 2010. Data mining with discrimination sensitive and privacy sensitive attributes. In *Proceedings of ISP 2010, international conference on information security and privacy*, 12–14, July 2010, Orlando, Florida.
Custers, B., T. Calders, B. Schermer, and T. Zarsky. 2013. *Discrimination and privacy in the information society; data mining and profiling in large databases*. Heidelberg: Springer.
Del Carmen, A. 2007. *Racial profiling in America*. Upper Sadle River: Prentice Hall.
Denning, D. 1983. *Cryptography and data security*. Amsterdam: Addison-Wesley.
Fayyad, U.-M., G. Piatetsky-Shapiro, P. Smyth, and R. Uthurusamy. 1996. *Advances in knowledge discovery and data mining*. Menlo Park: AAAI Press/The MIT Press.
Goldberg, I.A. 2000. *A pseudonymous communications infrastructure for the Internet*, dissertation, Berkeley: University of California at Berkeley.
Hornung, G. 2012. A general data protection regulation for Europe? Light and shade in the commission's draft of 25 January 2012. *SCRIPTed* 9(1): 64–81.
Kamiran, F., and T. Calders. 2009. Classification without discrimination. In *IEEE international conference on computer, control & communication (IEEE-IC4)*, 17–19 February 2009, Karachi, Pakistan.
Kamiran, F., and T. Calders. 2010. Exploiting independency constraints for classification. http://wwwis.win.tue.nl
Kuner, Chr. (2012) The European Commission's proposed data protection regulation: A Copernican Revolution in European Data Protection Law. *Privacy and security law report*, 6 February 2012.
Lindell, Y., and B. Pinkas. 2002. Privacy preserving data mining. *Journal of Cryptology* 15(3): 177–206.
Mannila, H., D. Hand, and P. Smith. 2001. *Principles of data mining*. Cambridge, MA: MIT Press.
Meeks, K. 2000. *Driving while black*. New York: Broadway Books.
Ohm, P. 2010. Broken promises of privacy: Responding to the surprising failure of anonymization. *UCLA Law Review* 57: 1701.
Pearl, D. 2009. *Causality: Models, reasoning, and inference*, 2nd ed. Cambridge: Cambridge University Press.
Pedreschi, D., R. Ruggieri, and F. Turini. 2008. Discrimination-aware data mining. In *Proceedings of the 14th ACM SIGKDD conference on knowledge discovery and data mining*. New York: ACM, pp. 560–568
Robinson, N., H. Graux, M. Botterman, and L. Valeri. 2009. *Review of the European data protection directive*. Cambridge: RAND Europe.
Schermer, B.W. 2007. *Software agents, surveillance, and the right to privacy: A legislative framework for agent-enabled surveillance*, PhD thesis, Leiden University.
Schermer, B.W. 2011. The limits of privacy in automated profiling and data mining. *Computer Law & Security Review* 27(7): 45–52.
Solove, D. 2004. *The digital person; technology and privacy in the information age*. New York: New York University Press.
van den Hoven, M.J. 1997. Privacy and the varieties of informational wrongdoing in an information age. *Computers and Society* 27(2): 33–37.
Vedder, A.H. 1999. KDD: The challenge to individualism. *Ethics and Information Technology* 1(4): 275–281.
Weitzner, D.J., H. Abelson, et al. 2006. *Transparent accountable data mining: New strategies for privacy protection*, MIT technical report. Cambridge: MIT.
Westin, A. 1967. *Privacy and freedom*. London: Bodley Head.
Withrow, B. 2006. *Racial profiling*. Upper Sadle River: Prentice Hall.
Zarsky, T.Z. 2003, Mine your own business! Making the case for the implications of the data mining of personal information in the forum of public opinion. *Yale Journal of Law and Technology* 5: 1–57.
Zarsky, T.Z. 2006. Chapter 12: Online privacy, tailoring, and persuasion. In *Privacy and technologies of identity, a cross disciplinary conversation*, ed. K. Strandburg and D. Stan Raicu, 209–224. New York: Springer.

Chapter 20
Military Robotics & Relationality: Criteria for Ethical Decision-Making

Lambèr Royakkers and Anya Topolski

Abstract In this article, we argue that the implementation of military robots must be preceded by a careful reflection on the ethics of warfare in that warfare must be regarded as a strictly human activity, for which human beings must remain responsible and in control and that ethical decision-making can never be transferred to autonomous robots in the foreseeable future, since these robots are not capable of making ethical decisions. Non-autonomous robots require that humans authorize any decision to use lethal force, i.e., they require a 'man-in-the-loop'. We propose a model of relationality for the moral attitude that is needed to confront the moral questions and dilemmas that will be faced by future military operations using robots. This model provides two minimal criteria for ethical decision making: non-binary thinking and reflexivity by means of rooting and shifting. In the second part of this article, we apply these criteria to today's human operators of non-autonomous military robots and secondly, to tomorrow's autonomous military robots, and ask whether robots are capable of relationality, and to what degree human operators make decisions on the basis of relationality. We then conclude with what we take to be a possible, albeit limited, role for robotics in the military with regard to both the current and the foreseeable future role of military robotics.

20.1 Introduction

In World War II, the Japanese navy equipped some of their submarines with the Kaiten, a manned torpedo which required a Kamikaze sacrifice from its pilot in exchange for a minute increase in accuracy. Today the military seems to be moving

L. Royakkers (✉)
School of Innovation Sciences, Eindhoven University of Technology, Eindhoven, The Netherlands
e-mail: L.M.M.Royakkers@tue.nl

A. Topolski
Institute for Philosophy, KU Leuven, Leuven, Belgium

J. van den Hoven et al. (eds.), *Responsible Innovation 1: Innovative Solutions for Global Issues*, DOI 10.1007/978-94-017-8956-1_20,

in another direction: unmanned vehicles make it possible to engage the enemy from a very safe distance – the pilots of Predators and Reapers, for instance, although wearing flight suits, do so without leaving their cubicle in Nevada (Singer 2009).[1] This use of unmanned systems is, on first sight at least, not very different (as long as there is 'a human in the loop') from using an aircraft to drop a bomb from a high altitude. The benefits of unmanned systems, however, are decreasing of the number of soldiers killed on the battlefield, gaining tactical and operational superiority, and reducing emotional and traumatic stress among soldiers (Veruggio and Operto 2008). To illustrate, almost 20 % of the soldiers returning from Iraq or Afghanistan have post-traumatic stress disorder or suffer from depression causing a wave of suicide, particularly among American veterans (Tanielian and Jaycox 2008). To further the goal of minimizing military casualties and stress-related casualties, Ronald Arkin (2007), the roboticist, has proposed equipping military robots with an artificial conscience that would suppress unethical lethal behaviour by adding an ethical dimension to these robots. This ethical dimension consists of prescriptive ethical codes which can govern its actions in a manner consistent with the Laws of War and Rules of Engagement. Arkin stated that "they [robot soldiers] can perform more ethically than humans are capable of"[2] (see also Sullins 2010). While Arkin's statement may seem like science-fiction to most, the fact is that the deployment of military robots or unmanned semi-autonomous vehicles is growing rapidly. American armed forces only operated about 100 Unmanned Arial Vehicles (UAVs) in 2001, and now have more than 5,000 ranging from the Learjet-size Global Hawk to the backpack-sized Raven (Krishnan 2009).

As if the tactical advantages brought by this technology was not enough, we now face the prospect of genuinely autonomous robot vehicles, those that involve 'artificial intelligence' and hence do not need human operators. The United States Air Force (2009), for example, expect the deployment of autonomous UAVs with "a fully autonomous capability" between 2025 and 2047. Though it is unclear what degree of autonomy these UAVs will have, "the eventual deployment of systems with ever increasing autonomy is inevitable" (Arkin 2007). A study by US Joint

[1]The Predator is an unmanned airplane which can remain airborne for 24 h and is currently employed extensively in Afghanistan. The Predator drones can fire Hellfire missiles and are flown by pilots located at a military base in the Nevada desert, thousands of miles away from the battlefield. Their successor, the Reaper, which may phase out the F-16, has already been spotted in Afghanistan. This machine can carry 5,000 pounds of explosive devices, Hellfire missiles, or laser directed bombs, and uses day-and-night cameras to navigate through a sheet of clouds. This unmanned combat aerial vehicle is operated by two pilots located at a ground control station behind a computer at a safe distance from the war zone.

[2]To make sure that military robots would perform reliably and lawfully they could be required to pass a military Turing Test "That means an autonomous system should be no worse than a human at taking decisions [about valid targets]" (Mick 2008). This test entails an assessment of the comparative morality pairs of descriptions of morally significant behaviour where one describes the actions of a human being in an ethical dilemma and the other the actions of a machine faced with the same dilemma. If the machine is not identified as the less moral pair significantly, then it has passed the test (Mick 2008).

Forces Command, Unmanned Effects: Taking the Human Out of the Loop, says that genuinely autonomous robots should be a reality by 2025.[3] The deployment of genuinely autonomous armed robots in battle, capable of making independent decisions as to the application of lethal force without human control, and often without any direct human oversight at all, would constitute a genuine military as well as moral revolution (Kaag and Kaufman 2009).

In this article, we would like to pause – something that rarely occurs in the race for technological progress in the military – and consider (1) what it means to make a decision about whether an action is ethical or not and, (2) consider whether robots are able to make such decisions. We argue that the implementation of military robots must be preceded by a careful reflection on the ethics of warfare in that warfare must be regarded as a strictly human activity, for which human beings must remain responsible and in control and that ethical decision-making can never be transferred to machines, since machines are not capable of making ethical decisions. Given the distinction between, on the one hand, non-autonomous military robotics *today*, in which – to differing degrees – human operators remain in the loop, and, on the other, the foreseeable *future* of military robotics which promise autonomous robots capable of ethical decision making, we will try to separate our analysis along these lines. We need to make a distinction for these autonomous robots between non-learning machines and learning machines. Learning military robots, based on neural networks, genetic algorithms and agent architectures, are able to decide on a course of action and to act without human intervention. The rules by which they act are not fixed during the production process, but can be changed during the operation of the robot, by the robot itself (Matthias 2004). The problem with these robots is that there will be a class of actions where no-one is capable of predicting the future behaviour of these robots any more. So, these robots would become a 'black box' for difficult moral decisions, preventing any second-guessing of their decisions. The control transfers then to the robot itself, but it is nonsensical to hold the robot responsible at the moment, since robots that will be built in the next two decades do not possess anything like intentionality or a real capability for agency.[4] The deployment of learning armed military robots will constitute a responsibility gap (Matthias 2004), since it would constitute the injustice of holding people responsible for actions of robots over which they could

---

[3]Singapore has announced its goal to build a military robot, named Urban Warrior, that can fight autonomously in urban environment like a human soldier, and conducted a contest in 2008. South Korea and Israel have been already deploying autonomous armed robot border guards. The South Korean system, the *Samsung Techwin SGR-A1* stationary sentry robot, is capable of interrogating suspects, identifying potential enemy intruders, and autonomous firing of its weapon (Krishnan 2009). The unmanned ground vehicle commissioned by the Israeli military, the *Guardium*, is an autonomous observation and target intercept system.

[4]Although we can state that the robot is causally responsible, but the robot is off the hook regarding moral responsibility. Some authors claim that fully autonomous robots can be considered as moral agents (Dennett 1997), but this discussion is beyond the scope of this article and such an attribution unnecessarily complicates the issue of responsibility ascription for immoral actions.

not have any control. These learning machines for military purposes seem, at least under present and near-term conditions, far from reasonable. This article is therefore about the non-autonomous and autonomous non-learning ones, that is to say, on the kind of armed military robots which are already deployed or that will be deployed in the foreseeable future. Non-autonomous robots require that humans authorize any decision to use lethal force, i.e., they require a 'man-in-the-loop'. This means that the decision to open fire, or the taking of any action that could threaten human life, is to be considered and approved by a person.

We will start by considering an age old philosophical debate, and one that is by no means new to the realm of military ethics – what are the standards for ethical judgment? In response to this question we propose a model of relationality for the moral attitude that is needed to confront the moral questions and dilemmas that will be faced by future military operations. This model provides two minimal criteria for ethical decision making: non-binary thinking and reflexivity by means of rooting and shifting. In the second part of this article, we apply these criteria to today's human operators of non-autonomous military robots and secondly, to tomorrow's autonomous military robots, and ask whether robots are capable of relationality, and to what degree human operators make decisions on the basis of relationality. We then conclude with what we take to be a possible, albeit limited, role for robotics in the military with regard to both the current and the foreseeable future role of military robotics.

20.2 Relationality: A Proposed Standard for Ethical Decision Making

While Clausewitz was by no means a post-modern military strategist, his account of the fog of war, its ambiguities and uncertainties are as true today as they were 200 years ago. In our times, neither peace nor war is without ambiguity. This is equally true of the technological advances introduced by military robotics and specifically those promised by autonomous systems. Designed to make the hard ethical choices that are often impossible for humans to make in the fog of war, autonomous military robots are intended to prevent military casualties by means of engaging from a safe distance, both physical and mental – the latter which brings with it greater 'objectivity' by minimising the ambiguities of affectivity combined with fear and stress. While there is no doubt that autonomous robotic systems offer a variety of military advantages, both in terms of ethics and efficacy, the question that cannot be so easily answered is – can robots, computer programmed deliberative and reactive systems, make ethical decisions? In order to answer this question, we must first define what an ethical decision is.

While it might be easier to reduce ethics to legal criteria, such as the laws of war, just war theory and military rule of engagement (Arkin 2007), ethics cannot so simply be reduced to a legal code. While we do not seek to deny the importance of military law, its ethical standards must be recognised as being exceptional in its justification of violence. This double standard of civil and military ethics is without a doubt problematic. This standard is not worthy of the name ethics as it only instrumentalises humanity's potential to improve itself and to learn from its mistakes (many of which occurred under the cloak of war). Rightfully ashamed of their colonial pasts, imperialist riches, and responsibility for genocide etc, taxpayers want no hand in murder. While one option is to weave a web of deceit which will eventually crumble, the other is simply to recognise that the legal standards of war are not a sufficient ethical standard for military decisions. We can no longer rely on traditional standards and norms. Responsibility in modernity entails a refusal to seek easy answers such as that of the just war tradition (Topolski 2010). Therefore, we propose a model of relationality, that we argue is the minimum standard for ethical decision-making and must be used when determining if human operators and autonomous military robots are capable of making ethical decisions.

Relationality is a social ontology that takes the relationships we have with others to be constitutive of the self. It is these relationships, implicit and explicit, that allow us to set shared standards for ethical decision making. As Hannah Arendt once wrote, reality is constructed by means of a web of relations (1968). What she meant by this is that there is no absolute certainty, there are no absolute guarantees. It is thus up to us, in relation with others, to determine our own standards. It is by means of a web of relations that we co-construct a shared notion of the world as well as a means to judge events in the world. It is also in this sense that relationality is a social ontology with the plurality of agents as its source. Relationality takes humanity to be not only constructed by actors, but also recognises that these actors themselves are constructed by means of their relationships to others.

The more important question to consider is – what does this mean in terms of ethics? Given that relationality seeks to be applicable in a military context, it must be suitable to complex and dynamic environments, its ethics cannot simply be a list of rules or norms as the situations that most often require ethical-decision making are exceptional cases where the standard rules or norms do not apply. Relationality is thus an ethical approach that emphasises the importance of a process of critical understanding. This differs greatly from approaches, like that of Arkin, who sees ethics as a military robotics specialist, and therefore in terms of information. They believe that applied ethics is essentially the application of theories to particular situations: "A machine (. . .) is able to calculate the best action in ethical dilemmas using ethical principles" (Anderson and Anderson 2007). This view is, however, problematic for a number of reasons (Van de Poel and Royakkers 2007). One reason is that no moral theory is universally accepted. Different theories might yield different judgments about a particular case. But even if there were one accepted theory, framework or set of principles, it is doubtful whether it could be straightforwardly applied to all particular cases. Theory development in ethics in general does not take place independent of particular cases. Rather, theory development is an attempt to systematize judgments over particular cases and to provide a rational justification for these judgments. So if we encounter a new case, we can of course try to apply the ethical theory we have developed until then to that case, but we should also be open to the possibility that the new case might sometimes reveal a flaw in the theory we have developed so far.[5]

Relationality seeks precisely to teach us to be prepared to think with and beyond the particular constraints of any one form of normative ethics. It forces us to consider our judgments not only in relation to theories and principles but also in terms of those with whom we interact, including the environment. The basic idea is that in a process of reflection different ethical judgments are contrasted with each other. This process is not so much about attaining the one absolute authoritative answer, as it is about eliminating problematic solutions as well as seeking creative options that appear by means of a comparison of different possibilities. This approach is inspired by an epistemology which:

> Recognises that from each positioning the world is seen differently, and thus that any knowledge based on just one positioning is 'unfinished' – which is not the same thing as saying it is 'invalid'. In this epistemology, the only way to approach 'the truth' is by dialogue between people of differential positioning (Yuval-Davies 1998).

Standpoint epistemology is perfectly suited to the current reality of military missions which are pursued by means of networked enabled operations (NEO) that involve a mixture of both military and non-military partners. Such networked environments are a means of communicating from a variety of perspectives, each of which helps all those on the network to have a better understanding of the complex and dynamic environment. In this manner, NEO brings together several partial perspectives for every situation. Each perspectives is thus incomplete and, as such, in need of the assessments of others to be as close to complete as possible. Relationality complements the epistemological presuppositions of standpoint epistemology in that it promotes an attitude of cooperation, communication and emphasizes that these, in addition to trust, are essential for both good judgments, and in the case of coordinated military operations – both efficiency and success.

20.3 The Capacity for Relationality in Human Operators and Autonomous Robots

Having defined the criteria for making an ethical decision as that of relationality, we now turn to its praxis, and propose that relationality should be the minimum standard for both humans and robots in modern warfare. The purpose is to consider in the

---

[5]If ethical theories do not provide moral principles that can be straightforwardly applied to get the right answer, what then is their role, if any, in applied ethics? Their role is, first, instrumental in discovering the ethical aspects of a problem or situation. Different ethical theories stress different aspects of a situation; consequentialism for example draws attention to how consequences of actions may be morally relevant; deontological theories might draw attention to the moral importance of promises, rights and obligations. And virtue ethics may remind us that certain character traits can be morally relevant. Ethical theories also suggest certain arguments or reasons that can play a role in moral judgments.

following section whether (a) human operators (Sect. 20.3.1), or (b) autonomous robots (Sect. 20.3.2), are capable of making decisions according to the minimal criteria of relationality. While these minimal criteria are by no means a guarantee of a good ethical outcome, we claim that there are no such guarantees in times of war, the criteria seek to make the decision making process itself, the process of gathering information, deliberating and judging it, as ethical as possible. This, we argue, is the best possible grounds for ethical decision making, leading to better discussion, deliberation, reflection and better outcomes.

Relationality, defined here in a minimal sense, provides us with the following two criteria for ethical judgments: first, one must consider all information and understand it in non-binary terms; second, all information must be understood in a reflexive manner by means of rooting– which includes an understanding of the situation of others with whom one is in relation. Let us consider each of these criteria in turn. Relationality calls for non-binary thinking by defining the self in terms of its relationships to others, that is by affirming that the other constitutes the self. A relational identity is one that deconstructs the distance we create within ourselves and between others. It thus denies us the ability to frame the world in terms of self-other or us-them. It is the refusal to see the self, others and the world in reductive or polarizing binary terms. The notion of relationality itself denies the possibility of binary, polarizing distinctions between people by emphasizing the self-constitutive nature of our relationships to other people. As a social ontology, relationality maintains that all distinctions are permeable and porous and, as such, there is neither an absolute 'I' nor an absolute 'us'. What is worth noting is that relationality does not deny that each and every relationship is immersed with power inequalities. The conscious acknowledgement of the power inequality itself is a precursor to an ethical interaction between people. Non-binary thinking is also, in this sense, a tool to deconstruct the normative frames such as us vs. them that allow one to create an artificial distance between oneself and the other.

The second criteria is that of reflexivity, which aims to demarcate the difference between information and understanding. More specifically, this is the ability to root oneself by means of a critical reflection. Critical reflexivity is the ability to reflect upon oneself, one's choices, one's actions, one's interactions with others. It means to challenge and question oneself. Reflexivity asks us to reflect on our location, our identity, and our values. One's location consists of a variety of factors, including one's social intersection in terms of gender, race, economics, class, sexuality, age etc. A second aspect of location is one's identity. While identities are often the product of a binary system, identities that arise from the critical reflexivity of rooting are relational identities developed by means of a narrative which weaves in and out of narratives of others with whom one interacts. The third aspect of reflexivity is a reflection on one's values. This is often the most difficult aspect of rooting. Unlike many of our social intersection and identity claims, our values are often not as easily identified without a serious commitment to reflexivity and dialogue, with oneself and others. Nonetheless, reflexivity asks us to consider what our values, beliefs, and attitudes are, and most importantly, why. It is in this threefold sense that rooting is a commitment to understanding oneself in terms of social intersection, identity and values.

Rooting is a precondition for shifting which is necessary for the development of a dynamic complex situational awareness. Rooting forces one to examine oneself before trying to imagine the 'roots' of the other by means of shifting. While the ability to imagine is essential to this praxis, one must neither sympathise with the 'roots' of the other nor have had a shared experience as in this case with empathy (although either can certainly not hurt). The praxis of rooting and shifting serves as a springboard for thinking, judging and acting out of relationality (rather than fear or egoism). Most importantly, it makes us aware of the co-constitutive nature of our relationship and hence responsibility to the other(s). It is in this sense that relationality also plays a structural role in promoting non-egoistic or binary based thinking and actions. The purpose of shifting is to attempt to understand the locations of those with whom one interacts. While there is no doubt that certain types of shifts are easier than others, as some locations are more reconcilable than others, the commitment to shifting itself already greatly facilitates one's ability to shift into difficult locations as does a thorough process of rooting. By means of shifting one is better able to engage the other in a dialogue that will lead to understanding and this remains the case if even only one of the members of the dialogue has engaged in relational thinking. One may be better able to avoid certain terms, topics or triggers that can prevent an open dialogue.

What is however a danger of shifting, one that rooting seeks to minimize but cannot eliminate, is the possibility of losing oneself in the other. While some, such as Levinas, view this as the ethical ideal, it is not ideal for our purposes here as it leads to the loss of self and hence another unique and valuable location. While the goal of relationality in terms of ethics and politics is one of solidarity, this should not come at the cost of the self. The other extreme, however, is that one affirms one's own roots too tightly. This can have the effect of preventing one from fully shifting and hence continues to reduce the other to oneself, or cause one to fail to recognise the differences among others of the same social group. Avoiding these extremes, shifting ideally leads one to a better understanding of the other in terms of their challenges, context, needs, etc, and most importantly, an awareness of the power dimensions involved in one's relationships to others. Perhaps the most important aspect of shifting in terms of the military is that it is a process that allows us to become aware of the importance of context whether social, political, cultural, historical etc. Shifting allows for a reflexive situational awareness. As one learns how critical context is by means of shifting into other locations, one begins to appreciate that judgments are not rooted in a universal standard but rather are very much tied to location, relationality and the particular context. What might be the best option in one's location may in fact be the worst option for another location. While this makes thinking, judging and acting much more difficult, it is precisely this difficulty that makes it more ethical.

Why are these criteria so essential for ethical decision-making? Why is relationality the foundation for responsible judgments? Perhaps the best way to answer this question is to consider a now infamous case of someone whose thoughtlessness, the by-product of failing to root and shift, led to radical evil – Adolf Eichmann (Arendt 2005, 2006). Eichmann went about his work – in a professional, efficient and thorough manner – making decisions based on information given to him by others, processing numbers, trains and costs. He worked diligently day and night in a machine-like manner, yet he never once considered who these numbers were, how he was related to them, and why it was not enough to desire to be as efficient as a machine. Quite simply, Eichmann's decisions were not ethical decisions, they were much more similar to the type of decisions made by robots – not deliberative ethical reflections. Eichmann could not have done what he did had he been thinking relationally, engaging in either rooting or shifting. It was nothing more than this thoughtlessness and egoism that led to his evil actions. While there is no doubt that Eichmann dreamt of being as efficient as a computer, it was precisely for this reason that he was incapable of making ethical decisions.

To help make the importance of rooting and shifting clear, both of which arise by means of understanding, let us consider a common scenario in places such as Afghanistan. The ISAF mission is composed of hundreds of different partners from different national militaries, several international NGOs, a variety of Afghani organizations and government institutions as well as several private, often corporate, partners. It is no surprise that given the complex nature of the relationships between these different partners that ISAF relies on networked enabled technologies to keep the mission, and all its communications between the partners, organized. As one can imagine, each of these relationships is unique. Each involves different levels of dependence, trust, security etc. As with all relationships, they are delicate, complex and dynamic. Such uncertainty and variations are not easily reduced to algorithms.

20.3.1 The Capacity for Relationality in Human Operators

With regard to human operators, who remain in control and responsible for the final decision (rather than simply playing a monitoring function, meaning that the human operator keep an eye on the process and only interferes if something goes wrong), relationality seeks to address two problems confronted by operators, that of moral disengagement and over-simplified situational awareness. While we argue that operators limited to a monitoring role are no longer acting relationally, operators who remain in control and are committed to both non-binary thinking and reflectivity are capable of acting relationally. As such, the question to be considered here is: what are the ethical benefits of relationality with regard to human operators? Given that one of the strongest arguments in favour of autonomous military robots is to avoid the ethical dangers of moral disengagement, we wish to make a plea for relationality as a means to both fight this tendency and keep humans-in-the-loop. The second criteria, that of reflexivity, which is realised by means of a continual process of shifting and rooting, aims to increase the operators ethical awareness by allowing for a complex, dynamic and relational approach to the situation. Once again, it is our contention that – while difficult – human operators in control (rather than monitoring) of all decisions are capable of reflexivity. While it is recognised that humans may chose not to act reflexively or only partially, it is the capacity to do so that is unique to human beings. By contrast, we will make the claim (in the following section), that autonomous robots are not capable of reflexivity and thus cannot act ethically according to the criteria established by relationality. While this may one day be otherwise, this seems unlikely at the present. If one considers that one becomes more relational by interacting with others by means of relationality, it is possible that robots – who also interact with humans – may slowly acquire traits of relationality.

Moral disengagement is a psychological process by which individuals detach themselves from a particular person(s) or distance themselves from a particular situation, a gradual process which allows them to 'turn off' their otherwise operational sense of morality. In terms of ethics, it is by means of moral disengagement that so many atrocities have occurred both in times of war and peace (McAlister 2001; Aquino et al. 2007). In terms of the physical and mental health of soldiers and civilians alike, moral disengagement often leads to an increased likelihood of a variety of mental difficulties including post traumatic stress disorder (Shalev 2002). As it is impossible to harm another, or in extreme scenarios kill another human being, without being personally harmed or traumatized (Bandura 2002), it is natural to seek to dehumanize the other to avoid further personal harm. Social psychologists point out that dehumanization, i.e., seeing people for something less than human, can open the door to more serious forms of unethical conduct (see, e.g., Bandura 1999, 2002; Mastroianni 2011; Moller and Deci 2010; Slim 2007). The use of military robots is a substantial cause for the increasing dehumanization of modern warfare.[6] Although fighting from behind a computer is not as emotionally potent as being on the battlefield, pushing a button to kill someone can still be a stressful job; various studies have reported physical and emotional fatigue and increased tensions in the private lives of military personnel operating the Predators in Iraq and Afghanistan (Donnelly 2005). The problem of 'residual stress' of human operators has led to proposals to diminish these tensions. In particular, the visual interface can play an important role in reducing stress; interfaces that only show abstract and indirect images of the battlefield will probably cause less stress than the more advanced real images (Singer 2009). From a technical perspective this proposal is a feasible one, since it will not be hard to digitally recode the war scene in such a way that it induces less moral discomfort with the war operator. Such 'photo shopping' of the war, however, raises some serious ethical issues in its own rite.

This need to dehumanize shows that prior to this moment there is a sense in which I, like this other, am human – we are related. It is only by denying this relationship that I am able to overcome the sense of shared humanity. Moral disengagement arises by silencing one's fundamental relationality. To do so, one must in fact distance oneself both internally and externally from others who have, throughout one's life, helped one to internalize this relationally produced voice. According to Bandura, this disengagement occurs by turning off one, or more,

---

[6]Besides the substantial cause, military robots also are an expression of, for example, the 'culture of fear' and increasing risk-averseness in our society.

of the following sanctions: (1) cognitive restructuring of conduct, (2) sanitizing language, (3) displacing responsibility, (4) disregarding or misrepresenting injurious consequences and (5) blaming the victim, or (6) dehumanization (Bandura 1999). This is perhaps clearest in terms of the process of dehumanization.[7] Dehumanization is the process of denying our relationality using the frame of us vs. them – a binary lens. This is often done by defining the out-group as dangerous beings (the enemy, racial impurity etc), non-human (i.e. animals, insects etc), or non-beings (i.e. garbage, machines etc) (Moshman 2007). This frame allows us, by means of repetition (and potential brainwashing) to view – and truly believe – that the other is a threat. This is also greatly aided by the use of euphemisms (for which the military is well known). Dehumanization thus allows us to limit our sphere of moral obligation to those included in our artificially constructed notion of 'us', leading to indescribable atrocities of the 'them'. As Mary Kaldor has shown, this is precisely the source of most casualties in 'new wars' (Kaldor and Vashee 1997; Kaldor 2004). Military robots, it goes without saying, neither have the potential to dehumanize another nor to 'lose it' under pressure. Part of the military's training, certainly since it became clear that soldiers often intentionally avoided shooting their 'enemies' (Grossman 1996), has been to teach soldiers to morally disengage by means of binary thinking in order to dehumanize the other, making it easier to kill. Drones, controlled by a human operator, are a further step in this same direction. Drones introduce an artificial asymmetry to warfare that allows one to avoid seeing the other face-to-face. Rather, the other appears as red dots/targets on a screen at a distance, much like in several video games designed by the US army for recruitment purposes. This depersonalization can even go so far that a soldier is no longer aware of the fact that she is actually involved in a real war. In the current situation it can already be hard to distinguish between a warfare video game and operating a drone. From a technological perspective there is only a small step between playing a computer game and destroying enemy 'avatars', and actually killing real people on the other side of the globe (Royakkers and Van Est 2010).

20.3.2 The Capacity for Relationality in Autonomous Robots

The most compelling argument, in terms of morality, in favour of autonomous military robots is that they decrease the personal pain and anxiety suffered by many soldiers who have to deliberate and choose whether or not to end a human life by allowing them to detach from the ethical responsibility of having to decide. By allowing such difficult ethical decisions to be made by an autonomous robot, military personnel seek to minimise and ideally avoid any mental discomfort. While

[7]One of the most cited arguments in support of drones is to avoid the practice of moral disengagement known as dehumanization which is a common cause of violence, murder and potentially genocide.

one can certainly understand how this is advantageous to the military, our contention is that this dehumanizing process comes at the cost of ethics. While we may be further removed from the decision and its immediate pain, such thinking leads to a variety of problems for soldiers who themselves have been dehumanized and are victims of a grave trauma. The same is true for society at large, which is not only responsible for helping these soldiers recover, but also for the atrocities committed by them.[8] Yet, as we argued above, relationality aims to keep humans responsible while also decreasing the pain caused by such detachment and difficult decisions. A further argument in favour of relationality, and against autonomous military robots, is that the latter are not capable of reflexivity which is absolutely necessary in situations of contemporary warfare and peacekeeping.

Within a military context, we argue that reflexivity is essential for ethical decision-making for a more fundamental reason. In order for moral judgments to be legitimate, they must be the result of a careful process of moral reflection. What this entails is that the determinations made by military robotics which are based on algorithms are not forms of moral deliberation and reflection. While it is clear that military robotics are capable of processing a greater amount of information at a much faster rate than human beings (which is the reason so many members of the military community are greatly in favour of drones), this ability is distinct from the ability to critically evaluate this information and to consider it when making difficult strategic decisions.[9] Knowledge, the process of transforming information into understanding, is a skill that only human beings are capable of. It is for this reason that drones are not capable of the type of shifts in awareness as the situation changes that arises from rooting. Robotics lack the ability to ask themselves questions about their own choices, actions and how their interactions affect those of their environment. As all military robots lack the ability to root, they have no basis upon which to reflect, that is they lack the understanding necessary for making ethical decisions in complex and changing environments characterized by different forms of relationality.

It is only by means rooting and shifting that human beings can understand the complex dynamics of each of these particular relationships, a process that is beyond the capacities of even the most advanced autonomous military robots. Military robots' inability to think, to reflect or to understand their complex situational

[8]Recent interviews with former members of the Taliban have also revealed the increase in PTSD and other war related mental disorders suffered by their fighters as well as the local population (Newsweek Dec 6, 2010) www.newsweek.com/2010/12/06/do-the-taliban-get-ptsd.html. The importance of setting standards for ethical decision making in the military is thus a shared responsibility as these standards affect not only those that are killed but also those doing the killing and those supporting them.

[9]As General Stanley McCrystal (the former commander of the ISAF mission) shared in an interview, while computers are able to process wealth of information they are not capable of understanding it. 3/11/10, www.idga.org/podcenter.cfm?externalid=826&mac=IDGA_OI_Featured_2010&utm_source=idga.org&utm_medium=email&utm_campaign=IDGAOptIn&utm_content=11/4/10.

environment has been demonstrated by the fact that they often miss important details or incorrectly interpret situations in a complex and dynamic military environment. Even the most excellent sensors can never compensate for a robot's deficient understanding of its environment (Krishnan 2009). Humans are better at discriminating targets, because they understand what a target is, and when and why to target something or somebody. Barring some major significant breakthrough in artificial intelligence research, situational awareness cannot be incorporated in software for lethal military robots (Gulam and Lee 2006; Fitzsimonds and Mahnken 2007; Kenyon 2006; Sharkey 2008; Sparrow 2007).

The ultimate goal, however, of autonomous military robots, according to United States Air Force (2009), is to create a military robot capable of making independent decisions as to the application of lethal force without human control. This development of autonomous non-learning machines with an additional ethical dimension is a newly emerging field of machine ethics. These robots, based on syntactic manipulation of linguistic symbols with the help of formal logic, are "able to calculate the best action in ethical dilemmas using ethical principles" (Anderson and Anderson 2007). It is thus once again assumed that it is sufficient to represent ethical theory in terms of a logical theory and to deduce the consequences of that theory. This view, analogous to the reduction of ethics to law or reflection to an algorithm, misunderstands the unique – non reducible – nature of ethical reflection. Arkin (2007) argues that some ethical theories, such as virtue ethics, do not lend themselves well by definition to a model based on a strict ethical code. While military robotic specialists claim that the solution is simply to eliminate ethical approaches that refuse such reduction, we would argue that this non-reducibility is the hallmark of ethics. While many ethical situations may be reducible, it is the ability to act ethically in situations that call for judgment – relational judgment – that are distinctly human. Furthermore, a consequence of this approach is that ethical principles themselves will be modified to suit the needs of a technological imperative: "Technology perpetually threatens to coopt ethics. Efficient means tend to become ends in themselves by means of the 'technological imperative' in which it becomes perceived as morally permissible to use a tool merely because we have it" (Kaag and Kaufman 2009).

20.4 Conclusions: Keeping Humans 'In-the-Loop' and Keeping Autonomous Robots Within the Limits of Military Ethics

On 22nd September 2010, a majority of the participants of the Expert Workshop on Limiting Armed Tele-Operated and Autonomous Systems in Berlin accepted a statement in which the following principle is included: "That it is unacceptable for machines to control, determine, or decide upon the application of force or violence in conflict or war. In all cases where such a decision must be made, at least one human being must be held personally responsible and legally accountable for the decision and its foreseeable consequences." In a footnote, it is made clear what is meant by a decision: "The decisions to which this principle should be applied include: The decision to kill or use lethal force against a human being; The decision to use injurious or incapacitating force against a human being; The decision to initiate combat or violent engagement between military units; The decision to initiate war or warfare between states or against non-state actors." Although we completely accept this statement, the statement does not explain why it is unacceptable for machines to make decisions upon the application of force or violence in conflict or war. What we hope to have made clear in this article is that autonomous military robots, while capable of a great many skills essential to the military, are not capable of making ethical decisions according to the criteria set forth by relationality. Autonomous robots, however, can serve those making ethical decisions, if they are used within particular limits. In order to make this argument, we wish here to introduce an important distinction between monitoring and controlling. The current situation is such that human operators maintain control over all final decisions, i.e., she provides or assigns tasks or brings changes and verifies the robot's execution to meet the requirements, which is the ideal ought to be required by military. The future role, however, may be restricted to monitoring, meaning that the human operator keeps an eye on the process and only interferes if something goes wrong which leads to unethical decision-making, since it does not fulfill the minimal standard for ethical decision-making. It is our contention that the path towards autonomous military robots also constitutes a shift away from controlling towards a paradigm of monitoring by human operators. Wallach and Allen (2008) express this concern that we have started on the 'slippery slope toward the abandonment of moral responsibility by human decision makers'. Treviño and Youngblood (1990) have argued that there is a link between the locus of control and ethical decision-making; those who see a clear connection between their own behaviour and its outcomes are more likely to accept responsibility for that behaviour.[10] Conversely, people who believe that they have little personal control in certain situations – such as monitoring – are more likely to go along with rules, decisions and situations even if they are unethical or have harmful effects (Detert et al. 2008).

While there is no doubt that the technological advances made by the military are to be admired, human progress is not always to be found in such artefacts. Progress, and certainly in the realm of ethics, often comes from restraint. It is precisely such restraint which we have argued in this article that is currently absent in discussions on military robots both human operated and autonomous. While this technology often allows us to accomplish a task with greatest efficiency – if ethics remains an important criteria, which we argue is in fact a criteria of ever increasing import to the military, than perhaps it is time to regain control of the situation. While the technology needed for autonomous military robotics races forward, it is the

[10]See also Levenson (1981), Rotter (1996).

task of ethicists to slow things down and to argue, as we have in this article, that robots are not capable of making ethical decisions. While technology may offer us the possibility of limiting our own casualties, the cost of this long-distance warfare is precisely our relationality, that is our ability to remain human and in relation to others. The most effective remedy for who wants to prevent unethical conduct consists of "humanization," a not so clearly defined concept that, however, includes the affirmation of relationality, instead of distancing oneself "from others or divesting them from human qualities" (Bandura 1999). Or, as Hugo Slim (2007) put it in his Killing Civilians, to be effective civilian immunity "requires that armed people find a fundamental identification with those called civilians and not an excessive distinction from them". Seeing people primarily as members of an enemy group is probably easiest "from an air force bomber or a computer screen that is miles away from the individuals one is killing".

In order to avoid reducing the reality of war to the virtuality of a video game, we must cherish our relationality at all costs, even the great costs of casualties and effectiveness. Relationality is the source of our ability to make ethical decisions, to think in non-binary terms and to reflect by means of rooting and shifting. While autonomous military robots are capable of rapid determinate decisions, ethical decisions are in fact indeterminate. The meaning of right or wrong in complex situations (such as in the fog of war) cannot be determined by a general metric, since the inherent controversy and ambiguity of moral judgment cannot be reduced to a logically consistent principle or set of laws, given the complex intuitions people have about right and wrong (Wallach and Allen 2008). It is for this reason that we have argued that autonomous military robots are not capable of ethical decision making,[11] and that relationality will not be attained by foreseeable autonomous military robots. Thus, instead of envisioning robots as idealized replacements for human soldiers, one might see the role of robotics as assisting human decision-making capacity. While robots excel at computational problems, humans are unsurpassed in what one might broadly called common sense, and 'unstructured' or 'open textured' decision-making required judgments rather than calculation (Clarke 1994). While we may continue to admire the technological powers of artefacts ranging from computers to military robots, perhaps it's worth taking a moment to consider our own human powers. While our powers may be less precise, human beings are capable of great wonders and our abilities, certainly when strengthened by relationality, ought not to be so hastily discharged.

Acknowledgments This research is part of the research program 'Moral fitness of military personnel in a networked operation environment', which is supported by the Netherlands Organization for Scientific Research (NWO) under grant number 313-99-110.

[11]However, military robots still controlled (as opposed to monitored) by human beings are indeed still able to effectuate ethical decisions made by a remote operator.

References

Anderson, M., and S.L. Anderson. 2007. Machine ethics: Creating an ethical intelligent agent. *AI Magazine* 28(4): 15–26.
Aquino, K., A. Reed, S. Thau, and D. Freeman. 2007. A grotesque and dark beauty: How moral identity and mechanisms of moral disengagement influences cognitive and emotional reactions to war. *Journal of Experimental Social Psychology* 43: 385–392.
Arendt, H. 1968. *The human condition*. New York: Schocken.
Arendt, H. 2005. *Responsibility and judgment*. New York: Schocken.
Arendt, H. 2006. *Eichmann in Jerusalem: A report on the banality of evil*. New York: Penguin Classics.
Arkin, R.C. 2007. *Governing lethal behavior: Embedding ethics in a hybrid deliberative/reactive robot architecture*, Technical report GIT-GVU-07-11. Atlanta: Georgia Institute of Technology.
Bandura, A. 1999. Moral disengagement in the perpetration of inhumanities. *Personality and Social Psychology Review* 3: 193–209.
Bandura, A. 2002. Selective moral disengagement in the exercise of moral agency. *Journal of Moral Education* 31(2): 101–119.
Clarke, R. 1994. Asimov's laws of robotics: Implications for information technology. *Computer* 27(1): 57–66.
Dennett, D.C. 1997. When Hal kills, Who's to blame? In *Hals legacy: 2001's computer as dream and reality*, ed. D. Stork, 351–365. Cambridge: MIT Press.
Detert, J.R., L.K. Treviño, and V.L. Sweitzer. 2008. Moral disengagement in ethical decision making: A study of antecedents and outcomes. *Journal of Applied Psychology* 93(2): 374–391.
Donnelly, S.B. 2005. Long-distance warriors. *Time Magazine*, 4 December 2005.
Fitzsimonds, J.R., and T.G. Mahnken. 2007. Military officer attitudes towards UAV adoption: Exploring institutional impediments to innovation. *Joint Force Quarterly* 46: 96–103.
Grossman, D. 1996. *On killing: The psychological cost of learning to kill in war and society*. New York: Little, Brown, and Company.
Gulam, H., and S.W. Lee. 2006. Uninhabited combat aerial vehicles and the law of armed conflicts. *Australian Army Journal* 3(2): 123–136.
Kaag, J., and W. Kaufman. 2009. Military frameworks: Technological know-how and the legitimization of warfare. *Cambridge Review of International Affairs* 22(4): 585–606.
Kaldor, M. 2004. *New & old wars: Organized violence in a global era*. Cambridge: Polity.
Kaldor, M., and B. Vashee (eds.). 1997. *New wars. Restructuring the global military sector*. London: Pinter.
Kenyon, H.S. 2006. Israel deploys robot guardians. *Signal* 60(7): 41–44.
Krishnan, A. 2009. *Killer robots. Legality and ethicality of autonomous weapons*. Farnham: Ashgate Publishing Limited.
Levenson, H. 1981. Differentiating among internality, powerful others, and chance. In *Research with the locus of control construct: Vol 1. Assessment methods*, ed. H.M. Lefcourt, 15–63. New York: Academic.
Mastroianni, G.R. 2011. The person–situation debate: Implications for military leadership and civilian–military relations. *Journal of Military Ethics* 10(1): 2–16.
Matthias, A. 2004. The responsibility gap: Ascribing responsibility for the actions of learning automata. *Ethics and Information Technology* 6: 175–183.
McAlister, A.L. 2001. Moral disengagement: Measurement and modification. *Journal of Peace Research* 38: 87–99.
Mick, J. 2008. Can robots commit war crimes? *Daily Tech*, 29 February 2008. http://www.dailytech.com/Can+Robots+Commit+War+Crimes/article10917.htm
Moller, A.C., and E.L. Deci. 2010. Interpersonal control, dehumanization, and violence: A self-determination theory perspective. *Group Processes and Intergroup Relations* 13(1): 41–53.
Moshman, D. 2007. Us and them: Identity and genocide. *Identity* 7(2): 115–135.

Rotter, J.B. 1996. *Generalized expectancies for internal versus external control of reinforcement*, Psychological monographs: General and applied, vol 80, 1–28. Washington: American Psychological Association.
Royakkers, L.M.M., and Q. van Est. 2010. The cubicle warrior: The marionette of digitalized warfare. *Ethics and Information Technology* 12: 289–296.
Shalev, A.Y. 2002. Acute stress reactions in adults. *Biological Psychiatry* 51: 532–543.
Sharkey, N. 2008. Cassandra or false prophet of doom: AI robots and war. *IEEE Intelligent Systems* 23(4): 14–17.
Singer, P.W. 2009. *Wired for war: The robotics revolution and conflict in the twenty-first century*. New York: The Penguin Press.
Slim, H. 2007. *Killing civilians: Method, madness and morality in war*. London: Hurst & Company.
Sparrow, R. 2007. Killer robots. *Journal of Applied Philosophy* 24(1): 62–77.
Sullins, J. 2010. RoboWarfare: Can robots be more ethical than humans on the battlefield? *Ethics and Information Technology* 12(3): 263–275.
Tanielian, T., and L.H. Jaycox (eds.). 2008. *Invisible wounds of war: Psychological and cognitive injuries, their consequences, and services to assist recovery*. Santa Monica: RAND Corporation.
Topolski, A. 2010. Peacekeeping without banisters: The need for new practices that go beyond just war. In *At war for peace*, ed. M. Forough, 49–56. Oxford: Inter-Disciplinary Press.
Treviño, L.K., and S.A. Youngblood. 1990. Bad apples in bad barrels: A causal analysis of ethical decision-making behavior. *Journal of Applied Psychology* 74: 378–385.
United States Air Force. 2009. *Unmannde aircraft systems flight plan 2009–2047*. Washington. http://www.unmanned.co.uk/unmanned-systems-special-reports/usaf-unmanned-aircraft-systems-flight-plan-2009-2047/.
Van de Poel, I.R., and L.M.M. Royakkers. 2007. The ethical cycle. *Journal of Business Ethics* 71(1): 1–13.
Veruggio, G., and F. Operto. 2008. Roboethics: Social and ethical implications of robotics. In *Springer handbook of robotics*, ed. B. Siciliano and O. Khatib, 1499–1524. Berlin: Springer.
Wallach, W., and C. Allen. 2008. *Moral machines*. New York: Oxford University Press.
Yuval-Davies, N. 1998. What is transversal politics? *Soundings* 12(Summer): 94–98.

Chapter 21
On Technology Against Cyberbullying

Janneke M. van der Zwaan, Virginia Dignum, Catholijn M. Jonker, and Simone van der Hof

21.1 Introduction

In 2008, 93 % of Dutch 6–17 year olds had access to the Internet, compared to 68 % in 2005/2006 (The Gallup Organisation 2008). On average, children spend 1–2 h a time on the Internet at home (Eurobarometer 2007). So, many children and adolescents spend a lot of time online. They use the Internet not only as an educational tool, but also for fun, games and to develop and maintain social contacts. One of the risks children and adolescents run online is to become a victim of cyberbullying. Cyberbullying can be defined as 'any behavior performed through electronic or digital media by individuals or groups that repeatedly communicates hostile or aggressive messages intended to inflict harm or discomfort on others' (Tokunaga 2010). Cyberbullying takes place via e-mail, instant-messaging programs, Internet chat rooms, multi-player online games, (social) websites and blogs. Recently, cyberbullying gained a lot of attention. With victimization rates ranging from 20 to 40 % (Tokunaga 2010), it is a common risk for children and adolescents. In addition, recent findings from the EU Kids Online II survey indicate that cyberbullying has a high impact on victims (Livingstone et al. 2010).

J.M. van der Zwaan (✉) • V. Dignum
Information and Communication Technology, Delft University of Technology, Jaffalaan 5, 2628 BX Delft, The Netherlands
e-mail: j.m.vanderzwaan@tudelft.nl; m.v.dignum@tudelft.nl

C.M. Jonker
Interactive Intelligence, Delft University of Technology, Delft, The Netherlands
e-mail: c.m.jonker@tudelft.nl

S. van der Hof
eLaw@Leiden, Universiteit Leiden, Leiden, The Netherlands
e-mail: s.van.der.hof@law.leidenuniv.nl

J. van den Hoven et al. (eds.), *Responsible Innovation 1: Innovative Solutions for Global Issues*, DOI 10.1007/978-94-017-8956-1_21,

Anti-social behavior such as cyberbullying can be regulated socially, legally, and/or technologically. Social norms play an important role in regulating behavior in general. Law may punish illegal behavior and regulate the enforcement of social norms. Technology can control or steer social behavior through functionalities in the software design (coined 'code as law' by Lessig (2000)) or through exerting social influence (persuasive technology Fogg 2002). Each regulatory modality can be more or less effective depending on the behavior and context involved. Cyberbullying is a complex problem that cannot be solved by measures from a single modality alone; better solutions may be found in a combination of measures from different modalities. Education and awareness, i.e., empowerment, of minors and adults is regarded a primary strategy for protection against online risks (Thierer 2009). However, the role of technology in addressing Internet safety issues is also recognized (Internet Safety Technical Task Force 2008).

This chapter focuses on using technology protect and empower children and adolescents against cyberbullying. So far, this topic has received little attention (exceptions are Internet Safety Technical Task Force 2008; Szwajcer et al. 2009; Mesch 2009). In 2008, the Technology Advisory Board of the Internet Safety Technical Task Force reviewed a total of 40 existing technology solutions for improving online safety of minors; none of which specifically target cyberbullying (Internet Safety Technical Task Force 2008). However, recently different initiatives have started to investigate the regulation of cyberbullying through technology, such as AMiCA (2010) and the project 'Evidenced-based ICT interventions against (cyber-)bullying amongst youngsters' (WISE Research Group 2010). Existing work (Szwajcer et al. 2009; Mesch 2009) seems to rely on the assumption that general Internet safety technologies can be used as protection against cyberbullying as well. In this chapter, we show that this assumption is unfounded and propose an alternative approach to addressing cyberbullying with technology.

This chapter is organized as follows. First, in Sect. 21.2, we provide a background on Internet safety technology and cyberbullying. In Sect. 21.3, this information is used to construct a framework of characteristics technology against cyberbullying should have to be able to protect against cyberbullying. In Sect. 21.4, we use the framework to discuss the expected effectiveness of existing Internet safety technologies against cyberbullying. The results indicate that these technologies are not effective against cyberbullying, mainly because they restrict online behavior that is not related to cyberbullying. The framework suggests that technology exerting social influence (persuasive technology) might be more effective. Therefore, in Sect. 21.5, we propose an alternative technology, that is, a virtual character acting as a supportive friend to victims of cyberbullying. We would like to emphasize that the proposed technology should be regarded as an additional channel for support rather than a 'miracle solution' for cyberbullying. Finally, in Sect. 21.6, we present our conclusions.

21.2 Background

21.2.1 Internet Safety Technology

Online safety of children and adolescents concerns risks such as harassment, bullying, sexual solicitation, exposure to problematic and illegal content (including pornography, hate speech, or violence), malicious software (for instance viruses), hackers, and online delinquency (for example identity theft or fraud). In their review of existing Internet safety technology, the Technology Advisory Board of the Internet Safety Technical Task Force (2008) distinguished the following functional goals:

- Limit harmful contact between adults and minors,
- Limit harmful contact between minors,
- Limit/prevent minors from accessing inappropriate content on the Internet,
- Limit/prevent minors from creating inappropriate content on the Internet,
- Limit the availability of illegal content on the Internet,
- Prevent minors from accessing particular sites without parental consent,
- Prevent harassment, unwanted solicitation, and bullying of minors on the Internet.

These goals show that Internet safety technology is restrictive; they clearly intend to restrict online behavior. This view on technology corresponds to the aforementioned 'code as law' perspective from Lessig. Web filtering software is an example of restrictive technology; a web filter blocks access to websites based on certain criteria.

Different types of Internet safety technologies can be distinguished, including (Internet Safety Technical Task Force 2008; Szwajcer et al. 2009):

- Content and behavior analysis,
- Filtering,
- Monitoring,
- Blocking undesirable contacts,
- Reporting,
- Age/identity verification, and
- Educational technology

Some technologies, such as age/identity verification, require storing personal data, which raises privacy concerns. Monitoring online behavior or automatically analyzing online communication might also invade privacy. In addition, restrictive technology could violate the right to freedom of expression. Children's privacy and their right to freedom of expression must be balanced against the potential benefits of Internet safety technologies. In some cases it might be appropriate to restrict behavior, for example to protect younger children, whereas for older children and adolescents protecting their privacy and/or freedom of expression might be more important.

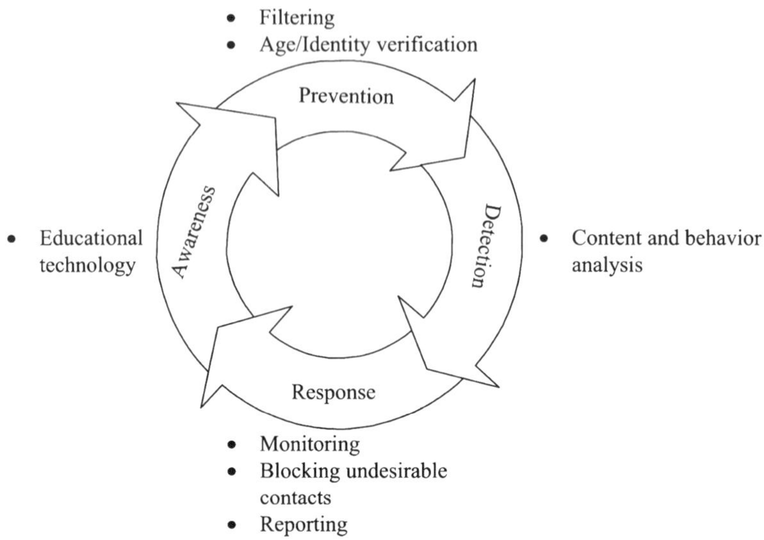

Fig. 21.1 Internet safety strategy cycle (Taken from Szwajcer et al. 2009)

Internet safety technologies can also be organized according to the solution strategy they follow. Solution strategies are awareness, prevention, detection, and response (Szwajcer et al. 2009). Figure 21.1 shows these strategies are connected. Increased awareness of a risk (presumably) prevents it. Prevented incidents diminish the need for detection. Detection of incidents allow for some kind of response (for example punishing the bully). And responding to incidents might lead to increased awareness.

21.2.2 Cyberbullying

Research on cyberbullying is still in the early stage. Little is known beyond prevalence, frequency among specific groups, and negative outcomes (Tokunaga 2010).

21.2.2.1 Compared to Traditional Bullying

Cyberbullying – by definition – is a type of bullying. According to Olweus, bullying is "characterized by the following three criteria: (1) it is aggressive behavior or intentional 'harm doing' (2) which is carried out 'repeatedly and over time' and (3) in an interpersonal relationship characterized by an imbalance of power" (Olweus 1999). In addition to these criteria, cyberbullying has some specific characteristics.

First, cyberbullies can remain anonymous relatively easy (Ybarra and Mitchell 2004; Patchin and Hinduja 2006; Kowalski and Limber 2007; Shariff 2008). Being bullied by an anonymous bully may be more distressing than being bullied by an acquaintance (Kowalski et al. 2008). In addition, it is difficult to punish an anonymous bully and to prevent him/her from bullying again (Patchin and Hinduja 2006). Another important difference is the lack of physical and social cues in online communication (Ybarra and Mitchell 2004; Patchin and Hinduja 2006; Kowalski and Limber 2007; Kowalski et al. 2008). This prevents the bully from being confronted with the consequences of the harassments (Kowalski et al. 2008). However, it could also lead to misinterpreting messages as cyberbullying when in fact they were not intended to be (Ybarra et al. 2007). A third difference is the 24/7-attainability provided by online communication (Patchin and Hinduja 2006; Kowalski and Limber 2007). Traditional bullying is usually characterized by a confined period of time during which bullies have access to their victims. In most cases, victims of traditional bullying are safe at home. However, physical separation is no limitation for cyberbullying. Other differences between traditional bullying and cyberbullying are the quick distribution of electronic messages to (potentially) infinite audiences (Kowalski and Limber 2007; Kowalski et al. 2008; Shariff 2008) and the permanent nature of information on the Internet (Shariff 2008).

21.2.2.2 Types, Media and Methods

Cyberbullying refers to bullying through electronic communication devices. Various types of behavior fall within the definition of cyberbullying (Vandebosch et al. 2006):

- *Outing* – Very personal information (pictures, home address, phone number, etc.) of individuals are broadcasted on the Internet without the victim's consent;
- *Trickery* – Individuals may be deceived into dispersing private information about themselves or others;
- *Impersonation* – The cyberbully acts as another person to deceive the victim;
- *Harassment* – The victim receives regular insulting and denigrating messages from the cyberbully;
- *Cyberstalking* – The victim is terrorised by frequent threatening and intimidating messages;
- *Denigration* – The cyberbully spreads online gossip and lies about a person;
- *Flaming* – The cyberbully provokes rude arguments against the victim;
- *Exclusion* – The victim is left out from participating in virtual groups or communities, e.g., by being blocked from buddy lists;

Cyberbullying happens through different media, such as e-mail, instant messenger applications, social networking websites, blogs, chat rooms, online games, virtual worlds, and mobile phones (sms).

In addition, cyberbullying can be communication-based or content-based. Methods used for online bullying include name-calling, gossiping, ignoring, threatening, hacking (breaking in to computers or online accounts), sending huge amounts of buzzers or winks (animated greetings and/or sound effects), spreading personal conversations, manipulating and spreading pictures, creating defamatory websites, humiliating someone in an open chat room, and sending sexual comments (Dehue et al. 2008; Vandebosch and Cleemput 2008).

21.2.2.3 Victims

Prevalence rates of cyberbullying victimization vary among studies. In a recent review of existing research, Tokunaga (2010) reports victimization rates of 20–40 %. Many studies explored demographic characteristics of victims, such as age and gender. There appears to be a correlation between age and the likelihood of victimization: victimization rates peak around ages 14–15 (Internet Safety Technical Task Force 2008; Tokunaga 2010). Reports regarding gender differences are inconclusive. Some studies report increased risk for females, other studies found no difference in gender with respect to cyberbullying victimization (Internet Safety Technical Task Force 2008; Tokunaga 2010).

Victims of cyberbullying tend to be heavier Internet users than youth that is not victimized (Smith et al. 2008). Victims of traditional bullying and those that bully others online are more likely to be cyberbullied (Ybarra et al. 2006; Li 2007). In addition, some online activities seem to be correlated with being cyberbullied. These activities include: having an active profile on social networking site (Mesch 2009), participating in public chat rooms (Ybarra et al. 2006; Mesch 2009), instant messaging (Ybarra et al. 2006), blogging (Ybarra et al. 2006) and participating in clip sharing networks (e.g. YouTube[1]) (Mesch 2009). In another study, Ybarra and Mitchell (2008) found that interpersonal online victimizations do not seem to occur to a greater degree in social networking sites. So, it remains unclear whether and to what extend cyberbullying victimization is correlated with specific online activities.

Of the online risks investigated in the EU Kids Online II survey, cyberbullying is most likely to upset children (Livingstone et al. 2010). About 60 % of the cyberbullying victims participating in this study reported to be very or fairly upset after being bullied. In another study, 40 % of victims reported to be emotionally affected by online bullying (Patchin and Hinduja 2006). Consequences of cyberbullying tend to be similar to those of traditional bullying. Victims of cyberbullying may experience physical, social, and psychosocial problems, such as serious depressive symptoms and stress (Finkelhor et al. 2000), feeling frustrated, angry, sad (Patchin and Hinduja 2006; Dehue et al. 2008), and not wanting to go to school (Dehue et al. 2008).

[1] http://www.youtube.com/

21.2.2.4 Bullies

Online bullies are typically the same age as their victims (Patchin and Hinduja 2006; Wolak et al. 2006, 2007; Kowalski and Limber 2007; Hinduja and Patchin 2009). And even though anonymity is often viewed as integral to cyberbullying, it seems that cyberbullying often takes place in the context of social groups and relationships (Mishna et al. 2009). Between 44 and 82 % of victims of cyberbullying know their bully or bullies offline (Wolak et al. 2006; Hinduja and Patchin 2009). So, online bullying has a strong connection with the offline world. Perpetrators of traditional bullying tend to bully others online (Li 2007). However, cyberbullies are not just traditional bullies, but also individuals that are afraid to bully others in real life (Mishna et al. 2009). Retaliation is also common. According to a study by Ybarra and Mitchell, four out of five harassers say their behavior was in response to online harassment initiated by someone else (Ybarra and Mitchell 2007). Most bullies cyberbully from their homes (85.6 %) and are alone when engaging in bullying (62.97 %) or with friends (24.6 %) (Dehue et al. 2008).

21.2.2.5 Other Stakeholders

Other stakeholders are parents and teachers. Parents are often unaware their child is engaged in cyberbullying, whether as a bully (5 % of parents knew, while 17 % of children admitted being a cyberbully) or as a victim (12 % of the parents knew, while 23 % of the children reported being bullied) (Dehue et al. 2008). Parents worry less about cyberbullying than about other online dangers, such as inappropriate contact with adults, accidental exposure to inappropriate material (such as pornography or violence), and visits to unapproved websites (Sharples et al. 2009). In addition, parents underestimate the negative online experiences of their children (Shariff 2008). Most parents set rules about the frequency with which their children are allowed to use the Internet (60 %) and about what they are and are not allowed to do on the Internet (80 %) (Dehue et al. 2008). According to a study by Mesch (2009), having rules in place about which websites are allowed to visit somewhat decreases the odds of being cyberbullied. However, other parental mediation techniques, such as installing filtering or monitoring software, checking the websites visited by children, and placing the computer in a shared location (e.g. the living room), did not affect the odds of being bullied online (Mesch 2009).

Teachers have started to include cyberbullying and other online safety issues in their lesson plans. Fifty-eight percent of 206 UK teachers surveyed by Sharples et al. instructed students about e-safety (Sharples et al. 2009). Cyberbullying often happens outside of school (Smith et al. 2008). For that reason, many teachers feel they can not (and should not) interfere (Shariff 2008). However, since cyberbullying may affect the victim's academic performance (Ybarra et al. 2007; Dehue et al. 2008), teachers should always try to help the victim (Shariff 2008). In addition, cyberbullying often starts in real life at school (Shariff 2008; Vandebosch and

Cleemput 2008). Some teachers become victims of cyberbullying themselves. There have been examples of students creating defamatory websites or popularity polls about teachers (Shariff 2008; Sharples et al. 2009).

21.2.2.6 Tackling Cyberbulling

Due to the recent nature of cyberbullying, validated approaches to stop or prevent it do not yet exist. However, some researchers made suggestions on how to tackle the problem.

Many studies stress the importance of education and awareness to reduce and prevent cyberbullying. Ybarra et al. (2006) support the idea to include cyberbullying prevention in conventional anti-bullying programs. It is important to educate both children and adults (e.g. teachers and parents) (Ybarra et al. 2006; Dehue et al. 2008). Educating parents and other adults might make it easier for children and adolescents to talk to them about their negative online experiences (Ybarra et al. 2006).

Teaching technological skills – again, both to children and adults – deserves special attention (Finkelhor et al. 2000; Smith et al. 2008), so children and adults know what can be done about certain situations. Technological skills include protecting private information, blocking or deleting contacts, contacting website moderators or Internet Service Providers, gathering evidence, and tracking bullies.

While educating potential victims and other stakeholders seems a good idea, Mishna et al. (2010) could not establish a link between and increase in online safety knowledge and a reduction in risky behavior. So, knowing about online dangers does not mean that children and adolescents show more careful behavior when they are online.

Ybarra et al. (2006) suggest that those who have trouble communicating are more likely to be involved in online harassment. In another study, Ybarra and Mitchell (2007) found that 80 % of online harassers said they retaliated in response to someone harassing them. Wright et al. (2009) found that cyberbullying often stems from 'misunderstandings' and 'mishearing stuff'. Therefore, a possible way to tackle cyberbullying is by improving interpersonal communication and conflict management skills (Ybarra et al. 2006; Ybarra and Mitchell 2007).

Some researchers consulted children and adolescents to find ways of tackling cyberbullying. Education, both of themselves and teachers and parents, is also commonly advised (Cassidy et al. 2009; Stacey 2009). Children and adolescents too recognize that 'working on creating a positive self-esteem in students' may help reduce cyberbullying (Cassidy et al. 2009). Finally, Stacey (2009) found that especially younger students could use some support when dealing with cyberbullying. Two types of support were discussed: practical advice on how to get rid of bullies, offensive material etc., and moral support. Younger students thought senior students would be their best resource in dealing with cyberbullying.

21.3 The Framework

In order to discuss the expected effectiveness of existing technology against cyberbullying, we constructed a framework consisting of desired characteristics of technology against cyberbullying. To identify these desired characteristics, we started with basic questions on what cyberbullying is and used the background knowledge from Sect. 21.2 the answers. The questions were: what are the online behaviors that can be characterized as cyberbullying?, who are the bullies?, and when do users need protection? Subsequently, we identified some risks associated with online technology in general.

Online behaviors that can be characterized as cyberbullying are diverse; different types, media and methods can be used to cyberbully others. Like traditional bullying, cyberbullying usually is communication-based (for example, name calling in chat conversations or sending threatening e-mails), but content-based cyberbullying also occurs (for example, creating a fake profile on a social network or posting manipulated pictures). Technology against cyberbullying should *take into account different types, media, and methods of cyberbullying and at least target online communication.*

Recent studies reveal that many of the online threats experienced by children and adolescents are perpetrated by peers, including sexual solicitation (Wolak et al. 2006) and online harassment (Smith et al. 2008; Hinduja and Patchin 2009). Although anonymity is often viewed as integral to cyberbullying, it seems that cyberbullying often takes place in the context of social groups and relationships (Mishna et al. 2009). Therefore, technology against cyberbullying should at least *take into account relationships with known and unknown peers.*

Cyberbullying can occur at any moment. This 24/7 attainability of cyberbullying is enabled by technology. Technology against cyberbullying should also be available at any moment and/or be able to intervene at any moment. In other words, technology against cyberbullying should *provide real-time support.*

Technology in general has some risks that might limit their suitability to protect against cyberbullying. For example, in Sect. 21.2 we observed that existing Internet safety technology always restricts users in some way. A disadvantage of restrictive technology is that it can be circumvented relatively easily by computer-savvy users. It is very hard to force people to use some technology. Therefore, it is suggested that technology against cyberbullying should *rely on voluntary use.* Victims (and potentially bystanders) are motivated to use some technology if they have something to gain (they want to stop the bullying), while cyberbullies are less likely to participate voluntarily, because bullying is an intentional act.

Additionally, technology might invade privacy and/or limit freedom of expression. Even though these issues are beyond the scope of this chapter, they are very important. Children's privacy and their right to freedom of expression should be balanced carefully against the potential benefits of technology against cyberbullying. Therefore, *protection of privacy and freedom of expression* are included in the framework.

Table 21.1 Desired characteristics for technology against cyberbullying

Characteristics
• Suitable for different types, media and methods
• Take peer contact into account
• Real-time
• Voluntary use
• Protecting the user's privacy
• Protecting the user's freedom of speech

The desired characteristics are summarized in Table 21.1. In the next section, this list will be uses to discuss the expected effectiveness of existing technology against cyberbullying.

21.4 Existing Internet Safety Technologies

This section reviews existing Internet Safety technologies and discusses their expected suitability against cyberbullying based on the framework proposed in the previous section. The following technologies are discussed: content and behavior analysis, filtering, monitoring, blocking undesirable contacts, reporting, age/identity verification, and educational technology. Most existing parental control applications, e.g., Net Nanny[2] or Cyber Patrol,[3] combine multiple technologies, such as content and behavior analysis, filtering and monitoring in one product. Below, we focus on the separate technologies, not complete applications.

21.4.1 Content and Behavior Analysis

Content and behavior analysis are about automatically extracting meaningful information from data, such as text, images, video material, and network traffic. Content analysis can be applied to detect inappropriate content. Potentially, these techniques can also be used to detect cyberbullying in text based conversations.

Preliminary results on related tasks show that it is rather difficult to automatically recognize different types of harassment. Pendar (2007) used a statistical approach to automatically distinguish between communication of sexual predators and victims. Classifier performance ranged from 40 to 95 %. Kontostathis et al. (2009) attempted to recognize sexual predation with a rule based approach and a model of the communication processes child sexual predators use in the real world. The resulting classifier correctly predicted predator speech 60 % of the time. These results seem

[2]http://www.netnanny.com/

[3]http://www.cyberpatrol.com/

promising, however, the studies reported have some limitations. First, the datasets used for the experiments were rather small (701 online conversations in the study by Pendar and 25 in the study by Kontostathis et al.[4]), especially when compared to the Reuters corpus (Lewis et al. 2004), which is a standard corpus for text classification experiments that contains about 810,000 news stories. Second, the classifiers tried to make a distinction between predators and victims, so the conversations used were known to be malicious. Most online conversations are not malicious. Data imbalance (data sets containing only a few objects that need to be detected) is a well known problem in machine learning that leads to suboptimal classifier performance (Chawla et al. 2004).

In 2009, the Content Analysis for the Web 2.0 Workshop (CAW2.0) offered a shared task on misbehavior detection.[5] Yin et al. (2009) trained classifiers to identify harassing messages in chat and online discussion forums. Harassing was defined as 'intended to annoy one or more persons', which is related to, but not the same as cyberbullying. Performance was between 25 and 40 %, so, there is much room for improvement.

In addition to analyzing the contents of text messages, statistical approaches can also be applied to determine whether a text was written by some known author (Abbasi and Chen 2008; Iqbal et al. 2008), and infer characteristics of users, such as gender (Kucukyilmaz et al. 2008) and potentially age (Szwajcer et al. 2009). Sudden changes in online behavior (log-in times, number of contacts, typing speed, etc.) could for example indicate identity theft (Szwajcer et al. 2009).

Automatically recognizing cyberbullying or other harmful content could be a first step in protecting children and adolescents against these threats (detection in Fig. 21.1). As mentioned before, most applications for parental control employ some form of content analysis. Content and behavior analysis can be used to detect different forms of cyberbullying, both communication-based and content-based. However, related work shows that detecting different types of harassment is not trivial and need to be improved before they can be used as (partial) protection against cyberbullying. The technology can be applied to all communication, including peer communication. In addition, the technology by itself does not rely on voluntary use. Content and behavior analysis can be applied in real-time. Because technology for content and behavior analysis stores and interprets online behavior which can be considered personal data, the privacy of users might be invaded. Depending on the actions taken after inappropriate data is recognized, the technology might also limit the freedom of expression.

---

[4]Pendar (2007) and Kontostathis et al. (2009) both used data made available by Perverted Justice (http://www.perverted-justice.com/).

[5]http://caw2.barcelonamedia.org/

21.4.2 Filtering

Web filtering software blocks access to websites with inappropriate content, such as pornography. Filtering techniques include white lists (lists of websites the user is allowed to visit), black lists (lists of websites the user is not allowed to visit) and content analysis (the content analysis algorithm decides whether the user is allowed to visit a website, e.g. based on the occurrence of certain key words). Common problems with web filtering are underblocking (fail to block access to websites with inappropriate material) and overblocking (block websites that do not contain inappropriate material). Hunter (2000) evaluated four commercial web filtering applications. He found the applications blocked inappropriate material 75 % of the time and appropriate material 21 % of the time.

Filtering is preventive (see Fig. 21.1). It does not specifically target communication, but filtering incoming and/or outgoing communication could limit or prevent harmful contact between minors and between minors and adults. However, automatically recognizing either communication-based or content-based cyberbullying is not a trivial task (see Sect. 21.4.1). Filtering technology does not exclude communication between peers. Because users do not get the choice to apply filtering or not before they go online, filtering does not rely on voluntary use. Filtering software may be circumvented. For example it is very easy to substitute terms that are filtered for unfiltered terms that are equally offending, for example 'loser' becomes 'l o s e r', 'L0S3R', 'looser', etc. Filtering software is real-time technology; websites are blocked and/or communication is filtered instantaneous. Since filtering software does not store personal data to block access to certain online resources, privacy is not at stake. However, blocking communication or preventing access to websites may affect freedom of expression.

21.4.3 Monitoring

Monitoring software informs parents about their children's online activities by recording websites addresses and online communication (for example instant messaging). Most parental control software allows monitoring online activities. A recent study found the use filtering and/or monitoring software does not correlate with less cyberbullying victimization (Mesch 2009).

Monitoring software is preventive (see Fig. 21.1) and works based on the assumption that users will adapt their behavior if they know their online activities are being watched. Because all online activity is stored, monitoring software theoretically targets all types, media, and methods of cyberbullying. In practice, however, cyberbullying incidents will have to be extracted by hand or automatically (see Sect. 21.4.1). Since cyberbullying might be hard to recognize and cyberbullying may only be a small part of all online activity, this is a tedious job. Because all online activities are recorded, peer communication is taken into account. Monitoring software does not rely on voluntary participation, users usually do not know or notice being monitored. Activities are recorded in real-time, however, action can be taken only after the records have been reviewed by an external party (for example a parent). For monitoring, privacy is an issue, because all online activities, which can be considered personal data, are recorded and stored for reviewing. Freedom of expression is not at stake.

21.4.4 Blocking Undesirable Contacts

Most instant messaging applications (e.g., Windows Live Messenger[6]), chat rooms, and social networking sites (e.g., Facebook[7] and MySpace[8]) give users the possibility to block other users, in order to prevent them from being contacted by these people. Many social networking sites also provide the possibility to restrict unknown users from contacting them and accessing their profile.

These blocking options are responsive (see Fig. 21.1) and limit harmful contact between minors and both minors and adults. Blocking contacts is suitable only for communication-based cyberbullying in applications where blocking options are available. It does take into account contact between peers. In fact, blocking bullies is a common advice for stopping cyberbullying.[9] Blocking is a voluntary act that allows users to control who can contact them. Users can block contacts whenever they want; in that sense blocking is real-time. Blocking users does not invade privacy or restrict freedom of expression.

Since incidents of cyberbullying often start or continue at school, the victim has to face the bully the next day anyway (Shariff 2008), so only blocking the bully online will not solve the problem. In addition, blocking does not guarantee the victim will not be contacted by the bully anymore, the bully may find other ways to contact and harass the victim (for example, by changing accounts and (anonymously) contacting the victim). Finally, blocking the bully (or bullies) could lead to social exclusion of the victim.

21.4.5 Reporting Content

Many social web applications (e.g., Facebook and MySpace) provide the possibility to report inappropriate and illegal content, for instance, by clicking a button labeled

---

[6]http://explore.live.com/windows-live-messenger

[7]http://www.facebook.com/

[8]http://www.myspace.com/

[9]See for example http://cybermentors.org.uk/, http://www.stopcyberbullying.org/, and http://www.cybersmart.gov.au/

'report abuse'. Reports are send to community moderators that manually review reported content and decide whether or not to remove it. Reporting is a response type of technology (see Fig. 21.1). Some social networking sites, chat rooms, online games, and forums also allow users to report others when they break the rules, for example, by cyberbullying. Moderators decide whether and how to punish offenders.

Reporting tools can be useful for limiting access to inappropriate material, including some forms of content-based cyberbullying (for instance happy slapping videos or fake profiles on social networking sites). Reporting communication-based cyberbullying is only possible if moderators are available in the application and communication records exist. Everybody can report content they feel is inappropriate, so this technology relies on voluntary use. Because moderators have to check reports manually, it may take some time before reported content is removed. Therefore, reporting is not real-time. Privacy is not at risk, since no personal data needs to be stored for reporting (reporters may be anonymous). Removing content might interfere with freedom of expression. Therefore, in the case of cyberbullying, content will only be removed in obvious and/or extreme cases of content-based cyberbullying.

Reporting (and subsequent removal by a moderator) can limit the consequences of the persistence of online content and stop the victim (and others) being confronted with it. However, since online content can be copied easily, removing the content from one site does not guarantee removal from the Internet. Another drawback is that moderators often can not follow up on every single report, so, a minimum number of reports is needed before action is taken. Possibly, the cyberbullying victim is unable to gather a sufficient amount of reports and the report will be ignored.

21.4.6 Age/Identity Verification

Age and/or identity verification technologies aim at restricting inappropriate contact between minors and adults as well as preventing minors to access inappropriate content. For example, in Second Life,[10] users must be 18 years old to view mature content. These technologies are preventive (see Fig. 21.1). Age and/or identity verification often use public or private databases containing information on either minors (for example school records) or adults (such as known sex offenders). People in the database (for instance minors or sex offenders) or people of certain ages (for example adults) are either allowed or not allowed to contact certain other people (such as minors) or access certain material (for example pornography).

Age and/or identity verification technologies do not target various forms of cyberbullying, such as content-based cyberbullying and harmful contact between

[10]http://secondlife.com/

peers. Age and/or identity verification may rely on voluntary participation (for example by becoming a member of a social network that applies age restrictions). In other cases participation may not be voluntary, for example if a school only allows its pupils to use the school's social networking website. This technology works online. However, since verifying age or identity requires gathering and storing personal data, the privacy of users might be at risk. Freedom of speech is not threatened.

21.4.7 Educational Technology

Education is another approach to improving the online safety of minors (awareness and prevention in Fig. 21.1). Since the topic of this chapter is technology, the discussion below is limited to educational technology, such as interactive computer games.

The first project described focused on traditional bullying instead of cyberbullying. FearNot! is an Intelligent Virtual Environment (IVE) in 3D, where synthetic characters act out bullying scenarios (Paiva et al. 2005). The application was designed for children 8–12 to witness the events from a third-person perspective. After a bullying episode, the victimized character turns to the user to ask for advice. The IVE offers children a safe environment that supports social and emotional learning. Evaluation of the system revealed that children care about and believe in the characters, not only in a single interaction but also over several separated interactions (Lynne et al. 2008). A controlled trial established a short-term effect of escaping victimization for a priory identified victims of bullying and a short-term overall prevention effect for UK children (Sapouna et al. 2010), demonstrating the potential of IVEs to support anti-bullying activities.

Other applications aimed at educating minors about online safety include Mr Ctrl[11] (not available anymore) and Internet Safety with Professor Garfield.[12] Mr Ctrl is a chatbot that answers questions about online safety. Internet Safety with Professor Garfield is a series of online interactive lessons about different topics concerning Internet safety. Both applications can be used individually, but also provide teaching material for classroom use. To the best of our knowledge, these applications have not been evaluated.

Because education is aimed at stimulating the right behavior in general, it basically targets all types, media and methods of cyberbullying. From the examples given here, it is not clear to what extend peer communication is explicitly taken into account. However, it would be easy to do so. Educational programs are usually mandatory, so there is no voluntary participation. Educational technology is designed to support traditional classroom teaching and not to protect or empower

[11] http://mrctrl.spaces.live.com/ (in Dutch).

[12] http://www.infinitelearninglab.org/

Table 21.2 Match between characteristics of existing technologies and the desired characteristics of technology against cyberbullying

	Different forms	Peer communication	Voluntary use	Real-time	Protect privacy	Protect freed. of expr.
Content & behavior analysis	±	+	−	+	±	+
Filtering	±	+	−	+	+	−
Monitoring	±	+	−	−	−	−
Blocking contacts	−	+	+	+	+	+
Reporting	−	+	+	−	+	−
Age/identity verification	−	−	±	+	−	+
Educational technology	+	?	−	−	+	+

+ good match, ± partial match, − no match, ? unknown, *n/a* not applicable)

pupils at the same time they use the Internet. Finally, privacy and freedom of expression are not at risk in normal educational settings.

Another concern regarding education and/or educational technology is its effectiveness. Mishna et al. (2010) performed a systematic review of interventions against cyber abuse of youth. The term 'cyber abuse' refers to online abusive interpersonal behaviors including online bullying, stalking, sexual solicitation, and exposure to problematic content, such as pornography. Three educational programmes were selected for the review. Mishna et al. concluded that participation in cyber abuse prevention and intervention strategies is associated with an increase in Internet safety knowledge, but changes to Internet risk attitudes and behavior are not significant. So, increased knowledge about safe Internet use does not necessarily correlate with less risk taking (or other behavior changes) online.

21.4.8 Summary

In this section we discussed the expected effectiveness of different existing Internet safety technologies against cyberbullying. The results of this discussion are summarized in Table 21.2. While all technologies satisfy at least some of the desired characteristics from the framework we proposed, we expect their effectiveness against cyberbullying to be limited. Technologies such as age/identity verification, filtering and monitoring, reporting, and blocking undesirable contacts have not been designed to protect against cyberbullying, but with other online risks in mind. Some of these technologies primarily target access to undesirable content. Their success in protecting against cyberbullying, which is mostly communication-based, is therefore limited.

One of the most salient features of existing Internet safety technology is its attempt to steer the behavior users by restricting them. While in certain cases restricting bullies and/or victims might be useful, teaching them to deal with cyberbullying incidents seems a promising approach. This viewpoint is supported by the literature. For example, Shariff (2008) argues that incidents of cyberbullying potentially are valuable learning experiences. This potential, however, is ignored by existing technologies. Additionally, Thierer (2009) claims education (media literacy) is the primary solution against online risks. In his view, the role of technology is to supplement (but never to supplant) education. Educational technology (Sect. 21.4.7) is a primary example of using technology to supplement education.

Our discussion was focussed on the separate existing Internet safety technologies. One might argue that combining multiple technologies, as is done in existing parental control software, might increase performance compared to individual technologies. However, the main issues, i.e., using technology that has been designed for other risks and restricting users instead of empowering them, will not be tackled by combining restrictive technologies.

Finally, we would like to emphasize that our discussion is limited to the expected effectiveness of technologies against cyberbullying. We do not claim the technologies discussed in this section should not be used; they might be very effective against other online risks, such as exposure to problematic and illegal content or identity theft. However, based on the characteristics proposed in the framework, we expect that the effectiveness of existing Internet safety technology against cyberbullying is limited.

21.5 Design for a Virtual Empathic Buddy Against Cyberbullying

In the previous section, we argued that technology that empowers users instead of restricting them might be more effective against cyberbullying. In fact, technology does not have to be restrictive to influence behavior. Persuasive technology steers behavior by exerting social influence (Fogg 2002). How can we use persuasive technology to empower victims of cyberbullying?

In many countries child helplines exist that allow children and adolescents to talk to counselors by telephone or chat about their problems, such as bullying. Additionally, peer support programs have been set up in schools to give pupils the opportunity to discuss their problems with trained peers. Research shows these peer support programs can be effective against traditional bullying (Cowie et al. 2002). However, helplines and peer supporters are not available 24/7, and one-on-one online counseling is very labor intensive. Automating this kind of support could help to reach more victims.

We propose to empower victims of cyberbullying by giving them access to a virtual empathic buddy. The buddy is a virtual character that interacts with users based on the principles of human face-to-face conversation. This means it

Fig. 21.2 Screenshot of the virtual empathic buddy

communicates with the user through text or speech, and displays emotions and other non-verbal behavior. The buddy's short term goal is to lower the victim's negative emotions (coping). On the long(er) term, the buddy aims at teaching the victim how to deal with cyberbullying. To achieve these goals the buddy exerts social influence by simulating humanlike communicative behavior.

The buddy 'lives' on the screen of potential cyberbullying victims. At the user's request, the buddy provides emotional support and practical advice on how to deal with the incident. Allowing users to decide when the buddy is activated, gives them a sense of being in control. We agree with Thierer (2009) that technology should only be used to supplement education and mentoring and not to replace it. Therefore, the buddy should be seen as an additional channel for victims to find support, not as a 'miracle solution' against cyberbullying. Ideally, the buddy should be an extension of an educational program, child helpline, or embedded in an anti-bullying training. However, the current focus of our research is on developing and testing the buddy as an autonomous agent. More details on the design can be found in van der Zwaan et al. (2010).

Figure 21.2 shows a screenshot of the buddy. A virtual character is displayed in the top left and communicates with the user through a chat interface. The look and feel of this character is still under investigation. The generic embodiment in the picture will be replaced with an appearance that appeals to the target audience. The user communicates her emotional state by manipulating the AffectButton (Broekens and Brinkman 2009) in the bottom left of the picture. The facial expression of the button can be changed by moving the mouse over it. When the expression matches the emotion the user wants to communicate, she clicks the button.

In the interaction with the child, the buddy has the social role of a peer. This role has been chosen for several reasons. First, it has been shown that peer support can be effective against traditional bullying (Cowie et al. 2002). Second, to decrease high expectations of intelligence, the agent is presented as a peer instead of an all-knowing expert. Finally, findings from the cyberbullying literature suggest that children do not like to talk to adults about their negative online experiences (Li 2007; Dehue et al. 2008; Mishna et al. 2009), therefore a peer is a more appropriate model for the buddy than an adult-like figure.

Interaction between the virtual buddy and the victim takes place in three stages:

1. Understand the victim's emotional state
2. Gather information about the current situation
3. Give advice

During the interaction, the buddy simulates skills that are expected from human peer supporters, such as being empathic, adopting a problem-solving approach, active listening and the ability to build up trust (Cowie et al. 2002). In order to be empathic, the buddy will adapt its response to the emotional state of the victim and display appropriate facial expressions. To decide which facial expressions are appropriate, the buddy needs some understanding of the user's emotional state. The problem-solving approach consists of gathering information about the current situation and giving practical advice. To interpret and reason about the situation, the buddy uses several knowledge bases, such as an advice module that maps situations to pieces of advice, an incident database that stores information about past incidents, and a user profile that contains personal information about the user. Active listening is implemented by showing appropriate listening behavior (such as nodding) and by asking follow up questions. It is expected that implementing these three skills will foster trust between the victim and the buddy. In addition, the social role of the buddy (peer) is chosen to increase trust. The appearance of the buddy will be designed to maximize perceived trustworthiness.

Since bullying refers to a series of events rather than isolated incidents (Olweus 1999), users are expected interact repeatedly with the virtual buddy. This gives the buddy the opportunity to (try to) establish a relationship. The buddy uses its knowledge about earlier (similar) events and bullies to give the user the sense he is being understood by the buddy. In order for the advice to be effective, the agent should (try to) persuade the victim to follow it's advice. Finally, the buddy requests explicit feedback about the way it handled situations, so it can adapt its approach to the preferences of the user.

The cyberbullying types, media, and methods the buddy can provide support for depends on the domain knowledge available. As mentioned before, the buddy's knowledge base contains a mapping between characteristics of bullying situations to specific pieces of advice. The buddy can only reason about situations based on its knowledge. This does not exclude any situation in advance, whether communication-based or content-based. In addition, the buddy's advice module can be extended with new knowledge if necessary. The goal of the virtual empathic buddy is to empower its users. It does in no way restrict the online behavior of its users, so peer contact is taken into account. The buddy provides support when requested by the user and aims at teaching the user to deal with future cyberbullying. In addition, the buddy's support is real-time. Since the buddy will collect and store personal data, the user's privacy is at risk. To protect the user's privacy, the agent will be implemented as a local application on the host computer; data will not be shared with any (social) web application used by the victim. However, additional measures to protect the privacy of users might be necessary. Since the buddy does not restrict user behavior, freedom of expression will not be violated.

While the virtual empathic buddy matches almost all the desired characteristics of technology against cyberbullying, some additional concerns arise. First, the buddy pretends to be a social being. It is important to make sure the victim does not become socially isolated as a result of interaction with the buddy. In addition, it should be made very clear that the buddy is not a replacement for professional help. There have been cases of severe (cyber)bullying that resulted in victims committing suicide. While these extreme cases are rare, care has to be taken not to harm users of the buddy. To address these concerns, the buddy will be included in a broader context of anti-cyberbullying measures, for example by employing the buddy within the context of a specialized helpline. If the buddy detects a case it cannot handle, it can refer the victim to the helpline or alert a human counselor that takes over the conversation.

21.6 Conclusion

This chapter makes three contributions. First, we presented a framework of desired characteristics of technology against cyberbullying based on literature on Internet safety technology and cyberbullying. Second, we discussed the expected effectiveness of existing Internet safety technologies based on this framework. The results indicate that existing Internet safety technology is not effective in protecting users against cyberbullying. Finally, we proposed an alternative technology that is aimed at empowering users instead of restricting them.

The framework consists of the following desired characteristics for technology against cyberbullying: it should be suitable for different types, media and methods of cyberbullying (at least communication-based cyberbullying), it should take into account peer contact, it should rely on voluntary use, it should be real-time, and user's privacy and freedom of expression should be balanced against restriction. This framework should be seen as a first step towards more formal criteria or requirements. The characteristics can be formalized and extended based on the results of experimental validation (for example by performing user studies and assessing the effectiveness of technologies in practice). The results of our review of existing technology indicates that prevention and detection of cyberbullying do not suffice. Five online safety task forces agree and conclude that empowerment, i.e. education and awareness, is a primary solution strategy to protect children and adolescents against online risks (Thierer 2009). Technology can be used to supplement education and awareness. We would like to emphasize that technology alone can never solve a complex problem such as cyberbullying. A combination of social, legal, and technological measures is required for best results.

Our discussion of existing Internet safety technologies shows that all of them satisfy at least some the characteristics from our framework. However, we conclude that the effectiveness of these technologies against cyberbullying still is limited. Technologies such as age/identity verification, filtering and monitoring, reporting, and blocking undesirable contacts have not been designed to protect against cyberbullying, but with other online risks in mind. Some of these technologies primarily target access to undesirable content. Their success in protecting against cyberbullying, which is mostly communication-based, is therefore limited. Additionally, apart from education, none of the existing online safety technologies discussed are designed to empower children and adolescents. Rather, the technologies either restrict the behavior of bullies and/or victims (filtering and monitoring, age/identity verification, blocking undesirable contacts). While in some cases restricting the behavior of bullies and/or victims might be useful, incidents of cyberbullying potentially can be valuable learning experiences (Shariff 2008), which are currently ignored by technology.

Instead of viewing technology from a restrictive 'code as law' perspective, this chapter proposes a socio-technological measure that does not restrict the possibilities of online communication and is aimed at empowering cyberbullying victims. The virtual empathic buddy is an animated character that interacts with users based on the principles of human face-to-face conversation. It provides emotional support and practical advice on how to deal with cyberbullying, and is available whenever a victim needs help.

Acknowledgements This work is funded by NWO under the Responsible Innovation (RI) program via the project 'Empowering and Protecting Children and Adolescents Against Cyberbullying'.

References

Abbasi, A., and H. Chen. 2008. Writeprints: A stylometric approach to identity-level identification and similarity detection in cyberspace. *ACM Transactions on Information Systems* 26(2): 1–29.

AMiCA. 2010. Automatic monitoring for cyberspace applications. http://www.clips.ua.ac.be/amica/. Accessed 6 July 2011 [Online].

Broekens, J., and W.P. Brinkman. 2009. Affectbutton: Towards a standard for dynamic affective user feedback. In *Affective computing and intelligent interaction (ACII)*, Amsterdam.

Cassidy, W., M. Jackson, and K.N. Brown. 2009. Sticks and stones can break my bones, but how can pixels hurt me?: Students' experiences with cyber-bullying. *School Psychology International* 30(4): 383–402.

Chawla, N.V., N. Japkowicz, and A. Kotcz. 2004. Editorial: Special issue on learning from imbalanced data sets. *SIGKDD Explorations Newsletter* 6(1): 1–6.

Cowie, H., P. Naylor, L. Talamelli, P. Chauhan, and P.K. Smith. 2002. Knowledge, use of and attitudes towards peer support: A 2-year follow-up to the Prince's Trust survey. *Journal of Adolescence* 25(5): 453–467.

Dehue, F., C. Bolman, and T. Völlink. 2008. Cyberbullying: Youngsters' experiences and parental perception. *Cyberpsychology & Behavior: The Impact of the Internet, Multimedia and Virtual Reality on Behavior and Society* 11(2): 217–223.

Eurobarometer. 2007. Safer internet for children, qualitative study in 29 European countries, national analysis: The Netherlands. http://ec.europa.eu/information_society/activities/sip/surveys/qualitative/index_en.htm.

Finkelhor, D., K.J. Mitchell, and J. Wolak. 2000. Online victimization: A report on the nation's youth. http://www.missingkids.com/missingkids/servlet/ResourceServlet?LanguageCountry=en_US&PageId=869.

Fogg, B.J. 2002. *Persuasive technology: Using computers to change what we think and do*. New York: ACM. Chap. Computers as persuasive social actors.

Hinduja, S., and J.W. Patchin. 2009. *Bullying beyond the schoolyard: Preventing and responding to cyberbullying*. Thousand Oaks: Corwin Press.

Hunter, C.D. 2000. Internet filter effectiveness (student paper panel): Testing over and underinclusive blocking decisions of four popular filters. In CFP'00: Proceedings of the tenth conference on computers, freedom and privacy, Toronto, 287–294. New York: ACM.

Internet Safety Technical Task Force. 2008. Enhancing child safety and online technologies: Final report of the Internet Safety Technical Task Force to the multi-state working group on social networking of state attorneys general of the United States. Tech. rep., http://cyber.law.harvard.edu/pubrelease/isttf/.

Iqbal, F., R. Hadjidj, B.C.M. Fung, and M. Debbabi. 2008. A novel approach of mining write-prints for authorship attribution in e-mail forensics. *Digital Investigation* 5(Supplement 1): 42.

Kontostathis, A., L. Edwards, and A. Leatherman. 2009. Chatcoder: Toward the tracking and categorization of internet predators. In *Proceedings of the 7th text mining workshop*, Sparks.

Kowalski, R., and S. Limber. 2007. Electronic bullying among middle school students. *Journal of Adolescent Health* 41(6, Supplement 1): 22–30.

Kowalski, R.M., S.P. Limber, and P.W. Agatston. 2008. *Cyber bullying: Bullying in the digital age*. Malden: Wiley-Blackwell.

Kucukyilmaz, T., B.B. Cambazoglu, C. Aykanat, and F. Can. 2008. Chat mining: Predicting user and message attributes in computer-mediated communication. *Information Processing & Management* 44(4): 1448–1466.

Lessig, L. 2000. *Code and other laws of cyberspace*. New York: Basic Books

Lewis, D.D., Y. Yang, T.G. Rose, and F. Li. 2004. RCV1: A new benchmark collection for text categorization research. *Journal of Machine Learning Research* 5: 361–397. http://dl.acm.org/citation.cfm?id=1005332.1005345.

Li, Q. 2007. New bottle but old wine: A research of cyberbullying in schools. *Computers in Human Behavior* 23(4): 1777–1791.

Livingstone, S., L. Haddon, A. Görzig, and K. Ólafsson. 2010. Risks and safety on the internet: The perspective of European children. Initial findings. http://www2.lse.ac.uk/media@lse/research/EUKidsOnline/EUKidsII%20%282009-11%29/home.aspx.

Lynne, H., A. Paiva, D. Wolke, K. Dautenhahn, E. Andre, and P. Rizzo. 2008. Virtual role-play in the classroom – Experiences with fearnot. In *eChallenges e-2008*, Stockholm.

Mesch, G.S. 2009. Parental mediation, online activities, and cyberbullying. *CyberPsychology & Behavior* 12(4): 387–393.

Mishna, F., M. Saini, and S. Solomon. 2009. Ongoing and online: Children and youth's perceptions of cyber bullying. *Children and Youth Services Review* 31(12): 1222–1228.

Mishna, F., C. Cook, M. Saini, M.J. Wu, and R. MacFadden. 2010. Interventions to prevent and reduce cyber abuse of youth: A systematic review. *Research on Social Work Practice* 21(1): 5–14.

Olweus, D. 1999. *The nature of school bullying: A cross-national perspective*, 7–27. New York: Routledge. Chap. Sweden.

Paiva, A., J. Dias, D. Sobral, R. Aylett, S. Woods, L. Hall, and C. Zoll. 2005. Learning by feeling: Evoking empathy with synthetic characters. *Applied Artificial Intelligence: An International Journal* 19(3): 235–266.

Patchin, J.W., and S. Hinduja. 2006. Bullies move beyond the schoolyard: A preliminary look at cyberbullying. *Youth Violence and Juvenile Justice* 4(2): 148–169.

Pendar, N. 2007. Toward spotting the pedophile telling victim from predator in text chats. In *ICSC'07: Proceedings of the international conference on semantic computing*, Irvine, 235–241. IEEE Computer Society, Washington, DC.

Sapouna, M., D. Wolke, N. Vannini, S. Watson, S. Woods, W. Schneider, S. Enz, L. Hall, A. Paiva, E. Andre, K. Dautenhahn, and R. Aylett. 2010. Virtual learning intervention to reduce bullying victimization in primary school: A controlled trial. *Journal of Child Psychology and Psychiatry* 51(1): 104–112.

Shariff, S. 2008. *Cyber-bullying: Issues and solutions for the school, the classroom and the home*. New York: Routledge.

Sharples, M., R. Graber, C. Harrison, and K. Logan. 2009. E-safety and web 2.0 for children aged 11–16. *Journal of Computer Assisted Learning* 25(1): 70–84.

Smith, P.K., J. Mahdavi, M. Carvalho, S. Fisher, S. Russell, and N. Tippett. 2008. Cyberbullying: Its nature and impact in secondary school pupils. *Journal of Child Psychology and Psychiatry* 49(4): 376–385.

Stacey, E. 2009. Research into cyberbullying: Student perspectives on cybersafe learning environments. *Informatics in Education* 8(1): 115–130.

Szwajcer, E., W. Ebbers, M. Oostdijk, C. Wartena, and B. Hulsebosch. 2009. Kinderen en nieuwe media – Technische and socio-technische oplossingsmogelijkheden voor gevaren in de online wereld. http://www.novay.nl/medialibrary/documenten/originelen/Eindrapportage_kinderen_en_nieuwe_media.pdf.

The Gallup Organisation. 2008. Towards a safer use of the Internet for children in the EU – A parents' perspectives. http://ec.europa.eu/information_society/activities/sip/surveys/quantitative/index_en.htm.

Thierer, A.D. 2009. Five online safety task forces agree: Education, empowerment & self-regulation are the answer. *Progress & Freedom Foundation Progress on Point Paper* 16(13).

Tokunaga, R.S. 2010. Following you home from school: A critical review and synthesis of research on cyberbullying victimization. *Computers in Human Behavior* 26(3): 277–287.

Vandebosch, H., and K.V. Cleemput. 2008. Defining cyberbullying: A qualitative research into the perceptions of youngsters. *CyberPsychology & Behavior* 11(4): 499–503.

Vandebosch, H., K. Van Cleemput, D. Mortelmans, and M. Walrave. 2006. Cyberpesten bij jongeren in vlaanderen, studie in opdracht van het viwta, brussel. Tech. rep.

van der Zwaan, J.M., M.V. Dignum, and C.M. Jonker. 2010. Simulating peer support for victims of cyberbullying. In *Proceedings of the 22nd Benelux conference on artificial intelligence (BNAIC 2010)*, Luxembourg.

WISE Research Group. 2010. Evidenced-based ICT interventions against (cyber-)bullying amongst youngsters. http://wise.vub.ac.be/content/evidenced-based-ict-interventions-against-cyber-bullying-amongst-youngsters. Accessed 6 July 2011 [Online].

Wolak, J., K. Mitchell, and D. Finkelhor. 2006. Online victimization of youth: Five years later. http://www.unh.edu/ccrc/pdf/CV138.pdf.

Wolak, J., K.J. Mitchell, and D. Finkelhor. 2007. Does online harassment constitute bullying?: An exploration of online harassment by known peers and online-only contacts. *Journal of Adolescent Health* 41(6): S51–S58.

Wright, V.H., J.J. Burnham, C.T. Inman, and H.N. Ogorchock. 2009. Cyberbullying: Using virtual scenarios to educate and raise awareness. *Journal of Computing in Teacher Education* 26(1): 8.

Ybarra, M.L., and K.J. Mitchell. 2004. Youth engaging in online harassment: Associations with caregiver-child relationships, internet use, and personal characteristics. *Journal of Adolescence* 27(3): 319–336.

Ybarra, M.L., and K.J. Mitchell. 2007. Prevalence and frequency of internet harassment instigation: Implications for adolescent health. *Journal of Adolescent Health* 41(2): 189–195.

Ybarra, M.L., and K.J. Mitchell. 2008. How risky are social networking sites?: A comparison of places online where youth sexual solicitation and harassment occurs. *Pediatrics* 121(2): 350–357.

Ybarra, M.L., K.J. Mitchell, J. Wolak, and D. Finkelhor. 2006. Examining characteristics and associated distress related to internet harassment: Findings from the second youth internet safety survey. *Pediatrics* 118(4): 1169–1177.

Ybarra, M.L., M. Diener-West, and P.J. Leaf. 2007. Examining the overlap in internet harassment and school bullying: Implications for school intervention. *Journal of Adolescent Health* 41(6, Supplement 1): 42–50.

Yin, D., Z. Xue, L. Hong, B.D. Davison, A. Kontostathis, and L. Edwards. 2009. Detection of harassment on web 2.0. In *CAW 2.0'09: Proceedings of the 1st content analysis in web 2.0 workshop*, Madrid.